Unexploded Ordnance Detection and Mitigation

NATO Science for Peace and Security Series

This Series presents the results of scientific meetings supported under the NATO Programme: Science for Peace and Security (SPS).

The NATO SPS Programme supports meetings in the following Key Priority areas: (1) Defence Against Terrorism; (2) Countering other Threats to Security and (3) NATO, Partner and Mediterranean Dialogue Country Priorities. The types of meeting supported are generally "Advanced Study Institutes" and "Advanced Research Workshops". The NATO SPS Series collects together the results of these meetings. The meetings are co-organized by scientists from NATO countries and scientists from NATO's "Partner" or "Mediterranean Dialogue" countries. The observations and recommendations made at the meetings, as well as the contents of the volumes in the Series, reflect those of participants and contributors only; they should not necessarily be regarded as reflecting NATO views or policy.

Advanced Study Institutes (ASI) are high-level tutorial courses intended to convey the latest developments in a subject to an advanced-level audience

Advanced Research Workshops (ARW) are expert meetings where an intense but informal exchange of views at the frontiers of a subject aims at identifying directions for future action

Following a transformation of the programme in 2006 the Series has been re-named and re-organised. Recent volumes on topics not related to security, which result from meetings supported under the programme earlier, may be found in the NATO Science Series.

The Series is published by IOS Press, Amsterdam, and Springer, Dordrecht, in conjunction with the NATO Public Diplomacy Division.

Sub-Series

A.	Chemistry and Biology	Springer
B.	Physics and Biophysics	Springer
C.	Environmental Security	Springer
D.	Information and Communication Security	IOS Press
E.	Human and Societal Dynamics	IOS Press

http://www.nato.int/science
http://www.springer.com
http://www.iospress.nl

Series B: Physics and Biophysics

Unexploded Ordnance Detection and Mitigation

Edited by

Jim Byrnes
Prometheus Inc.,
Newport, RI, USA

Springer

Published in cooperation with NATO Public Diplomacy Division

Proceedings of the NATO Advanced Study Institute on Unexploded Ordnance Detection and Mitigation
Il Ciocco, Italy
20 July–2 August 2008

Library of Congress Control Number: 2008938812

ISBN 978-1-4020-9252-7
ISBN 978-1-4020-9251-0
ISBN 978-1-4020-9253-4 (eBook)

Published by Springer,
P.O. Box 17, 3300 AA Dordrecht, The Netherlands.

www.springer.com

Printed on acid-free paper

This work relates directly to Department of the Navy Grant N00014-08-1-1073 issued by Office of Naval Research Global. The United States Government has a royalty-free license throughout the world in all copyrightable material contained herein.

Preface

The chapters in this volume were presented at the July–August 2008 NATO Advanced Study Institute on Unexploded Ordnance Detection and Mitigation. The conference was held at the beautiful Il Ciocco resort near Lucca, in the glorious Tuscany region of northern Italy. For the ninth time we gathered at this idyllic spot to explore and extend the reciprocity between mathematics and engineering. The dynamic interaction between world-renowned scientists from the usually disparate communities of pure mathematicians and applied scientists which occurred at our eight previous ASI's continued at this meeting.

The detection and neutralization of unexploded ordnance (UXO) has been of major concern for very many decades; at least since the First World war. UXO continues to be the subject of intensive research in many fields of science, including mathematics, signal processing (mainly radar and sonar) and chemistry. While today's headlines emphasize the mayhem resulting from the placement of improvised explosive devices (IEDs), humanitarian landmine clearing continues to draw significant global attention as well. In many countries of the world, landmines threaten the population and hinder reconstruction and fast, efficient utilization of large areas of the mined land in the aftermath of military conflicts.

Current estimate state that there are about 110 million unexploded mines in more than 60 countries, and that roughly 30,000 people per year, a large percentage of whom are innocent civilians, are killed or maimed globally. Moreover, the injury rate among those searching for and attempting to disarm mines, even outside war zones, is as high as one casualty per 100 mines.

The combination of basic ideas in mathematics, radar, sonar, and chemistry with ongoing improvements in hardware and computation, as well as very new advances in multisensor data fusion, offers the promise of more sophisticated and accurate UXO detection and identification capabilities than currently exist. Coupled with the dramatic rise in the need for surveillance in innumerable aspects of our daily lives, brought about by hostile acts deemed unimaginable only a few short years ago, the time was ripe for scientists in these usually diverse fields to join together in a concerted effort to combat both the new brands of terrorism and the long-standing existence of UXOs throughout the world. We envisage this ASI as one important step.

To encompass the diverse nature of the subject and the varied backgrounds of the participants, the ASI involved two broadly defined but interrelated areas:

- Mathematical, computer science, chemical and signal processing technologies for automatic detection and identification
- Robotic and other methods for safe neutralization and removal of UXOs

A deep understanding of these topics and of their interdependency, is clearly crucial to meet the challenges resulting from both the widespread existence of UXOs and the increasing sophistication of those who wish to do us harm. The authors whose works appear in this volume include many of the world's leading experts in these areas.

The ASI brought together world leaders from academia, Government and industry, with extensive multidisciplinary backgrounds evidenced by their research and participation in numerous workshops and conferences. This created and interactive forum for initiating new and intensifying existing efforts aimed at furthering the required interdisciplinary approach to the automatic identification and mitigation of UXOs. The forum provided opportunities for young scientists and engineers to learn more about these problem areas, and the vital role played by new mathematical and scientific insights, from recognized experts in this crucial and growing area of both pure and applied science.

The talks and following chapters were designed to address an audience consisting of a broad spectrum of scientists, engineers, and mathematicians involved in these fields. Participants had the opportunity to interact with those individuals who have been on the forefront of the ongoing intense work in UXO detection and mitigation, to learn firsthand the details and subtleties of this important and existing area, and to hear these experts discuss in accessible terms their contribution and ideas for future research. This volume offers these insights to those unable to attend.

The cooperation of many individuals and organizations was required in order to make the conference the success that it was. First and foremost I wish to thank NATO, and especially Dr. F. Pedrazzini and his most able assistant, Ms. Alison Trapp, for the initial grant and subsequent help.

Very generous financial support was also received from:

- Office of the U.S. Deputy Assistant Secretary of the Army (Environment, Safety and Occupational Health), Mr. Tad Davis, Mr. Hermann Spitzer and Mr. Bob Lavelle
- U.S. Defense Advanced Research Projects Agency, Drs. Tony Tether and Bob Leheny
- U.S. Army Research office, Dr. Russ Harmon
- U.S. Office of Naval Research, Dr. Tom Swean and Mr. Brian Almquist
- U.S. Office of Naval Research Global, Mr. Dave Marquis
- U.S. Army Space and Missile Defense Technical Center Office, Drs. Rodney Roberston, Bob McMillan and Pete Kirkland
- Prometheus Inc.

This additional support is gratefully acknowledged.

For their assistance in obtaining very substantial U.S. Department of Defense funding. I sincerely thank the Rhode Island Congressional Delegation:

- Congressman Patrick Kennedy
- Congressman Jim Langevin
- Senator Jack Reed
- Senator Sheldon Whitehouse

In this regard I especially thank Mr. Dan Murphy of Congressman Kennedy's office, who arranged for this invaluable Congressional support.

I note that United States Government support does not necessarily reflect the position or the policy of the United States Government and no official endorsement should be inferred.

I wish to express my sincere appreciation to my assistants Marcia Byrnes and Seda Vural for their invaluable aid.

Finally, my heartful thanks to the Il Ciocco staff, especially Bruno Giannasi, for offering an ideal setting, not to mention the magnificent meals, that promoted the productive interaction between the participants of the conferences. All of the above, the speakers, and the remaining conferees, made it possible for our Advanced Study Institute, and this volume, to fulfill the stated NATO objectives of disseminating advanced knowledge and fostering international scientific contacts.

Il Ciocco, Italy
August 4, 2008

Jim Byrnes

Contents

Wavelet Decomposition of Measures: Application to Multifractal Analysis of Images

Patrice Abry[1], Stéphane Jaffard[2], Stéphane Roux[1], Béatrice Vedel[1] and Herwig Wendt[1]

Abstract We show the relevance of multifractal analysis for some problems in image processing. We relate it to the standard question of the determination of correct function space settings. We show why a scale-invariant analysis, such as the one provided by wavelets, is pertinent for this purpose. Since a good setting for images is provided by spaces of measures, we give some insight into the problem of multifractal analysis of measures using wavelet techniques.

Keywords: Fourier transform, function spaces, fractals, fractional integration, Hölder regularity, image classification, image processing, measures, multifractal analysis, scaling function, scale invariance, spectrum of singularities, wavelets, wavelet leaders

1 Introduction

The detection of UXO (Unexploded Ordnance) uses sensor technologies, such as: GPR (Ground Penetrating Radar), where electromagnetic waves penetrate the ground and are reflected by layers with electrically different natures; IR (Infrared sensors), based on the different thermal properties of different layers of the ground; and Ultrasound sensors, which use ultrasound waves as a probe. In each case, one faces difficult signal or image processing problems. Indeed, ill-posed inverse problems have to be solved in the presence of noise. Note however that these problems are related to similar technological challenges which have been extensively studied in the past. For instance, oil detection can be performed by studying the reflections of vibrations emitted at the surface of the earth. Similarly, the deep structure of

[1]CNRS UMR 5672 Laboratoire de Physique, ENS de Lyon, 46, allée d'Italie, F-69364 Lyon cedex, France, e-mail: { patrice.abry, sroux, herwig.wendt}@ens-lyon.fr, beatrice.vedel@u-picardie.fr
[2]Laboratoire d'Analyse et de Mathématiques Appliquées, CNRS UMR 8050, Université Paris Est, 61 Avenue du Général de Gaulle, 94010 Créteil Cedex, France, e-mail: jaffard@univ-paris12.fr

J. Byrnes (ed.), *Unexploded Ordnance Detection and Mitigation,*

the mantle of the earth is studied by such methods, but the (much more powerful) vibrations used actually are earthquakes.

The resolution of such ill-posed problems in the presence of noise usually necessitates preprocessing which involves denoising, deblurring, and then the inversion of operators which are of pseudo-differential type. In order to be numerically stable, these operations require the choice of a function space which

- Supplies a proper mathematical setting for the resolution
- Is a realistic framework for the kind of signals or images considered

While the first problem has attracted a lot of attention among mathematicians, the second one is usually disregarded. However, in a completely independent way, this question has been addressed since the 1940s, initially by physicists working to determine the function space regularity of fully developed turbulence. Their motivation was, first, the fundamental comprehension of the physical phenomena at work, but they also wanted to use this information as a classification tool in order to select among the many turbulence models that have been proposed. Mutifractal analysis is now used in a large number of problems in signal and image processing, but still retains this initial motivation of a classification tool based on function space regularity.

Images are often stored, denoised, and transmitted using their wavelet coefficients. In particular, due to the success of wavelet techniques in the 1990s, the JPEG 2000 benchmark is based on wavelet decompositions. Therefore, it is relevant to analyze images directly using their wavelet coefficients instead of starting from the pixel values, and many image processing techniques are now based directly on the wavelet coefficients of the image. Multifractal analysis is one example of such a situation. It was introduced in signal processing in the mid-1980s (but relies on insights developed as early as the 1940s by N. Kolmogorov), and can be interpreted as the determination of the smoothness index of the signal analyzed inside some families of function spaces. This smoothness index is stored through a one-dimensional family of parameters, the *scaling function*, which is based on the computation of p-order averages of local quantities (such as oscillations) of the signal. Initially introduced as a tool for the study of fully developed turbulence, it turned out to be also pertinent in order to study signals of many different origins and has lead to new methods of classification and identification.

In Section 2 we start by describing wavelet bases and some of their properties; a particularly relevant one is that by construction, their algorithmic form implies that they are fitted to the dectection of scale-invariance properties in signals and images. Another important property is that wavelets allow simple characterizations of function spaces.

In Section 3 we give a short overview of the use of function spaces in image modeling and image processing; indeed, it has become a key issue in many algorithms, such as denoising, inpainting or texture classification.

In Section 4 we introduce the wavelet scaling function and give its most important properties. We show that the information supplied by function space regularity is encapsulated in this scaling function, and that wavelet techniques yield numerically simple algorithms for the determination of this scaling function.

In Section 5 we recall the basics of multifractal analysis: we show that the scaling function can be given an alternative interpretation in terms of the pointwise smoothness of the signal. This interpretation has proved particularly important for several reasons: It has allowed the introduction of other scaling functions, which are better suited for that purpose, and it also allowed to extend the scaling function to negative values of p, see [9], which proved particularly important for some classification problems, where the difference between several possible models can only be drawn for negative p's. We will focus on the *wavelet leader scaling function* which now plays a key-role in several fields of applications because it is mathematically well understood, numerically stable, and can be coupled with powerful statistical tests.

In Section 6 we show that this method cannot be directly used in image processing because it assumes that the function studied is bounded, and such a requirement is usually not a valid framework in image analysis. Therefore, one has to perform first a preprocessing which associates to the image another bounded function; this association should be one-to-one in order to lose no information, and should retain as much as possible the relevant features of the image. A standard way to solve this problem is to perform a *fractional integration* of large enough order. However, in practice, this is difficult to realize; therefore, we introduce the notion of *pseudo-fractional integration* which is numerically simple, and retains the same qualitative properties. We investigate how this affects the multifractal properties of the image, and we give a general condition, which is usually met in mathematical models, under which these properties can be exactly determined.

2 Wavelet Bases

Recall that $L^2(\mathbb{R}^d)$ is the space of square-integrable functions, i.e. of functions satisfying

$$\int_{\mathbb{R}^d} |f(x)|^2 dx < \infty.$$

It is endowed with the norm

$$\| f \|_2 = \left(\int_{\mathbb{R}^d} |f(x)|^2 dx \right)^{1/2}.$$

Historically, the first wavelet basis was introduced by A. Haar in 1909. He noticed that, if $\psi = 1_{[0,1/2)} - 1_{[1/2,1)}$, then the collection of the function 1 and the $\psi_{j,k} = 2^{j/2}\psi(2^j x - k)$ for $j \geq 0$ and $k = 0, \cdots, 2^j - 1$ form an orthonormal basis of $L^2([0,1])$, and this irregular basis (its elements have discontinuities) nonetheless displays some better properties than the trigonometric system: If f is a continuous function, then the partial sums of the reconstruction converge uniformly to f. The next wavelet basis, which has the same simple algorithmic form, was introduced by J. Strömberg in the 1980s: he constructed functions ψ, which can be arbitrarily smooth, and so that the wavelet basis generated allows to decompose functions of arbitary smoothness, or, by duality, distributions. An important feature noticed by Strömberg, and which will play a key role in the following, is that therefore the same wavelet basis can be used in order to analyse functions or distributions, without any a priori assumption on their regularity, and on the function spaces to which they belong. The "rule of thumb" is that the wavelet expansion of f will converge in "most" function spaces that actually contain f, if the wavelets are smooth enough. This is particularly important in signal and image processing, where smoothness properties can vary significantly from one type of image to another, and therefore the analysis tool should not imply unnecessary a priori assumptions on the data, since their regularity is unknown (actually, one of our purposes will precisely be to determine regularity indices in scales of function spaces).

We will now recall the algorithmic form of wavelet bases, in particular in several dimensions. We refer to [5, 10, 11] for detailed expositions of the construction of such bases.

Orthonormal wavelet bases on $\mathbb{R}^d$ are of the following form: There exists a function $\varphi(x)$ and 2^d-1 functions $\psi^{(i)}$ with the following properties: The functions $\varphi(x-k)$ $(k \in \mathbb{Z}^d)$ and the $2^{dj/2}\psi^{(i)}(2^jx-k)$ $(k \in \mathbb{Z}^d,\ j \in \mathbb{Z})$ form an orthonormal basis of $L^2(\mathbb{R}^d)$. This basis is r-smooth if φ and the $\psi^{(i)}$ have partial derivatives up to order r and if the $\partial^\alpha\varphi$, and the $\partial^\alpha\psi^{(i)}$, for $|\alpha| \le r$, have fast decay.

Therefore, $\forall f \in L^2$, we have the following decomposition

$$f(x) = \sum_{k\in\mathbb{Z}^d} C_k\varphi(x-k) + \sum_{j=0}^{\infty}\sum_{k\in\mathbb{Z}^d}\sum_{i} c^i_{j,k}\psi^{(i)}(2^jx-k); \tag{1}$$

the $c^i_{j,k}$ are the wavelet coefficients of f:

$$c^i_{j,k} = 2^{dj}\int_{\mathbb{R}^d} f(x)\psi^{(i)}(2^jx-k)dx, \tag{2}$$

and

$$C_k = \int_{\mathbb{R}^d} f(x)\varphi(x-k)dx. \tag{3}$$

Note that, in (1), we do not use the L^2 normalisation for the wavelets, but a normalisation which is better fitted to the definition of the wavelet leaders that we will give below.

Formulas (2) and (3) make sense even if f does not belong to L^2; indeed, if one uses smooth enough wavelets, they can be interpreted as a duality product betweeen smooth functions (the wavelets) and distributions.

We will use more compact notations for indexing wavelets. Instead of using the three indices (i,j,k), we will use dyadic cubes. Since i takes 2^d-1 values, we can assume that it takes values in $\{0,1\}^d-(0,\ldots,0)$; we introduce:

- $\lambda\ (=\lambda(i,j,k))\ = \dfrac{k}{2^j} + \dfrac{i}{2^{j+1}} + \left[0, \dfrac{1}{2^{j+1}}\right)^d.$

- $c_\lambda = c^i_{j,k}$
- $\psi_\lambda(x) = \psi^{(i)}(2^j x - k)$.

The wavelet ψ_λ is essentially localized near the cube λ; more precisely, when the wavelets are compactly supported

$$\exists C > 0 \quad \text{such that} \quad \forall i, j, k, \qquad supp(\psi_\lambda) \subset C \cdot \lambda$$

(where $C \cdot \lambda$ denotes the cube of same center as λ and C times wider). Finally, Λ_j will denote the set of dyadic cubes λ which index a wavelet of scale j, i.e. wavelets of the form $\psi_\lambda(x) = \psi^{(i)}(2^j x - k)$.

Among the many families of wavelet bases that exist, two will prove particularly useful:

- Lemarié-Meyer wavelets: φ and $\psi^{(i)}$ both belong to the Schwartz class, see [11].
- Daubechies wavelets: the functions φ and $\psi^{(i)}$ can be chosen arbitrarily smooth and with compact support, see [5].

Finally, note that in practice one never needs to compute integrals in order to determine the wavelet coefficients of a signal or a function. There exist fast decomposition and reconstruction algorithms which allow to compute the coefficients via discrete convolutions (filtering algorithms). These algorithms were discovered by S. Mallat: They are a consequence of the method of construction of wavelet bases, see [5, 10].

3 Image Processing: The Function Space Approach

Image processing often requires a priori assumptions, which amount to deciding that the image considered belongs to a given function space.

A standard approach consists of assuming that the relevant information in an image can be modeled by a "cartoon", which is composed of piecewise smooth parts separated by discontinuities along piecewise smooth curves. This is typical of photographs taken inside buildings, when no texture is involved. Note that natural images rarely follow this assumption, since most objects are textured and often have "fractal" edges (e.g. trees, clouds, mountains,...). However, the assumption of discontinuities along (not necessarily smooth) lines is mandatory in image processing, because of the *occlusion phenomenon*: one object can be partially hidden behind another; therefore, this "cartoon model" is the smoothest one we can expect in practice. It is easy to associate a function space to such a model. Indeed, the gradient of a cartoon will be smooth, except along the lines of discontinuities, where Dirac masses will appear along those lines. Therefore the gradient will be a *bounded measure*.

The space of functions whose gradient is a bounded measure is called BV (for "bounded variation"). Note however that modeling using the space BV does not entirely recapture the essence of the cartoon model, since a cartoon necessarily is a bounded function and, in dimension 2, a function in BV can be unbounded: The reader will easily check that singularities which behave locally like $|x - x_0|^{-\alpha}$ for $\alpha < 1$ can occur. Therefore the alternative space $BV \bigcap L^\infty$ is often proposed (recall that L^∞ is the space of bounded functions).

Real-life images never are cartoons, since they always contain some parts with either rough boundaries, textures or noise. A standard assumption is that they can be modeled as a sum of a function $u \in BV$ and another term v which will model the noise and texture parts. There is much less consensus on which regularity should be assumed for the second term v. The first "$u+v$ model" (introduced by Rudin, Osher and Fatemi in 1992 [14]) assumed that this part belongs to L^2; however, the very strong oscillations displayed by some textures have suggested that such components do not have a small L^2 norm, but might have a small norm in spaces of negative regularity index (i.e. spaces of distributions). Therefore the use of spaces such as divergences of L^∞ functions (or divergences of L^2 functions) were proposed (note that, here again, derivatives have to be taken in the sense of distributions), initially by Y. Meyer, see [12], and then by several other authors, see [4, 13] and references therein. More sophisticated models also aim to separate the noise from the texture, and therefore propose to split the image into three components ($u+v+w$ models, see [4]). All these methods are minimization algorithms based on the assumption that each of these components belongs to a different function space.

The Rudin-Osher-Fatemi algorithm proposed to extract the cartoon component u by minimizing the functional

$$J(u) = \| u \|_{BV} + t \| f - u \|_2^2,$$

where f is the initial image, and t is a scale parameter which has to be tuned.

In 2001, Y. Meyer proposed to minimize the alternative functional

$$J(u) = \| u \|_{BV} + t \| f - u \|_G,$$

where

$$\| f \|_G = \inf_{g:\, f = \nabla \cdot g} \| g \|_\infty .$$

More recently, in 2003, Osher, Solé and Vese proposed another model which recaptures the same fundamental idea but uses for the texture and noise component a space of distributions easier to handle, the Sobolev space H^{-1}, generated by partial derivatives of order 1 of L^2 functions. The corresponding functional is

$$J(u) = \| u \|_{BV} + t \| f - u \|_{H^{-1}}^2 .$$

Several alternatives have been more recently proposed, based on the same fundamental ideas, but using other function spaces. However the relevance of one particular function space is usually advocated using either theoretical arguments derived from functional analysis, or practical arguments motivated by the algorithmic implementation. The fundamental problem of determining to which function spaces a given image (or a part of a given image) belongs has been rarely considered. (See however [7] where the authors question the fact that natural images belong to BV, and actually answer in the negative.) The resolution of this problem is justified by several reasons. A first motivation rises implicitly from the short review we just performed: The function spaces used in modeling should fit the data. Another motivation is that, if these function spaces depend strongly on the image that is considered, then this information might prove useful in image classification. This second motivation is at the origin of multifractal analysis. Before describing the functional information supplied by multifractal analysis, we turn to another fundamental question in function-space modeling: Can one find a "natural" function space which a priori contains all images?

Without any assumption, we can of course safely adopt the widest possible mathematical setting, which is supplied by distributions. However, the physical procedure through which an image is captured tells us that it is a local average of the light intensity, and therefore is a nonnegative quantity. Therefore an image is a positive distribution; but a famous theorem of L. Schwartz asserts that positive distributions necessarily are bounded measures. Therefore the setting supplied by bounded measures seems to be a conservative option for the choice of a "universal" space that would contain all possible natural images.

4 The Wavelet Scaling Function

The first seminal ideas that led to mutifractal analysis were introduced by N. Kolmogorov, in the field of fully developed turbulence. Let f be a function $\mathbb{R}^d \longrightarrow \mathbb{R}$. N. Kolmogorov associated to f its *scaling function* which is defined as follows.

Let $p \geq 1$, and assume that, when $h \to 0$,

$$\int |f(x+h) - f(x)|^p dx \quad \sim \quad |h|^{\eta_f(p)}, \tag{4}$$

then $\eta_f(p)$ is the scaling function of f. It can be given a function space interpretation with the help of the Lipschitz spaces $\mathrm{Lip}(s,L^p)$: Let $s \in (0,1)$, and $p \in [1,\infty]$; f belongs to $\mathrm{Lip}(s,L^p(\mathbb{R}^d))$ if $f \in L^p$ and

$$\exists C > 0,\ \forall h, \qquad \| f(x+h) - f(x) \|_p \leq C|h|^s. \tag{5}$$

It follows from this definition that, if $\eta_f(p) < p$,

$$\eta_f(p) = \sup\{s : f \in \mathrm{Lip}(s/p, L^p(\mathbb{R}^d))\}. \tag{6}$$

The initial definition given by Kolmogorov is difficult to use in practice, and suffers from mathematical restrictions. An obvious one is that we have to assume the precise scaling law (4); we also have to assume that f is a function, and we saw that we actually want to analyze larger classes of mathematical objects (spaces of measures, and distributions); finally, we want to derive the scaling function from the wavelet coefficients of f, through a simple formula. One solution is to extend the characterization of the scaling function by using Besov spaces instead of Lipschitz spaces. The easiest way to define Besov spaces is through their wavelet characterization (we assume that the wavelet basis used is smooth enough).

Let $p \in (0,\infty)$; a function f belongs to the Besov space $B^s_p(\mathbb{R}^d)$ (also referred to as $B^{s,\infty}_p(\mathbb{R}^d)$) if and only if $(C_k) \in l^p$ and

$$\exists C, \forall j, \qquad \sum_{\lambda \in \Lambda_j} \left[2^{(s-d/p)j} |c_\lambda| \right]^p \leq C. \tag{7}$$

We will pay special attention to the case $p = +\infty$: f belongs to $B^s_\infty(\mathbb{R}^d)$ if and only if $(C_k) \in l^\infty$ and

$$\exists C, \forall \lambda, \qquad |c_\lambda| \leq C 2^{-sj}. \tag{8}$$

The spaces B^s_∞ coincide with the uniform Lipschitz spaces $C^s(\mathbb{R}^d)$; for instance, if $0 < s < 1$, an equivalent definition is given by: $f \in L^\infty$ and

$$\exists C, \ \forall x,y \qquad |f(x) - f(y)| \leq C|x-y|^s.$$

The *uniform Hölder exponent* of f is

$$H^{min}_f = \sup\{s : \ f \in C^s(\mathbb{R}^d)\}; \tag{9}$$

it yields an additional parameter for image processing and classification that will prove important in the following.

The embeddings between Besov and Lipschitz spaces imply that, if f is an L^1 function such that $\eta_f(p) < p$, then its scaling function can be defined indifferently using the Besov or Lipschitz scales:

$$\eta_f(p) = \sup\{s : f \in B^{s/p}_p\}. \tag{10}$$

Let

$$S_f(p,j) = 2^{-dj} \sum_{\lambda \in \Lambda_j} |c_\lambda|^p$$

then

$$\eta_f(p) = \liminf_{j \to +\infty} \frac{\log\left(S_f(p,j)\right)}{\log(2^{-j})}, \tag{11}$$

which follows immedialtely from (10). This formula has practical implications: it allows to compute the scaling function through a linear regression on a log-log plot. Figure 1 (top right) shows an example of a wavelet scaling function for a real-world image.

Note that the uniform Hölder exponent of f can be derived from the scaling function

$$H^{min}_f = \lim_{p \to +\infty} \eta'_f(p);$$

it can also be derived directly from the wavelet coefficients of f; indeed, it follows from (9) and the wavelet characterization of the Besov spaces B^s_∞ that, if

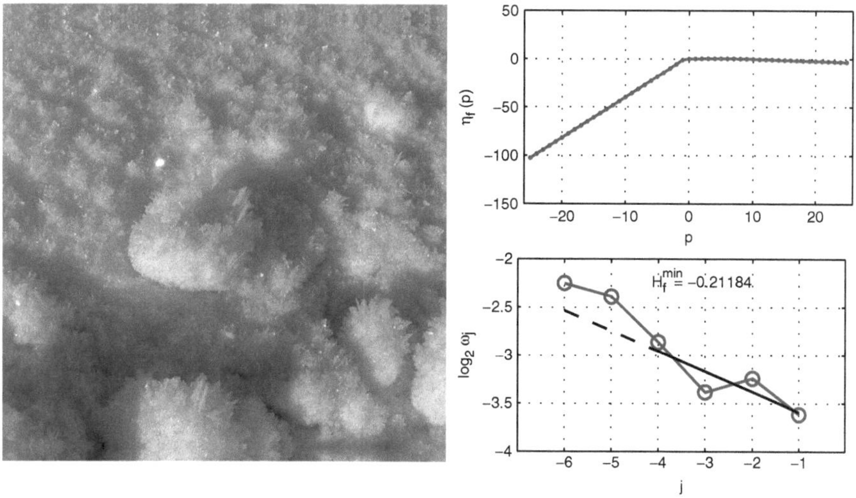

Fig. 1 Image of snow (*left*), wavelet scaling function $\eta_f(p)$ (*top right*) and uniform Hölder exponent H_f^{min} (*bottom right*). Their respective estimated values are $\eta_f(1) = 0.254$, $\eta_f(2) = 0.412$ and $H_f^{min} = -0.212$.

$$\omega_j = \sup_{\lambda \in \Lambda_j} |c_\lambda|,$$

then

$$H_f^{min} = \liminf_{j \to +\infty} \frac{\log(\omega_j)}{\log(2^{-j})}. \tag{12}$$

This is illustrated in Figure 1 (bottom right).

The derivation of the scaling function through (11) has several advantages:

- Since Besov spaces are defined for $p > 0$, it makes sense for $p \in (0,1)$ whereas Lipschitz spaces are not defined for $p < 1$. This yields an additional useful range of values for classification.
- It does not make any a priori assumption of the regularity of f, which can be a measure or even a distribution.
- It allows for an easy numerical implementation.

The knowledge of the scaling function allows to settle the issues we raised concerning the function spaces which contain a given image. For instance, the embeddings between the Besov spaces and the other classical function spaces have the following consequences:

Proposition 1. *Let f be a distribution defined on $\mathbb{R}^2$. The values taken by the scaling function at 1, 2 and $+\infty$ have the following implications:*

- *If $\eta_f(1) > 1$, then $f \in BV$, and if $\eta_f(1) < 1$, then $f \notin BV$*
- *If f is a measure, then $\eta_f(1) \geq 0$, and, if $\eta_f(1) > 0$, then f belongs to L^1.*
- *If $\eta_f(2) > 0$, then $f \in L^2$ and if $\eta_f(2) < 0$, then $f \notin L^2$.*

- *If* $\eta_f(2) > -2$, *then* $f \in H^{-1}$ *and if* $\eta_f(2) < -2$, *then* $f \notin H^{-1}$.
- *If* $H_f^{min} > 0$, *then* f *is bounded and continuous, and if* $H_f^{min} < 0$, *then* $f \notin L^\infty$.
- *If* $H_f^{min} > -1$, *then* $f \in G$ *and if* $H_f^{min} < -1$, *then* $f \notin G$.
- *If* f *is a measure, then* $H_f^{min} \geq -2$.

Most of these statements are easy consequences of standard function space embeddings. The second one is particularly important for the validation of many models. Indeed, in several fields of applications, models which are singular measures are used. Since they are measures, it follows that $\eta_f(1) \geq 0$, and since they are not L^1 functions, $\eta_f(1) \leq 0$. It follows that they must necessarily satisfy $\eta_f(1) = 0$, a sharp requirement which has the widest range of validity (it is completely non-parametric, i.e. does not make the assumption that the measure has a particular form) and it can be checked on real-life data in order to validate those models.

We only prove the first assertion which concerns measures because of the particular importance of this result (the other assertions have similar proofs). It is a direct consequence of the following lemma.

Lemma 1. *Let* μ *be a bounded measure on* $\mathbb{R}^d$*; then its wavelet coefficients* $\mu_{j,k}$ *satisfy*

$$\exists C\, \forall j, \qquad 2^{-dj} \sum_{\lambda \in \Lambda_j} |c_\lambda| \leq C. \tag{13}$$

Conversely, if μ *satisfies the slightly stronger requirement*

$$\exists C \qquad \sum_j 2^{-dj} \sum_{\lambda \in \Lambda_j} |c_\lambda| \leq C, \tag{14}$$

then μ *is an* L^1 *function.*

Proof of Lemma 1: Recall that a bounded measure μ is a linear form on the space of continuous bounded functions, i.e. satisfies

$$|\langle f | d\mu \rangle| \leq C \parallel f \parallel_\infty$$

for any continuous bounded function f.

Denote by c_λ the wavelet coefficients of μ, and by ε_λ their signs (with the convention that $sign(x) = 0$ if $x = 0$). Let

$$f_j = \sum_{\lambda \in \Lambda_j} \varepsilon_\lambda \psi_\lambda.$$

On one hand,

$$\langle f_j | d\mu \rangle = \sum_{\lambda \in \Lambda_j} \varepsilon_\lambda c_\lambda 2^{-dj} = 2^{-dj} \sum_{\lambda \in \Lambda_j} |c_\lambda|;$$

but, on the other hand,

$$\langle f_j | d\mu \rangle \leq C \parallel f_j \parallel_\infty \leq C',$$

it follows that (13) holds.

Conversely, suppose that (14) holds. Then

$$\| \sum_j \sum_{\lambda \in \Lambda_j} c_\lambda \psi_\lambda \|_1 \leq \sum_j \sum_{\lambda \in \Lambda_j} |c_\lambda| \, \| \psi_\lambda \|_1 \leq C \sum_j \sum_{\lambda \in \Lambda_j} |c_\lambda| 2^{-dj} < +\infty.$$

So that the wavelet series of f converges normally in L^1, so that $f \in L^1$.

Using a wavelet formula for the obtention of the scaling function has additional advantages. Up to now, we implicitly assumed that images are functions (or perhaps distributions) defined on $\mathbb{R}^2$ (or a subset of $\mathbb{R}^2$ such as a square or a rectangle). Of course, this is an idealization that we used because it is convenient for mathematical modeling. However, real-life images are sampled and given by a finite array of numbers (usually of size $1,024 \times 1,024$). This practical remark has an important consequence: The problem that we just raised is ill-posed. Indeed, given any "classical" space of functions defined on a square, and such an array of numbers, one can find a function in this space that will have the preassigned values at the corresponding points of the grid. In other words, paradoxically, any function space could be used. Let us however show extreme consequences of this simple remark.

Recall that the Fourier transform of a function $f(x_1,x_2)$ is defined by

$$\hat{f}(\xi_1,\xi_2) = \int_{\mathbb{R}^2} f(x_1,x_2) e^{-i(x_1\xi_1+x_2\xi_2)} dx_1 dx_2.$$

One can, for instance, assume that images are *band-limited* which means that their Fourier transforms vanish outside a ball centered at 0, and whose radius is proportional to the inverse of the sampling width (according to Shannon's theorem); note that this assumption is often made, in particular in deblurring and denoising algorithms. This assumption implies that the model used is composed of C^∞ functions; however it would lead to incompatibilities, for instance if we want to use a realistic model which includes discontinuites along edges (which, as we saw, is a natural requirement).

Another commonly met pitfall is that an image is given by grey-levels, and thus takes values in $[0,1]$. Therefore, it may seem appropriate to use a modeling by bounded functions, and this is indeed a classical assumption (note that the "cartoon model" clearly implies boundedness). We will see that the wavelet techniques we introduced allow to discuss this assumption, and show that it is not satisfied for most images.

The resolution of the paradox we raised in this section requires the use of *multiscale techniques* such as the one supplied by wavelet analysis. Let us consider for instance the last example we mentioned: Starting with a discrete image, given by an array of $1,024 \times 1,024$ numbers all lying between 0 and 1, how can we decide if it can be modeled or not by a bounded function? It is clear that, if we consider the image at only one scale (the finest scale in order to lose no information), then the answer seems to be affirmative. However, as mentioned earlier, any other space would also do. One way to solve the difficulty is to consider the image at all the scales available (in theory, there are ten of them, since $1,024 = 2^{10}$) and inspect if certain quantities behave **through this range of scales** as is the case for a bounded function. If not, we can give an unexpected negative answer to our problem, but this negative answer should however be understood as follows:

The image considered is a discretization at a given scale of a "hidden function" defined on a square (to which we have no access) and, if the scaling properties of this "hidden function" are, at all scales, the same ones as we observe in the range of scales available, then it is not bounded.

The recipe in order to settle this point is the following: one uses (12) in order to determine numerically the value of H_f^{min},which is done by a regression on a log-log plot, and using Proposition 1, it follows that, if $H_f^{min} < 0$, then the image is not bounded, and if $H_f^{min} > 0$, then the image is bounded. Of course, if the numerical value obtained for H_f^{min} is close to 0 (i.e. if 0 is contained in the confidence interval which can be obtained using statistical methods, see [15, 16]) then the issue remains unsettled.

The same method holds for the other classical function spaces, as a consequence of Proposition 1. More generally, it allows to determine if the image belongs to a given function space A_p^s, as soon as this space has "close embeddings" with Besov spaces, see [2, 15]; this means that

$$\forall \varepsilon > 0, \qquad B_p^{s+\varepsilon} \hookrightarrow A_p^s \hookrightarrow B_p^{s-\varepsilon}.$$

This includes for instance Sobolev spaces, Hardy spaces or Triebel-Lozorkin spaces. Note that, of course, one can consider spaces with non-integer integrability exponent p and non-integer smoothness index.

5 The Leader Scaling Function

In the mid-1980s, two physicists, U. Frisch and G. Parisi proposed an interpretation of the scaling function in terms of the pointwise Hölder singularities of the function considered, see [6]; this interpretation had a wide amount of consequences, see [2,3] and references therein: It gave a deep insight into the understanding of the information contained in the scaling function, and it led to the introduction of new scaling functions which are better fitted for that purpose. The one we will describe in this section is the only one which meets the two following requirements: Its mathematical properties are well understood and its numerical implementation is easy, in any space dimension, see [1, 8].

We start by recalling the mathematical definitions related to pointwise Hölder regularity.

Definition 1. Let f be a bounded function $\mathbb{R}^d \to \mathbb{R}$, $x_0 \in \mathbb{R}^d$ and let $\alpha \geq 0$; f belongs to $C^\alpha(x_0)$ if there exist $C > 0$ and a polynomial P of degree less than α such that

$$|f(x) - P(x - x_0)| \leq C|x - x_0|^\alpha.$$

The Hölder exponent of f at x_0 is

$$h_f(x_0) = \sup\{\alpha : \; f \in C^\alpha(x_0)\}.$$

The isohölder sets are

$$E_H = \{x_0 : \quad h_f(x_0) = H\}.$$

Note that Hölder exponents met in signal processing often lie between 0 and 1, in which case the Taylor polynomial $P(x-x_0)$ boils down to $f(x_0)$ and the definition of the Hölder exponent means that, heuristically,

$$|f(x) - f(x_0)| \sim |x - x_0|^{h_f(x_0)}.$$

U. Frisch and G. Parisi suggested that the scaling functions yield information concerning sizes of the isohölder sets. These sizes are measured with the help of *Hausdorff dimensions*, which we recall.

Definition 2. Let $E \subset \mathbb{R}^d$ and $\alpha > 0$. Let us introduce the following quantities : Let $n \in \mathbb{N}$; if $L = \{l_i\}_{\, i \in \mathbb{N}}$ is a countable collection of dyadic cubes of width smaller than 2^{-n} which forms a covering of E, then let

$$\mathscr{H}_n^\alpha(E,L) = \sum_{i \in \mathbb{N}} diam\,(l_i)^\alpha, \quad \text{and} \quad \mathscr{H}_n^\alpha(E) = \inf\left(\mathscr{H}_n^\alpha(E,L)\right),$$

where the infimum is taken over all possible coverings of E by dyadic cubes of scales at least n. The α-dimensional Hausdorff measure of E is

$$\mathscr{H}^\alpha(E) = \lim_{n \to +\infty} \mathscr{H}_n^\alpha(E).$$

The Hausdorff dimension of E is

$$dim\,(E) = \sup\{\alpha > 0\,;\; \mathscr{H}^\alpha(E) = +\infty\} = \inf\{\alpha > 0\,;\; \mathscr{H}^\alpha(E) = 0\}\;.$$

If E is empty then, by convention, $dim_H\,(E) = 0$.

If f is bounded, the function $H \to dim(E_H)$ is called the *spectrum of singularities* of f.

A *uniform Hölder function* is a function satisfying $H_f^{min} > 0$. In particular, it is continuous. One can prove the following relationship between the scaling function of a function and its pointwise Hölder singularities, see [8].

Theorem 1. *Let $f : \mathbb{R}^d \to \mathbb{R}$ be a uniform Hölder function. Then*

$$dim(E_H) \leq \inf_{p > p_0} \left(d + Hp - \eta_f(p)\right),$$

where p_0 is such that $\eta_f(p_0) = dp_0$.

We will introduce an alternative scaling function for which a stronger relationship with the spectrum of singularities can be proved. Its definition is similar to the wavelet scaling function, except that wavelet coefficients have to be replaced by wavelet leaders, which are defined as follows.

Let λ be a dyadic cube; 3λ is the cube of same center and three times wider. If f is a bounded function, the *wavelet leaders* of f are the quantities

$$d_\lambda = \sup_{\lambda' \subset 3\lambda} |c_{\lambda'}|.$$

Let $x_0 \in \mathbb{R}^d$; $\lambda_j(x_0)$ is the dyadic cube of width 2^{-j} which contains x_0; and

$$d_j(x_0) = d_{\lambda_j(x_0)} = \sup_{\lambda' \subset 3\lambda_j(x_0)} |c_{\lambda'}|.$$

It is important to require f to be bounded; otherwise, the wavelet leaders of f can be infinite. The reason for introducing wavelet leaders is that they give information on the pointwise Hölder regularity of the function. Indeed, one can show that (see [8] and references therein) if f is a uniform Hölder function, then

$$h_f(x_0) = \liminf_{j \to +\infty} \left(\frac{\log(d_j(x_0))}{\log(2^{-j})} \right).$$

Therefore, it is clear that a scaling function constructed with the help of wavelet leaders will incorporate pointwise smoothness information. For any $p \in \mathbb{R}$, let

$$T_f(p,j) = 2^{-2j} \sum_{\lambda \in \Lambda_j} |d_\lambda|^p.$$

The *leader scaling function* is defined by

$$\forall p \in \mathbb{R}, \qquad \zeta_f(p) = \liminf_{j \to +\infty} \frac{\log(T_f(p,j))}{\log(2^{-j})}.$$

An important property of the leader scaling function is that it is "well defined" for $p < 0$, which is not the case for the wavelet scaling function. By "well defined", we mean that it has the following robustness properties if the wavelets belong to the Schwartz class (they still partly hold otherwise, see [2, 8]):

- ζ_f is independent of the wavelet basis.
- ζ_f is invariant under the addition of a C^∞ perturbation.
- ζ_f is invariant under a C^∞ change of variable.

Note that the wavelet scaling function does not possess these properties when p is negative.

The leader scaling function can also be given a function-space interpretation for $p > 0$. Let $p \in (0, \infty)$; a function f belongs to the *Oscillation space* $\mathcal{O}_p^s(\mathbb{R}^d)$ if and only if $(C_k) \in l^p$ and

$$\exists C, \forall j, \qquad \sum_{\lambda \in \Lambda_j} \left[2^{(s-d/p)j} d_\lambda \right]^p \leq C.$$

Then

$$\zeta_f(p) = \sup\{s : f \in \mathcal{O}_p^{s/p}.$$

Properties of oscillation spaces are investigated in [2, 8].

We denote by $\mathscr{L}u$ the Legendre transform of a concave function u, i.e.

$$(\mathscr{L}u)(H) = \inf_{p\in\mathbb{R}} (d + Hp - u(p)).$$

The *leader spectrum* of f is defined through a Legendre transform of the leader scaling function as follows

$$L_f(H) = (\mathscr{L}\zeta_f)(H).$$

Of course, the leader spectrum of f has the same robustness properties as the leader scaling function.

Theorem 2. *If f is uniform Hölder then,*

$$\forall H, \qquad dim(E_H) \leq L_f(H).$$

We already saw that the cartoon assumption implies that $f \in BV \cap L^\infty$. We can actually get a sharper result which yields the exact scaling functions of cartoons for $p > 0$.

Lemma 2. *Let f be a piecewise smooth function with discontinuities along piecewise smooth curves. Then its wavelet and leader scaling functions are given by*

$$\forall p > 0, \qquad \eta_f(p) = \zeta_f(p) = 1.$$

This result gives a numerically sharp and simple way to decide if the cartoon assumption is satisfied for an image.

Proof: We use compactly supported wavelets, and we first compute the contribution of the wavelet coefficents such that the support of the wavelet intersects the curves of dicontinuities. There are $\sim C2^j$ such coefficients, and the size of these coefficients are $\sim C$. It follows that

$$2^{-2j} \sum |c_\lambda|^p \sim C2^{-j}.$$

The contribution of the other wavelet coefficients is negligible, because they decay faster than 2^{-Aj} for any $A > 0$.

It also follows that the wavelet leaders are of the same order of magnitude. Hence the lemma holds.

As stated above, we can use wavelet leaders only if the function considered is bounded, and the mathematical results we mentioned only hold under the slightly stronger property that the function considered is uniform Hölder. Note however that we do not expect this assumption to be usually satisfied for images, since it implies continuity, an assumption which, as already stated, is not realistic in image processing. Recall however that the condition $H_f^{min} > 0$ (which is the definition of uniform hölderianity) can be practically checked, and inspection of image databases shows that, indeed, images quite often have negative H_f^{min}, which shows the necessity of a modification of the computation of the leader-based scaling function for practical purposes.

6 Multifractal Formalism for Unbounded Functions and Measures

In order to be able to use the wavelet leader-based method described above, one has to associate to the image a bounded function, in a one-to-one way in order to lose no information; furthermore, this association should retain as much as possible the relevant features of the image. For instance, it should keep the locations of the Hölder singularities, and transform the wavelet scaling function in a simple way. In one dimension, the simplest way to solve this problem is to perform an integration of the function. If one starts with a bounded measure, it is clear that one will obtain in this way a bounded function; thus, at most two integrations will be sufficient in order to obtain a uniform Hölder function. In dimension larger than one, the natural substitute is given by fractional integration, which we now describe. Note that, even in dimension 1, the tool supplied by fractional integration can prove useful, since it allows to tune the order of integration, which need not be an integer.

In dimension 1, taking a derivative of order $s \in \mathbb{N}$ amounts to multiplying the Fourier transform of the function by $(i\xi)^s$; therefore, the inverse operator (integration of order n) amounts to dividing the Fourier transform by $(i\xi)^s$. This may pose a problem if the Fourier transform does not vanish at the origin, therefore, one prefers to use the alternative operator, I^s defined by

$$\widehat{I^s(f)} = (1+|\xi|^2)^{-s/2}\hat{f}(\xi);$$

indeed, it has the same behavior at high frequencies, but does not have the drawback we mentioned; another advantage of this definition is that it immediately extends to non-integer values of s. The operator I^s is the *fractional integration* of order s.

Let us recall a few simple properties of I^s which show that it is relevant for our purpose.

First, the uniform regularity exponent H_f^{min} is always shifted exactly by s:

$$\forall f, \qquad H_{I^s(f)}^{min} = H_f^{min} + s.$$

This simple property shows a possible strategy we can follow in order to perform the multifractal analysis of an image which is not bounded: First determine its exponent

H_f^{min}, then, if $H_f^{min} < 0$, perform a fractional integration of order $s > -H_f^{min}$; it follows that the uniform regularity exponent of $I^s(f)$ is positive, and therefore its leader scaling function is well defined. This is essentially the strategy we will follow except for a slight modification which will allow us to eliminate the numerical computation of the fractional integration.

The pointwise Hölder exponent of a function f is shifted by an amount larger than or equal to s under a fractional integration of order s:

$$\text{if} \quad s > 0, \qquad h_{I^s(f)}(x_0) \geq h_f(x_0) + s.$$

We usually expect this Hölder exponent to be exactly shifted by s. This is the case for Hölder singularities of *cusp-type*, i.e. such that

$$|f(x) - f(x_0)| \sim |x - x_0|^{\alpha}.$$

However, this is not the case if the singularity has strong oscillations near x_0, such as for the *chirp functions*

$$|x - x_0|^{\alpha} \sin\left(\frac{1}{|x - x_0|^{\beta}}\right).$$

We will give a simple sufficient condition under which the function has no chirp and the fractional integrals satisfy

$$\forall x_0, \ \forall s > 0, \qquad h_{I^s(f)}(x_0) = h_f(x_0) + s.$$

The wavelet scaling function is always tranformed in a simple way under the action of a fractional integration:

$$\forall p > 0, \qquad \eta_{I^s(f)}(p) = \eta_f(p) + sp.$$

Note that such a transformation is easier to check on the Legendre tranforms, since it implies that

$$\mathscr{L}(\eta_{f^{-s}})(H) = \mathscr{L}(\eta_f)(H - s)$$

(the spectrum is shifted under fractional integration). Such simple formulas do not exist for the leader scaling function. In particular, the shape of its Legendre transform can be modified (it is not just shifted) under a fractional integration. This is both an advantage and a drawback; indeed, on one side, it shows that the scaling functions of all fractional integrals contain non-redundant information. On the other hand, there is no canonical way to pick a particular order of fractional integration in order to perform the multifractal analysis.

However, numerically, a fractional integration in a bounded domain is difficult to realize; In practice, it is equivalent to perform a *pseudo-fractional integration* which is numerically simple, and retains the same properties: its scaling functions and pointwise exponents are the same as for a fractional integral. Let us first define this transform.

Let f be a function, or a distribution, with wavelet coefficients c_λ, and let ψ_λ be a given wavelet basis. The *pseudo-fractional integral* of f of order s, denoted by $\tilde{I}^s(f)$, is the function whose wavelet coefficients on the same wavelet basis are

$$\tilde{c}_\lambda = 2^{-sj} c_\lambda.$$

Therefore, one obtains the pseudo-fractional integral by just multiplying the wavelet coefficients of f by 2^{-sj}.

Theorem 3. *The following properties hold for any function or distribution f:*

- *For any $s \in \mathbb{R}$, the wavelet scaling functions of $I^s(f)$ and $\tilde{I}^s(f)$ coincide.*
- *If $s > -H_f^{min}$ then, the leader scaling functions of $I^s(f)$ and $\tilde{I}^s(f)$ coincide.*
- *If $s > -H_f^{min}$ then*

$$\forall x_0, \qquad h_{I^s(f)}(x_0) = h_{\tilde{I}^s(f)}(x_0).$$

The strategy in order to perform a multifractal analysis of a distribution is, this is illustrated in Figure 2 First determine its uniform Hölder exponent H_f^{min}, then compute the leader scaling function associated to $\tilde{I}^s(f)$ for an $s > -H_f^{min}$, i.e. based on the "pseudo-leaders"

$$\tilde{d_\lambda} = \sup_{\lambda' \subset 3\lambda} 2^{-sj'} |c_{\lambda'}|,$$

finally, compute the Legendre transform of this scaling function. If the function considered has cusp singularities only, then we expect that

$$\mathscr{L}(\zeta_{\tilde{I}^s(f)})(H) = \mathscr{D}_f(H - s), \tag{15}$$

for a certain function D_f which is independent of s. This allows to define a "canonical" spectrum $\mathscr{D}_f(H)$. If it is not the case, then retaining all this collection of trans-

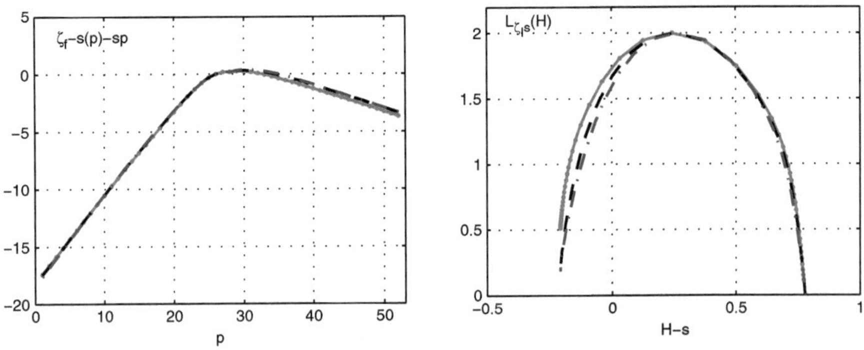

Fig. 2 Leader scaling function (*left*) of the image in Figure 1, obtained with $s = 0.5$. Superposition of $\mathscr{L}(\zeta_{\tilde{I}^s})(H)$ (*right*), obtained from the image in Figure 1 with $s = 0.5$, $s = 0.75$ and $s = 1$.

forms for all values of (large enough) s, yields exhaustive information on the oscillations of f.

We now give a simple condition under which a function has only cusp-type singularities, and therefore (15) holds.

Theorem 4. *Let f be a bounded function. Let $M(\lambda)$ denote the scale j' where the supremum is attained in the definition of the wavelet leaders*

$$d_\lambda = \sup_{\lambda' \subset 3\lambda} |c_{\lambda'}|. \tag{16}$$

If

$$\sup_{\lambda \in \Lambda_j} (M(\lambda) - j) = o(j)$$

then (15) holds, and

$$\forall x_0, \ \forall s > 0, \qquad h_{\tilde{I}^s(f)}(x_0) = h_f(x_0) + s.$$

Proof: Let $\lambda'(\lambda)$ denote the cube where the supremum is attained in (16), and denote by j' its scale. It follows that

$$j \le j' \le j + \omega(j), \qquad \text{where} \qquad \omega(j) = o(j).$$

Let

$$d_\lambda^s = \sup_{\lambda' \subset 3\lambda} |2^{-sj'} c_{\lambda'}|.$$

Since $s > 0$ and $j' \ge j$,

$$d_\lambda^s \le 2^{-sj} \sup_{\lambda' \subset 3\lambda} |c_{\lambda'}| = 2^{-sj} d_\lambda.$$

Let $\varepsilon > 0$. For j large enough, $\omega(j) \le \varepsilon j$, so that

$$d_\lambda^s \ge |2^{-sj'} c_{\lambda'(\lambda)}| = 2^{-sj'} d_\lambda \ge 2^{-s(j+\varepsilon j)} d_\lambda;$$

therefore:

$$2^{-s(j+\varepsilon j)} d_\lambda \le d_\lambda^s \le 2^{-sj} d_\lambda. \tag{17}$$

Since

$$h_{\tilde{I}^s(f)}(x) = \liminf_{j \to +\infty} \frac{\log(d_\lambda^s)}{\log(2^{-j})},$$

it follows from (17) that

$$\forall \varepsilon > 0, \qquad h_f(x) + s \le h_{\tilde{I}^s(f)}(x) \le h_f(x) + s + \varepsilon;$$

so that the second assertion of the theorem follows.

It also follows from (17) that

$$\forall p > 0, \quad 2^{-dj}2^{-spj}\sum(d_\lambda)^p \le 2^{-dj}\sum(d_\lambda^s)^p \le 2^{-dj}2^{-sp(j+\varepsilon j)}\sum(d_\lambda)^p.$$

Therefore

$$\zeta_f(p) + sp \le \zeta_{I^s(f)}(p) \le \zeta_f(p) + sp(1+\varepsilon)$$

and the first assertion of the theorem follows.

References

1. P. Abry, S. Jaffard and B. Lashermes. Wavelet Analysis and Applications, T. Qian et al. eds., *Applied and Numerical Harmonic Analysis Series*, 201–246 Springer, New-York 2006.
2. P. Abry, S. Jaffard,S. Roux, B. Vedel and H. Wendt. The contribution of wavelets in multifractal analysis. Proceedings of the Zuhai Summer School on Wavelets and Applications, *Preprint*, 2008.
3. A. Arneodo, B. Audit, N. Decoster, J.-F. Muzy and C. Vaillant. Wavelet-Based Multifractal Formalism: Applications to DNA Sequences, Satellite Images of the Cloud Structure and Stock Market Data, A. Bunde, J. Kropp, H.J. Schellnhuber Eds., *The Science of Disasters. Springer Berlin.* 27–102, 2002.
4. J.-F. Aujol and A. Chambolle. Dual norms and image decomposition models. *Int. J. Comput. Vis.* 63, 85–104, 2005.
5. I. Daubechies. Ten Lectures on Wavelets. *SIAM.*, 1992.
6. U. Frisch and G. Parisi. Fully developed turbulence and intermittency. *Proc. Int. Summer School on Turbulence and Predictability in Geophysical Fluid Dynamics and Climate Dynamics* North Holland, Amsterdam 84–88, 1985.
7. Y. Gousseau and J.-M. Morel. Are natural images of bounded variation? *SIAM J. Math. Anal.* 33, 634–648, 2001.
8. S. Jaffard. Wavelet techniques in multifractal analysis. *Fractal Geometry and Applications: A Jubilee of Benoît Mandelbrot,* M. Lapidus et M. van Frankenhuijsen eds., *Proceedings of Symposia in Pure Mathematics, AMS*, 91–152, 2004.
9. B. Lashermes, S. Roux, P. Abry and S. Jaffard. Comprehensive multifractal analysis of turbulent velocity using wavelet leaders. *Eur. Phys. J. San Diego, B,* 61(2), 201–215, 2008.
10. S. Mallat. *A Wavelet Tour of Signal Processing. Academic CA*, 1998.
11. Y. Meyer. Ondelettes et Opérateurs. *Hermann.* 1992.
12. Y. Meyer. Oscillating patterns in image processing and nonlinear evolution equations. *University Lecture Series 22*, AMS, 2001.
13. S. Osher, A. Solé and L. Vese. Image decomposition and restoration using total variation minimization and the L^1 norm. *Multiscale Model Simul.*, 1,349–370, 2003.
14. L. I. Rudin, S. Osher and E. Fatemi. Nonlinear total variation based noise removal algorithms. *Physica D*, 60:259–268, 1992.
15. H. Wendt, P. Abry and S. Jaffard. Bootstrap for Emperical Multifractal Analysis. *IEEE Signal Proc. Mag.* 24, 38–48, 2007.
16. H. Wendt, S. Roux P. Abry and S. Jaffard. Bootstrapped wavelet leaders for multifractal analysis of images. *Preprint*, 2008.

Volatile Compounds Detection by IR Acousto-Optic Detectors

Arnaldo D'Amico[1,2], Corrado Di Natale[1,2], Fabio Lo Castro[1], Sergio Iarossi[1], Alexandro Catini[2] and Eugenio Martinelli[2]

Abstract Many important gasses and liquids, including the aggressive or anomalous ones for which our attention is higher, have strong absorption lines in the near and mid infrared spectral range. Infrared sensors exploit the fact that most gasses and liquids present unique infrared signatures in the 2–14 μm wavelength region. Due to this uniqueness infrared sensors provide conclusive identification and measurement of the target sample with little interference from other unwanted volatile compounds. Infrared sensors have the characteristics of being highly accurate, reliable, and, in general, low noise devices. In this chapter we will consider the most important infrared sources and sensors as well as the absorption techniques employed in this context. Furthermore the acousto-optic principle will be presented and discussed in some detail as the promoter of a multi-wavelength infrared generator. Finally system performance and data on gas detection will also be introduced and commented upon.

Keywords: Infrared spectra, near, medium, far infrared intervals, infrared sources, infrared detectors, acousto-optic devices, volatile compounds infrared absorption

1 Optical Radiation

The International Commission on Illumination (CIE) recommended the division of optical radiation into the following three bands: [14, 15].

- IR-A: 700–1,400 nm
- IR-B: 1,400–3,000 nm
- IR-C: 3,000–1 mm

[1]C.N.R. Institute of Acoustics (IDAC), Via del Fosso del Cavaliere 100, 00133 Rome, Italy, email: fabio.locastro@idac.rm.cnr.it, sergio.iarossi@idac.rm.cnr.it
[2]University of Rome "Tor Vergata", Department of Electronic Engineering, Via del Pollitecnico 1, 00133 Rome, Italy, email: damico@eln.uniroma2.it, dinatale@uniroma2.it, catini@ing.uniroma2.it, martinelli@ing.uniroma2.it

J. Byrnes (ed.), *Unexploded Ordnance Detection and Mitigation*,

A commonly used sub-division scheme is:

- Near-infrared (NIR, IR-A DIN): 0.75–1.4 μm in wavelength, defined by the water absorption, and commonly used in fiber optic telecommunication because of low attenuation losses in the SiO_2 glass (silica) medium. Image intensifiers are sensitive to this area of the spectrum. Examples include night vision devices such as night vision goggles.
- Short-wavelength infrared (SWIR, IR-B DIN): 1.4–3 μm, water absorption increases significantly at 1,450 nm. The 1,530–1,560 nm range is the dominant spectral region for long-distance telecommunications.
- Mid-wavelength infrared (MWIR, IR-C DIN) also called intermediate infrared (IIR): 3–8 μm. In guided missile technology the 3–5 μm portion of this band is the atmospheric window in which the homing heads of passive IR "heat seeking" missiles are designed to work, homing on to the IR signature of the target aircraft, typically the jet engine exhaust plume.
- Long-wavelength infrared (LWIR, IR-C DIN): 8–15 μm. This is the "thermal imaging" region, in which sensors can obtain a completely passive picture of the outside world based on thermal emissions only and requiring no external light or thermal source such as the sun, moon or infrared illuminator. Forward-looking infrared (FLIR) systems use this area of the spectrum. Sometimes also called the "far infrared".
- Far infrared (FIR): 15–1,000 μm.

NIR and SWIR is sometimes called reflected infrared while MWIR and LWIR is sometimes referred to as thermal infrared. Due to the nature of the blackbody radiation curves, typical "hot" objects, such as exhaust pipes, often appear brighter in the MW compared to the same object viewed in the LW.

Another scheme is based on the response of various sensors [30]:

- Near Infrared (NIR): from 0.7 to 1.0 μm (from the approximate end of the response of the human eye to that of silicon).
- Short-Wave Infrared (SWIR): from 1.0 to 3 μm (from the cut off of silicon to that of the MWIR atmospheric window). *InGaAs* covers to about 1.8 μ meters; the less sensitive lead salts cover this region.
- Mid-Wave Infrared (MWIR): from 3 to 5 μm (defined by the atmospheric window and covered by Indium antimonide [*InSb*] and $HgCdTe$ and partially by lead selenide [*PbSe*]).
- Long-Wave Infrared (LWIR): from 8 to 12, or from 7 to 14 μm: the atmospheric window (Covered by $HgCdTe$ and microbolometers).
- Very-Long Wave Infrared (VLWIR): from 12 to about 30 μm, covered by doped silicon.

These divisions are justified by the different human responses to this radiation: near infrared is the region closest in wavelength to the radiation detectable by the human eye, mid and far infrared are progressively further from the visible regime. Other definitions follow different physical mechanisms (emission peaks, vs. bands, water absorption) and the newest follow technical reasons (The common silicon sensors are sensitive to about 1,050 nm, while *InGaAs* sensitivity starts around 950 nm and ends between 1,700 and 2,600 nm, depending on the specific configuration). Unfortunately, international standards for these specifications are not currently available. The boundary between visible and infrared light is not precisely defined. The human eye is markedly less sensitive to light above 700 nm wavelength, so shorter frequencies make insignificant contributions to scenes illuminated by common light sources. But particularly intense light (e.g., from lasers, or from bright daylight with the visible light removed by colored gels) can be detected up to approximately 780 nm, and will be perceived as red light. The onset of infrared is defined (according to different standards) at various values typically between 700 and 800 nm.

2 Infrared Sources

2.1 Thermal infrared sources

Thermal sources are resistors of various sorts heated by applying an electric current. Some can be electrically modulated by interrupting the current flow. Others have a larger thermal mass and cannot be modulated effectively at a frequency suitable for most analytical instruments. Black Bodies (B.B.) usually belong to both categories. In fact for many years it has been possible to design and fabricate small B.B. made by thin wires heated at high temperature while, after the advent of micromachining engineering, integrated B.B. have been fabricated and successfully tested. Traditional B.B. have the possibility to deliver high power that can offer modulated B.B. energy through external choppers able to operate mechanically up to a frequency of 5 kHz. Electro-optical choppers can allow an even higher frequency to be reached for particular applications. A B.B. can be considered as a passive structure able to adsorb any radiation frequency and emit a radiation spectra related to its average temperature. Its radiant emittance, or radiance, can be expressed as $R = \epsilon\sigma T$, where ϵ is the emissivity, σ is the Stefan-Boltzman constant equal to: $\sigma = 5.67 \cdot 10^{-8}$ W/(m$^2 \cdot$ K^4) and T is the absolute temperature value.

2.2 IR Light Emitting Diodes (LEDs)

2.2.1 Tunable laser diodes

Laser diodes generate light by a single photon being emitted when a high energy electron in the conduction band recombines with a hole in the valence band. The energy of the photon and hence the emission wavelength of laser diodes is therefore determined by the band gap of the material system used. Different diode lasers are available for specific applications in the range over which tuning is to be performed. These lasers can be also tuned by either adjusting their temperature or by changing injection current density into the gain medium. While temperature changes allow tuning over 100 cm^{-1}, it is limited by slow tuning rates (a few hertz), due to the thermal inertia of the system. On the other hand, adjusting the injection current can provide tuning at rates as high as $\sim$10 GHz, but it is restricted to a smaller range (about 1–2 cm^{-1}) over which the tuning can be performed. The typical laser linewidth is on the order of $10^{-3}cm^{-1}$ or smaller.

2.2.2 Quantum Cascade Lasers

Quantum cascade lasers (QCLs) are semiconductor lasers that emit in the mid- to far-infrared portion of the electromagnetic spectrum. A QCL however does not use bulk semiconductor materials in its optically active region. Instead it comprises a periodic series of thin layers of varying material composition forming a so called superlattice. The superlattice introduces a varying electric potential across the length of the device, meaning that there is a varying probability of electrons occupying different positions over the length of the device. This is referred to as one-dimensional multiple quantum well confinement and leads to the splitting of the band of permitted energies into a number of discrete electronic subbands. By a suitable design of the layer thicknesses it is possible to engineer a population inversion between two subbands in the system which is required in order to achieve laser emission. Since the position of the energy levels in the system is primarily determined by the layer thicknesses and not by the material, it is possible to tune the emission wavelength of QCLs over a wide range in the same material system. In quantum cascade structures, electrons undergo intersubband transitions and photons are emitted. The electrons tunnel to the next period of the structure and the process repeats. Additionally, in semiconductor laser diodes, electrons and holes are annihilated after recombining across the band gap and can play no further part in photon generation. However in a unipolar QCL, once an electron has undergone an intersubband transition and has emitted a photon in one period of the superlattice, it can tunnel into the next period of the structure where another photon can be emitted. This process of a single electron causing the emission of multiple photons as it goes through the QCL structure gives rise to the name *cascade* and yields a quantum efficiency greater than unity, which leads to higher output powers than semiconductor laser diodes.

3 Sensor Types and Related Materials

3.1 Detection mechanisms

The most important types of infrared sensors can be classified on the basis of their intrinsic working mechanisms as illustrated below [25].

3.1.1 Quantum sensors

They can be divided in three categories:

(a) Photon sensors

(a.1) Photoconductive(intrinsic) $Hg_{(1-x)}$ Cd_xTe, $Pb_{(1-x)}Sn_xTe$, $Pb_{(1-x)}Sn_xSe$. For each x-value a different cut-off frequency is obtained.
(a.2) Photoconductive(extrinsic) *Ge*, *Ge* – *Si*, *InSb*, *GaAs* are the most interesting materials. Germanium can be doped by different materials and the following performances can be obtained as far as the cut-off frequency is concerned. $Ge : Au(9\ \mu m)$, $Ge : Hg(14\ \mu m)$; $Ge : Cd(23\ \mu m)$; $Ge : In(105\ \mu m)$; $Ge : Sb(125\ \mu m)$. *Types of photoconductivity:*
(a.3) Photovoltaic (photodiodes).
(a.4) Superconducting sensors (based on the Josephson effect).

(b) Thermal sensors (operating at room temperature)

(b.1) Golay cell
(b.2) Thermistor bolometer
(b.3) Thermocouple
(b.4) Thermopile
(b.5) Pyroelectric

(c) Thermal sensors (operating at cryogenic temperature)

(c.1) Carbon bolometer
(c.2) Ge bolometer
(c.3) Si bolometer
(c.4) *InSb* free carrier-absorption bolometer
(c.5) Superconducting bolometer

Another kind of classification concerns the presence or not of an external power supply during the signal detection procedure. In this context we have:

1. Active sensors which are those not requiring any external supply, such as

(1.a) Photovoltaic
(1.b) Thermocouple and thermopiles
(1.c) Pyroelectric

2. Passive sensors requiring external DC bias

(2.a) Photoconductive
(2.b) All kind of bolometers

3.2 Photon sensors

Photon sensors are based on the absorption of long-wavelength radiation as a result of a specific quantum event, such as the photoelectric emission of electrons from a surface, or electronic transitions in semiconductor materials. Therefore, the output of photon sensors depends on the photon's absorption rate and not directly on photon energy. They normally require to be cooled to cryogenic temperatures in order to get rid of excessive dark current, but have high performance, with larger detectivities and smaller response times. They respond only to photons whose energy $h\nu$ is equal to or larger than the energy gap or than the ionization energy. The rate of carrier generation due to a given incident power P is given by:

$$G(s^{-1}) = \eta P/h\nu \tag{1}$$

for $h\nu \geq Eg$, and $G = 0$ for $h\nu \leq Eg$. If τ does represent the lifetime of the carriers in a photoconductor we have $\Delta n = G\tau$, while in photovoltaic sensors the current is given by $I = qG$.

Photon sensors can be further subdivided into *photoconductive* and *photovoltaic* devices. The function of photoconductive sensors are based on the photogeneration of charge carriers (electrons, holes or electron-hole pairs). These charge carriers increase the conductivity of the device material. Detector materials possible to be utilized for photoconductive sensors are:

- Indium Antimonide ($InSb$)
- Quantum Well Infrared Photodetector (QWIP)
- Mercury Cadmium Telluride ($Hg_xCdTe_{(1-x)}$)
- Lead Sulfide (PbS)
- Lead Selenide ($PbSe$)
- Lead Tin Telluride ($Pb_xSnTe_{(1-x)}$)

Photovoltaic devices (Figure 1a) require an internal potential barrier which derives from the presence of a built-in electric field in order to be able to separate the photo-generated electron-hole pair. Such potential barriers can be created by the use of p-n junctions or Schottky barriers. While the current-voltage characteristics of photoconductive devices are symmetric with respect to the polarity (if we neglect small deviation due to the presence of delocalized space charge regions at the contacts or inside the non perfect material) of the applied voltage, photovoltaic devices exhibit rectifying behavior. Photon sensors may also be classified on the basis of whether the photo-transitions take place across the fundamental band gap of the infrared sensitive material, or from impurity states to either of the valence or the conduction band. In the first case they are denoted *intrinsic*, in the latter case *extrinsic*. The quantum well type of detector discussed below is however not easily classified according to this criterion. (See Figure 2).

In most cases photon sensors need to be cooled to cryogenic temperatures, i.e. down to 77 K (liquid nitrogen) or 4 K (liquid helium). In some favorable cases thermoelectric cooling down to 200 K is sufficient (e. g. 3–5 μm wavelength mer-

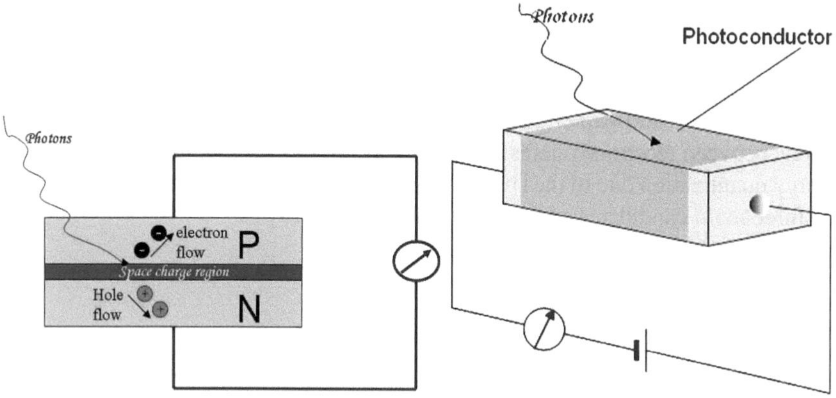

(a) Photovoltaic device. (b) Photoconductive device.

Fig. 1 Photon sensors.

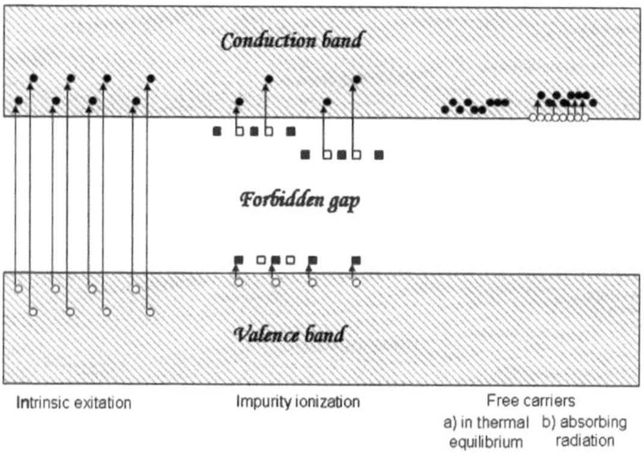

Fig. 2 Photoconductivity processes showing the electron transitions in four different cases.

cury cadmium telluride). The main workhorse in the field of photon sensors is mercury cadmium telluride ($HgCdTe$), and to a less extent indium antimonide (*InSb*). Vigorous work has been done on cadmium telluride both in the US and Europe since its discovery in 1959 and work is still being done. Cadmium telluride is used both for the 3–5 μm (MWIR) and 8–12 μm (LWIR) atmospheric transmission windows, whereas indium antimonide is only used for the 3–5 μm range. Platinum silicide (*PtSi*) Schottky barrier sensors also work in the MWIR domain. Large (512 × 512 pixels) *PtSi* focal plane arrays have been fabricated, they are compatible with silicon CCD/CMOS technology, and show high performance, due to the extremely good pixel to pixel uniformity, in spite of the very low quantum efficiency. As regards FPAs for the 3–5 μm window, both cadmium telluride, *InSb* and *PtSi* materials pose no major technological problems and are considered to be a finished product. In contrast, to date, no photon sensors FPAs operating in the 8–12 μm window exhibit sufficient performance to be operated at 77–80 K. In the course of only the last five years, sensors based on low-dimensional structures have evolved as viable candidates for FPAs (focal plane arrays), especially in the LWIR region. These so called band-gap engineered sensors operate on account of electronic transitions between electronic states arising as a result of size quantization, i.e. electron energy quantization due to the small layer dimensions in the growth direction. There are three main candidates of interest for IR sensor arrays:

(i) The $AlGaAs/GaAs$ quantum wells
(ii) The strained $SiGe/Si$ superlattices (SL)
(iii) The strained InAs/GaInSb SLs and others

Among them the most mature is the $AlGaAs/GaAs$ quantum well (QW) structure, which is a spin-off from *GaAs* technology. This sensor type is generally named Quantum Well Infrared Photoconductor or QWIP. Here special grating structures are necessary in order to achieve a high quantum efficiency of the detector. QWIP FPAs need operating temperatures around 70–75 K in order to work properly, temperatures which are easily achievable by miniature Stirling coolers. The main advantages of SiGe/Si QWs are the compatibility with silicon technology and that grating structures are not necessary. The cooling requirements seem, however, to be more extensive than for $AlGaAs/GaAs$ quantum wells. InAs/GaInSb so called type II SLs in theory offer the possibility of high sensitivity and operating temperatures of an intrinsic detector. In addition, the materials processing and uniformity are expected to be superior to that of III–VI materials such as cadmium telluride. However, presently the maturity of the sensor technology is far from being comparable to cadmium telluride sensors.

3.3 Thermal sensors

In contrast to photon sensors, the operation of thermal sensors is straightforward . The absorption of infrared radiation in these sensors raises the temperature of the device, which in turn changes some temperature-dependent parameter such as electrical conductivity, gas pressure or thermal polarizability. All these kinds of thermal sensors show a remarkably flat response of the detectivity versus the wavelength (see Figure 3). Thermal sensors may be thermopile (Seebeck effect), bolometer, Golay cell sensors, thermopile or pyroelectric sensors ($LiTaO_3$). This last sensor deserves a remark; in fact the current generated in a given load resistor is proportional to the average temperature rate and for this reason it is sometimes called a derivative temperature detector. If p represents the pyroelectric coefficient and A is the detector area, the current is given by:

$$I = p \cdot A \cdot \frac{d(<T>)}{dt}. \qquad (2)$$

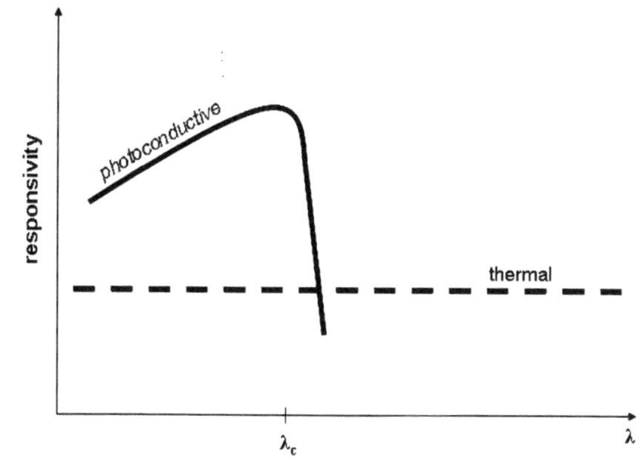

Fig. 3 Theoretical response of thermal and photosensors.

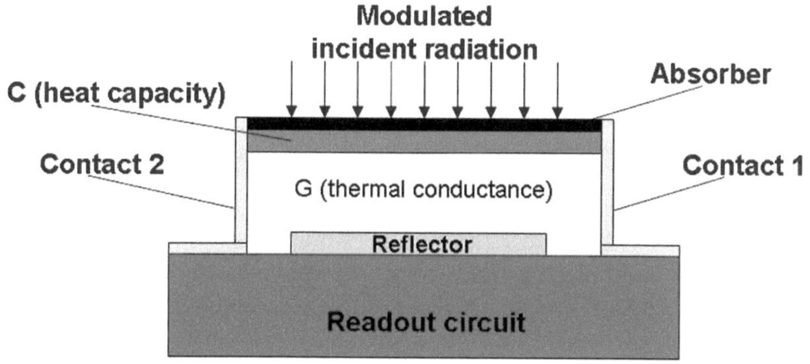

Fig. 4 Thermal detector.

The most relevant advantage of thermal sensors is that they can operate at room temperature. However, the detectivity is lower and the response time longer than for photon sensors. This makes thermal sensors suitable for focal plane array operation, where the latter two properties are less critical.

A thermal detector is conveniently divided into three functional parts:

- Absorber for infrared radiation
- Membrane or other structure for achieving thermal insulation
- Temperature detector

The *absorber* can be a finely subdivided metal such as platinum black, or be based on an interferometric structure. A simple model useful to understand the operating principle of a thermal detector is shown in Figure 4. From this figure it is possible to derive the heat equation of this device which can be expressed as follows:

$$C \cdot \frac{d(\Delta T)}{dt} + G \cdot \Delta T = Re[W \cdot e^{j\omega t}] \quad (3)$$

where:

- C is the heat capacity of the detector.
- G is the thermal conductance between the sensor and the heat sink at temperature T_0.
- W is the adsorbed peak power.
- ω is the angular modulation frequency.

Leaving out the mathematical details the solution of the heat equation rewritten as:

$$\frac{d(\Delta T)}{dt} + \frac{\Delta T}{\tau} = Re\left[\frac{W}{C} \cdot e^{j\omega t}\right] \tag{4}$$

which considers the modulated response is given by:

$$\Delta T = \left(\frac{W}{C} \cdot e^{j\omega t}\right) \cdot (1 + j\omega t) \tag{5}$$

where $\tau = \frac{C}{G}$ is the thermal time constant of the device. The real time dependence of the temperature change versus the frequency is then given by:

$$\Delta T = \frac{W}{G} \cdot \frac{1}{\sqrt{1+\omega^2\tau^2}}. \tag{6}$$

If $\omega\tau$ is less than 1 the temperature rise does not depend on the heat capacity C which is usually minimized for achieving fast responses. In order to obtain high sensitivity it is of utmost importance that the detector element is thermally insulated from the detector substrate. Therefore, when fabricating thermal detector arrays it is common to make thin membranes using micro-mechanical processing techniques. The material may be silicon nitride or silicon dioxide, which both are compatible with silicon processing techniques.

The *temperature detector* is usually integrated into a suitable membrane, and utilized to detect the usually minute temperature change resulting from exposure to infrared radiation from a room-temperature scene and subsequent absorption. Thermal sensors are conveniently classified according to their means of detecting this temperature change:

- A resistive bolometer contains a resistive material, whose resistivity changes with temperature. To achieve high sensitivity the temperature coefficient of the resistivity should be as large as possible and the noise resulting from contacts and the material itself should be low. Resistive materials could be metals such as platinum, or semiconductors (thermistors). Metals usually have low noise but have low temperature coefficients (about 0.2%/K), semiconductors have high temperature coefficients (1–4%/K) but are prone to be more noisy. Semiconductors used for infrared sensors are e.g. amorphous, polycrystalline silicon, or vanadium oxide.

- A thermoelectric device (thermocouple or thermopile) is based on the presence of one or several junctions between two materials. The junctions properly arranged and connected develop a *thermo-emf* that changes with temperature, the so-called Seebeck effect. In order for the sensitivity to be high the Seebeck coefficient should be as high as possible. Certain alloys containing antimony and bismuth have very high Seebeck coefficients of 150 μV/K. The CMOS compatible combination aluminum/polycrystalline silicon gives about 65 μV/K.
- A pyroelectric sensor is based on the fact that certain dielectric materials of low crystal symmetry exhibit spontaneous dielectric polarization. When the electric dipole moment depends on temperature the material becomes pyroelectric. Usually a capacitor is fabricated from the material and the variation of charge on it is sensed. Common pyroelectric materials used for infrared sensors are *TGS* (triglycine sulphate), $LiTaO_3$ (lithium tantalate), *PZT* (lead zinc titanate) and certain polymers.A *dielectric bolometer* makes use of pyroelectric materials operated in a way to detect the change of the dielectric constant with temperature. A suitable material for this application is *SBT* (Strontium Barium Titanate).
- The Golay detector is based on the volume or pressure change of an encapsulated gas with temperature. The volume change is measured e.g. by the deflection of light rays resulting from the motion of properly positioned mirrors fastened to the walls of the gas container.

3.4 Infrared imaging

There are two basic types of infrared imaging systems: mechanical *scanning systems* and systems based on detector arrays without a scanner. It should be mentioned that *detector arrays* as well are used for scanning systems, but the number of detector elements (picture elements – pixels) generally is smaller in this case.

A mechanical *scanner* utilizes one or more moving mirrors to sample the object plane sequentially in a row-wise manner and project these onto the detector. The advantage is that only one single detector is needed. The drawbacks are that high precision and thus expensive opto-mechanical parts are needed, and the detector response time has to be short. As mentioned above, detector arrays are also used for this application. For example, a long linear detector array can be used to simultaneously sample one column of the object plane. By using a single moving mirror the whole focal plane can be sampled. In contrast, when a single detector is used, two mirrors moving in two orthogonal directions must be used, one of them moving at high speed, the other one at lower speed.

Detector arrays operated as *focal plane arrays* (FPA) (or *staring arrays*) are located in the focal plane of a camera system, and are thus replacing the film of a conventional camera for visible light. The advantage is that no moving mechanical parts are needed and that the detector sensitivity can be low and the detector slow. The drawback is that the detector array is more complicated to fabricate. However, rational methods for semiconductor fabrication yield economic advantages, provided that production volumes are large. The general trend is that infrared camera systems will be based on FPAs, except for special applications.

The spatial resolution of the image is determined by the number of pixels of the detector array. Common formats for commercial infrared sensors are 320 × 240 pixels (320 columns, 240 rows), and 640 × 480. The latter format (or something close to it), which is nearly the resolution obtained by standard TV, will probably become commercially available in the next few years. Today, for example, indium antimonide and platinum silicide sensors are commercially available in the 320 × 240 pixels format. Typical pitches between pixels are in the range 20–50 μm.

Detector arrays are more complicated to fabricate, since besides the detector elements with the function of responding to radiation, electronic circuitry is needed to multiplex all the detector signals to one or a few output leads in a serial manner. The output from the array is either in analogue or digital form. In the former case analogue to digital conversion is usually done external to the detector array. The electronic chip used to multiplex or read out the signals from the detector elements are usually called simply *readout integrated circuit* (ROIC) or (analogue) multiplexer.

The ROIC is usually made using silicon CCD (charge coupled device) or CMOS technology. However, the detector elements must often be fabricated from more exotic materials as discussed above. The exceptions are e. g. platinum silicide or micro-bolometers which can be based on silicon technology. In the former case a hybrid approach is most common, in which case all the detector pixels are fabricated from a separate detector chip. This detector chip is then *flip-chip bonded* to the ROIC chip. Flip-chip bonding involves the processing of metal bumps onto contact holes, one per pixel, of both the detector chip and the ROIC . Using special equipment, the two chips are first aligned to each other. Then the chips are put in contact, while applying heat and/or mechanical force. During this process the two chips become electrically connected to each other via the metal bumps. Usually indium is used for the bumps due to its excellent low temperature properties.

Uniformity of the detector elements across the array is a key issue for obtaining high performance. In fact, individual pixel response characteristics differ considerably across an array in most cases. Therefore so called pixel correction has to be done prior to the presentation of the final image. This amounts to calibrating each individual pixel, by exposing the array to calibrated surfaces of known temperature.

IR sensor can be divided in two broad categories: incoherent and coherent. Incoherent infrared sensors can be seen as sensors sensitive to the photon energy. Examples are: photomultiplier, photoconductors, bolometers, etc. For all of them in the detection process, information of phase and frequency is lost. Coherent IR sensors are all those that can maintain frequency and phase information; examples are linear amplifiers heterodine sensors (mixers).

4 Figure of Merit of Incoherent IR Sensors (I.I.R.S.)

The most important figures of merit for I.I.R.S. are:

4.1 Noise Equivalent Power (NEP)

It is defined as the r.m.s. value of a sinusoidally modulated radiant power falling on the sensor able to determine signal to noise ratio equal to unity.

The NEP (or P_N) depends on the (S/N) noise bandwidth of the preprocessing circuit.
The smaller this bandwidth, the lower the NEP. The NEP is given with reference to 1 Hz bandwidth (W/$\sqrt{\text{Hz}}$).
The NEP should be written as follows:
NEP (500 K, 900, 1) where, as an example, 500 K represents the Black Body temperature, 900 is the chopping frequency, and 1 Hz is the bandwidth.
P_N can be experimentally estimated by the following relationship:

$$P_N(orNEP) = I \cdot A_S \cdot \frac{V_n/V_s}{\sqrt{\Delta f}} \tag{7}$$

where I is the irradiance falling on the sensor area A_S, (V_n/V_s) is the noise to signal ratio evaluated in the bandwidth (Δf).

4.2 Responsivity

It can be represented by the electrical output of a sensor divided by the power (P) of the radiation striking it. Since electrical output can be either voltage or current, one distinguishes between voltage and current responsivity.

$$R = \frac{V_s}{P} = \frac{V_s}{I \cdot A_s}. \tag{8}$$

The responsivity unit is $[V/W]$.

It is worth underlining that the responsivity is linked to the NEP and to the D^* as follows:

$$R = \frac{V_n}{P_N \cdot \sqrt{\Delta f}} = \frac{D^* \cdot V_n}{\sqrt{A_s \cdot \Delta f}}. \tag{9}$$

It is relevant also to mention that the responsivity is frequency dependent (see Figure 5) and in most cases its behaviour can be expressed as follows:

$$R(f) = \frac{R_0}{\sqrt{1+\omega^2\tau^2}}. \tag{10}$$

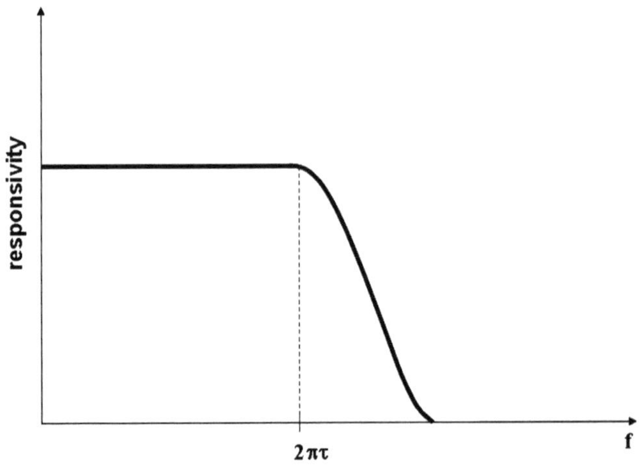

Fig. 5 Frequency dependence of the responsivity.

Responsivity can be measured for monochromatic radiation, in which case the responsivity is called spectral responsivity. Alternatively, a blackbody source kept at a fixed temperature can be used. In this case one talks about black body responsivity. Spectral responsivity plotted versus wavelength is often used for presentation of a detector's spectral response properties.

4.3 Detectivity D^*

Whereas responsivity takes into account the detector's signal properties only, the detectivity or D^* value is a measure of its signal to noise properties. The D^* value is normalized with respect to detector area (provided that the signal to noise ratio increases with the square root of the detector area, which is often the case, at least for photon sensors). It is defined as:

$$D^* = \frac{R}{V_N} \cdot \sqrt{A_S \cdot \Delta f} = \frac{R}{I_N} \sqrt{A_S \cdot \Delta f} \tag{11}$$

where V_N and I_N is the noise voltage and current, respectively, R is the responsivity, A_S the detector area and Δf the noise bandwidth.

4.4 Temperature resolution (NETD) and other important parameters

NETD is an abbreviation for Noise Equivalent Temperature Difference and is a measure of the smallest object temperature difference that can be detected by an IR camera.

When dealing with I.R. sensors also the following parameters are important:

- *Operating temperature:* temperature of the I.R. sensor.
- *Cut-Off wavelength:* wavelelength above which the response goes to zero.
- D_{λ_p}*:* which is similar to the normal D^* apart from the fact that here the input power is related to a small $\Delta\lambda$ generated, for instance, by a monochromator.
- *Response time:* time which goes from 0.1 to 0.9 of the overall response.
- *Noise mechanism:* which describes the kind of noise (shot, thermal, flicker, burst).
- *Resistance/squares:* representing a measure of a thin film of sensitive material.
- *Mode of operation:* describes how the device is applied and how the output voltage (current) is taken.

Figure 6 gives a broad representation of the most important infrared sensors in terms of detectivity (D^*) versus wavelengths. Only a few sensors, namely pyroelectric, Golay cell, thermopile, thermistor (letters S-T-U-V), have a flat response from the visible to above 20 μm. Other sensors (letters A-B-C-D-E-F-G-H-R-I-J-L-R-O) have a band pass behaviour in the 1–5 μm region, others (K-N-M-Q) show broad response covering many microns of range. The envelope of all the curves are located below the ideal limit of both photoconductive and photovoltaic sensors.

When IR sensors are taken into consideration it is worth keeping in mind the following radiation terms:

- Radiant power: P (Watts).
- Spectral radiant power: P_λ (W/μm).

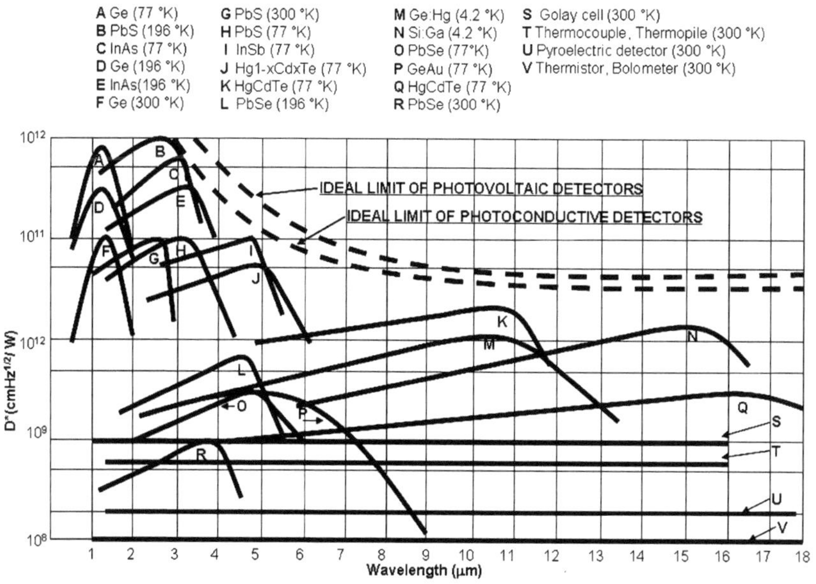

Fig. 6 Detectivity versus wavelength in the infrared region of different sensors.

- Radiant intensity: I (W/steradiants).
- Spectral radiant intensity: I_λ (Watts/steradiant. μm).
- Radiant emittance: R_e (W/m^2).
- Spectral radiant emittance: $R_{e\lambda}$ (W/m$^2 \cdot$ μm).
- Radiance: R_Ω (W/m$^2 \cdot$ steradiant).
- Spectral radiance: $R_{\Omega\lambda}$ (W/m$^2 \cdot$ steradiant $\cdot$ μm).
- Irradiance: I_R (W/m^2).
- Reflectivity (R); Absorptivity(α); Transmittivity(T); Emissivity(E): which are all expressed by numbers, each less than one, in practical cases.

After this brief introduction to IR sensors we will talk about the acousto-optic technique which seems to be suitable for the detection of aggressive volatile compounds in the 8–14 μm region. Below we list a number of aggressive compounds to which a great deal of attention is paid:

- **Nerve Agents:** Sarin, Ciclosarin, Tabun, Soman, VX
- **Blister Agents:** Nitrogen Mustards (HN-1, HN-2, HN-3), absorption @ 14 μm
- **Pulmonary Agents:** Cloropicrina, Perfluoro-Isobutilene (PFIB), Fosgene
- **Blood Agents:** CNCl

5 Acousto-Optic Devices

The early studies on acousto-optic phenomena, i.e. the optical interactions with acoustic waves, go back to around the 1920s, as pointed out by C.F. Quate and M. Born in their articles [2, 37]. The acousto-optic interactions occur in an optical media (usually solid, sometimes liquid, rarely a gas) when an optic wave and an acoustic wave are present in the same place and time. A possible scenario is shown in Figure 7. The strain of the medium due to the pressure fluctuation of the acoustic

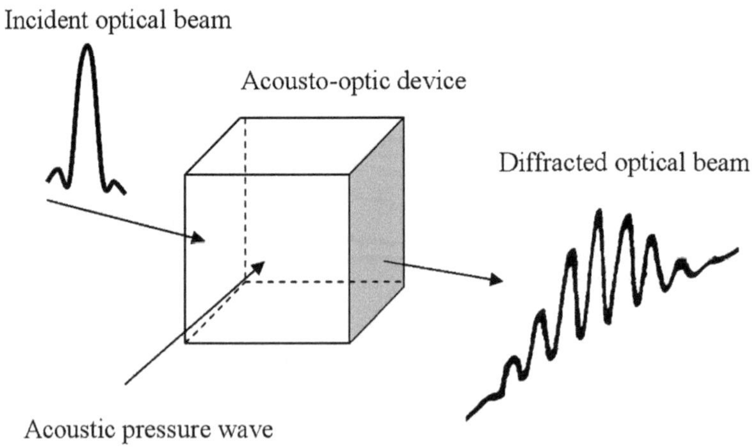

Fig. 7 Scenario.

wave causes diffraction, refraction, interference and reflection of the optical wave [38]. After the advent of the laser acousto-optics research was especially dedicated to the development of devices able to modulate and deflect the laser beam [7,12,20]. Here we develop a bit of theory, taken from the literature, related to the acousto-optic mechanism and give some explanations of some related devices, essential for the generation of power in relatively small $\Delta\lambda$, to be used in volatile compound absorption based detection techniques.

5.1 Acousto-optic interaction theory

5.1.1 The elasto-optical effect

Let $\mathbf{E}(\mathbf{r},t)$ and $\mathbf{H}(\mathbf{r},t)$ be respectively the electric field and the magnetic field of the light beam. Thus Maxwell's equation within the medium can be written as

$$\nabla \times \mathbf{E} = -u_0 \frac{\partial}{\partial t}\mathbf{H} \tag{12}$$

$$\nabla \times \mathbf{H} = \frac{\partial}{\partial t}(\varepsilon \mathbf{E}) \tag{13}$$

$$\mathbf{D} = \varepsilon \mathbf{E} \tag{14}$$

$$\nabla \cdot \mathbf{D} = \rho = 0 \tag{15}$$

by Gauss' law. The optical properties of a medium are completely characterized by the electric impermeability tensor $\beta = \varepsilon_0 \varepsilon^{-1}$ (not to be confused with the impedance of the medium), where ε^{-1} is the inverse of the tensor ε. So Equation (14) can be inverted and rewritten as

$$\varepsilon_0 \mathbf{E} = \beta \mathbf{D}. \tag{16}$$

In the directions for which $\mathbf{E}$ and $\mathbf{D}$ are parallel, both tensor ε and β share the same principal axes and are represented by a diagonal matrix. Thus the principal values of β are

$$\frac{\varepsilon_0}{\varepsilon_1} = \frac{1}{n_1^2}; \frac{\varepsilon_0}{\varepsilon_2} = \frac{1}{n_2^2}; \frac{\varepsilon_0}{\varepsilon_3} = \frac{1}{n_3^2} \tag{17}$$

where $\varepsilon_1 = \varepsilon_{11}$, $\varepsilon_2 = \varepsilon_{22}$ and $\varepsilon_3 = \varepsilon_{33}$.

The quadratic representation of the electric impermeability tensor β

$$\sum_{ij} \beta_{ij} x_i x_j = 1 \qquad i,j=1,2,3 \tag{18}$$

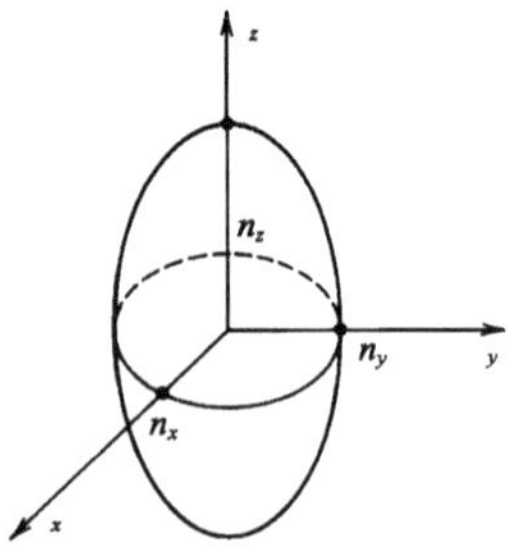

Fig. 8 The index ellipsoid.

Table 1 Lookup table for the index m or n that represents the pair of indices (i,j) or (k,l).

j	i:1	2	3
1	1	6	5
2	6	2	4
3	5	4	3

is called the index ellipsoid or optical indicatrix (Figure 8). Using principal axes as the coordinate system the quadratic form is described by

$$\frac{x^2}{n_x^2} + \frac{y^2}{n_y^2} + \frac{z^2}{n_z^2} = 1. \tag{19}$$

For example, if the medium is isotropic then the indicatrix will describe a spherical surface or, if it is a uniaxial crystal, the surface will be an ellipsoid of revolution [26].

When the medium is perturbed by a sound wave $\mathbf{S}(\mathbf{r},t)$, the compression and refraction waves change the local density and the resulting strain of the component atoms and molecules of the scattering medium change the optical polarization [12].

As a consequence the permittivity tensor ε changes its coefficients, and hence β too according to the equation

$$\beta_{ij} = \beta_{0_{ij}} + \Delta\beta_{ij}, \tag{20}$$

where

$$\Delta\beta_{ij} = p_{ijkl} S_{kl} \qquad i,j,k,l=1,2,3. \tag{21}$$

$\Delta\beta_{ij}$ is the variation due to the perturbation, $\beta_{0_{ij}}$ is the indicatrix coefficient before the perturbation and p_{ijkl} are the elastooptic or photoelastic constant coefficients of the strain-optic tensor of fourth rank. They reflect a particular symmetry due to symmetry both of $\Delta\beta_{ij}$ and of S_{kl}, in particular $p_{ijkl} = p_{jikl} = p_{ijlk} = p_{jilk}$ [22, 41]. In this way it is possible to contract the pair of indices (i,j) to a single index m $= 1,2,\ldots,6$ using the Nye's notation reported at Table 1 [42]. Also the pair of indices (k, l) can be contracted to the index n $= 1,2,\ldots,6$ in the same way. For example the pair (k, l) $= (1,2)$ is labeled n $= 6$.

Moreover the symmetry of the crystal adds other constraints on the coefficients. For example the matrix pmn of a cubic crystal [32, 40, 42] has the structure

$$pmn = \begin{bmatrix} p_{11} & p_{12} & p_{12} & 0 & 0 & 0 \\ p_{12} & p_{11} & p_{12} & 0 & 0 & 0 \\ p_{12} & p_{12} & p_{11} & 0 & 0 & 0 \\ 0 & 0 & 0 & p_{44} & 0 & 0 \\ 0 & 0 & 0 & 0 & p_{44} & 0 \\ 0 & 0 & 0 & 0 & 0 & p_{44} \end{bmatrix}.$$

Furthermore if the medium material is isotropic

$$p_{44} = \frac{1}{2}(p_{11} + p_{12}). \tag{22}$$

Other photoelastic matrices for different types of crystal have been published by Mason [27], Nye [34] and Krishnan [24].

Table 2 reports some elastoplastic coefficients. More detailed tables can be found in [8, 46].

For example, let us now consider a longitudinal acoustic wave characterized by a displacement $u_x = A_0 \sin(\omega_{ac}t - k_{ac}x)$, $u_y = 0$ and $u_z = 0$, traveling into an isotropic cubic crystal ($n_{0x} = n_{0y} = n_{0z}$) along the x direction. Thus the strain tensor S_{ij}, defined as

$$S_{ij} = \frac{1}{2}\left(\frac{\partial u_i}{\partial x_j} + \frac{\partial u_j}{\partial x_i}\right) \tag{23}$$

where $i, j = 1, 2, 3$ denotes the coordinate (x, y, z), which has all components vanishing except

$$S_{11} = S_1 = S_0 \cos(\omega_{ac}t - k_{ac}x) = -k_{ac}A_0 \cos(\omega_{ac}t - k_{ac}x). \tag{24}$$

By substituting (24) and (20) into (21) we find

$$\beta_{11} = \frac{1}{(n_x)^2} + p_{11}S_1 \tag{25}$$

Table 2 Elastoplastic coefficients and refractive index.

Material	l (mm)	n	P_{11}	P_{12}	P_{44}	P_{31}	P_{13}	P_{33}	P_{41}	P_{14}	P_{66}
Fused quartz	0.63	1.46	0.121	0.270	−0.075						
GaP	0.63	3.31	−0.151	−0.082	−0.074						
GaAs	1.15	3.37	−0.165	−0.140	−0.072						
TiO_2	0.63	2.58	0.011	0.172		0.0965	0.168	0.058			
$LiNbO_3$	0.63	2.20	0.036	0.072		0.178	0.092	0.088	0.155		
$LiTaO_3$	0.63	2.18	0.0804	0.0804	0.022	0.086	0.094	0.150	0.024	0.031	
KDP	0.63	1.51	0.251	0.249		0.225	0.223	0.246			0.058
H_2O		1.33	0.31								

$$\beta_{22} = \beta_{33} = \frac{1}{(n_x)^2} + p_{12}S_1 \tag{26}$$

$$\beta_{ij} = 0,\ i \neq j. \tag{27}$$

Equations (25), (26), (27) show that the initial isotropic crystal has become a uniaxial crystal (two refractive index are equal) and the quadratic form of the optical indicatrix represents an ellipsoid of revolution whose axes $n_o = n_2 = n_3$ and $n_e = n_1$ are given by

$$\frac{1}{(n_o)^2} = \frac{1}{(n_{0_x})^2} + p_{12}S_1 \tag{28}$$

$$\frac{1}{(n_e)^2} = \frac{1}{(n_{0_x})^2} + p_{11}S_1 \tag{29}$$

where n_o and n_e represent the ordinary and extraordinary refractive index.
Using the approximation

$$\frac{1}{\sqrt{1+a}} \approx 1 - \frac{1}{2}a \quad with \quad a \ll 1 \tag{30}$$

it is possible to write (28) or (29), with $i = e, o$ as

$$n_i = \frac{n_{0_x}}{\sqrt{1 + (n_{0_x})^2 p_{mn} S_m}} = n_{0_x}\left[1 - \frac{1}{2}(n_{0_x})^2 p_{mn} S_m\right]$$
$$= n_{0_x} - \frac{1}{2}(n_{0x})^3 p_{mn} S_m \tag{31}$$

and hence

$$\Delta n_i = n_i - n_{0_x} = -\frac{1}{2}(n_{0_x})^3 p_{mn} S_m. \tag{32}$$

The variation of the reflective index Δn is negative compared with the positive strain perturbation S. Other than the longitudinal wave, also, the transverse share wave is very common in acousto-optic devices. Here the displacement wave, $u_x = 0$, $u_y = 0$ and $u_z = A_0 \sin(\omega_{ac}t - k_{ac}x)$, travels in an isotropic cubic crystal along the x direction but vibrates orthogonally in the z direction. Thus the strain tensor has all components zero except

$$S_{13} = S_{31} = S_5 = S_0\cos(\omega_{ac}t - k_{ac}x) = -\frac{1}{2}k_{ac}A_0\cos(\omega_{ac}t - k_{ac}x) \tag{33}$$

and the crystal will become biaxial (the three principal indices are different from each other). Reassembling (32) and using (33), the principal refractive index will be given by

$$n_x = n_{0_x} - \frac{1}{4}(n_{0_x})^3 k_{ac}A_0\cos(\omega_{ac}t - k_{ac}x) \tag{34}$$

$$n_y = n_{0_x} \tag{35}$$

$$n_z = n_{0_x} + \frac{1}{4}(n_{0_x})^3 k_{ac} A_0 \cos(\omega_{ac} t - k_{ac} x). \tag{36}$$

If we want to consider the variation of the permitivity ϵ related to the photoelastic effect we can write [23, 44]

$$\epsilon(\mathbf{r},t) = \varepsilon_0(1 + C\mathbf{S}(\mathbf{r},t)) = \varepsilon_0 + \epsilon'(\mathbf{r},t), \tag{37}$$

where $\epsilon(\mathbf{r},t) = \varepsilon_0 C\mathbf{S}(\mathbf{r},t)$ is the time-varying permitivity and C is a constant dependent on the medium material.

Furthermore, if we assume that the acoustic wave is a planar traveling wave with sinusoidal vibration, then the relationship between the variation of the refractive index D*n(x,t)* and the acoustic strain wave can be written as [21]

$$\Delta n(x,t) = \frac{n}{2} CS(x,t) \tag{38}$$

and

$$C = -n^2 p. \tag{39}$$

Note that the above (39) defines the constant C using the refractive index n as a scalar, and hence (39) can usually be applied to a liquid [40].

5.1.2 Raman-Nath diffraction

Let us consider now a progressive sinusoidal perturbation wave **S**(**r**,t) propagating into an optical medium large L and width W, characterized by a permeability μ_0 and a permitivity ϵ, and the electric field **E**(**r**,t) incident at an angle θ_0 upon the acoustic beam as shown in Figure 9a and b. In order to model the acousto-optic interaction we have to consider the Maxwell equation stated above considering ϵ as a function of *x, y, z* coordinates and time *t*. Eliminating **H**, we obtain [31]

$$u_0 \varepsilon_0 \frac{\partial^2 \mathbf{D}}{\partial t^2} = \nabla^2 \mathbf{E} + \frac{\nabla \epsilon}{\epsilon} \nabla \times \mathbf{E} + \left(\frac{\nabla \epsilon}{\epsilon} \nabla \right) \mathbf{E} + (\mathbf{E}\nabla) \frac{\nabla \epsilon}{\epsilon}. \tag{40}$$

Since $\omega_{ac} << \omega_{op}$

$$|\nabla \varepsilon| \lambda << 1. \tag{41}$$

Thus (40) can be written as

$$\nabla^2 \mathbf{E} - \mu_0 \varepsilon_0 \frac{\partial^2}{\partial t^2} (\epsilon \mathbf{E}) = 0 \tag{42}$$

Hence, assuming that the frequency of time variation of ϵ is very small compared to that of **E**, (42) is reduced to

$$\nabla^2 \mathbf{E} = \mu_0 \epsilon \frac{\partial^2}{\partial t^2} (\mathbf{E}) \quad or \quad \nabla^2 \mathbf{E} = \frac{\mathbf{n}(\mathbf{r},t)^2}{c^2} \frac{\partial^2}{\partial t^2} (\mathbf{E}) \tag{43}$$

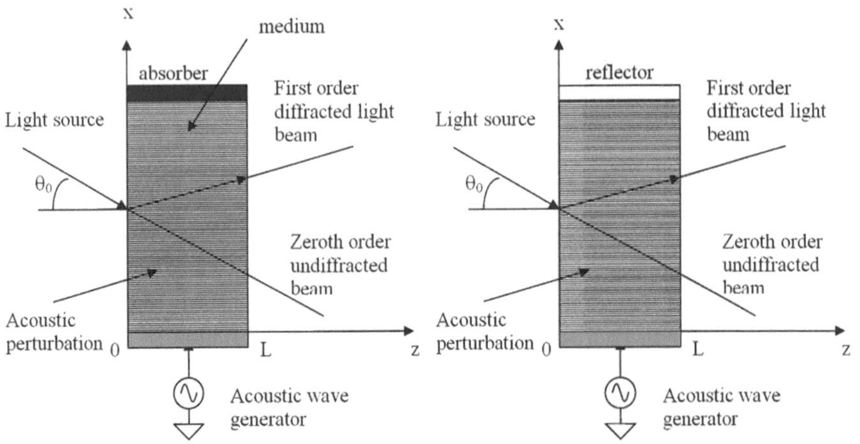

(a) Propagation acoustic wave. (b) Standing acoustic wave.

Fig. 9 Schematic of acousto-optic devices.

where c is the light velocity in free space and $\mathbf{n}(\mathbf{r},\mathrm{t}) = \frac{\mathbf{c}}{c_{medium}(\mathbf{r},t)}$ is the refracting index of the medium [46].

Let $\mathbf{S}(\mathbf{r},\mathrm{t})$ be the sinusoidal perturbation propagating along the x axis, so $\mathbf{S}(\mathbf{r},\mathrm{t})$ can be written as

$$S(x,t) = \frac{1}{2} S_0 \exp\left[j(\omega_{ac}t - k_{ac}x)\right] + c.c. \tag{44}$$

and let $\mathbf{E}(\mathbf{r},t)$ be defined as

$$\mathbf{E}(\mathbf{r},t) = \frac{1}{2} E_{op}(r) \exp\left[j(\omega_{op}t - \mathbf{k_{op}} \cdot \mathbf{r})\right] + c.c. \tag{45}$$

where S_0 is the amplitude of the perturbation, E_{op} is the amplitude in free space and $c.c.$ is the complex conjugate. Furthermore let $n(x,t)$ be the refractive index of the medium perturbed in time and space by the pressure acoustical wave, given by

$$n(x,t) = n_0(x) + \Delta n(x,t) \tag{46}$$

where $\Delta n(x,t)$ can be written according to (32) as

$$\Delta n(x,t) = -\frac{1}{2}(n_0)^3 p S_0 \sin(\omega_{ac}t - k_{ac}x). \tag{47}$$

As the scattering process is essentially lossless, or reactive, the wave energy-momentum conservation principles are applicable. In the case of a plane monochromatic optical and acoustic wave propagated in a medium that is optically inhomogeneous, nonmagnetic and isotropic, the energy-momentum relations are given by

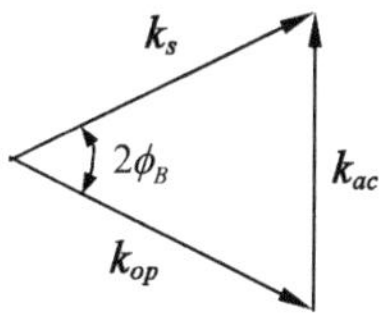

Fig. 10 Momentum scattering of a plane and monochromatic optical and acoustic waves in isotropic medium.

$$\hbar k_s = \hbar k_{op} + \hbar k_{ac}$$
$$\hbar \omega_s = \hbar \omega_{op} + \hbar \omega_{ac}$$

where the subscript s means scattered, $\hbar$ is Planck's constant h divided by 2π, k_s, k_{ac} and k_{op} are respectively the scattered, acoustic and optic wave vector, and k_{ac} and k_{op} are defined as $k_{ac} = \frac{\omega_{ac}}{v_{ac}}$ and $k_{op} = \frac{\omega_{op}}{v_{op}} = \frac{\omega_{op} n_0}{c}$.

Figure 10 shows the vector representation of the moment conservation. Along the interaction area, the perturbed optical field $\mathbf{E}$ can be written as [5, 18, 46]

$$E(x,z,t) = \frac{1}{2} \sum_{m=-\infty}^{+\infty} E_m(z) \exp\left[j(\omega_m t - \mathbf{k}_m \cdot \mathbf{r})\right] + c.c. \tag{48}$$

where

$$\omega_m = \omega_{oc} + m\omega_{ac} \tag{49}$$

$$\mathbf{k}_m = k_{op} + mk_{ac} \tag{50}$$

$$\mathbf{k}_m \cdot \mathbf{r} = k_{op}[z\cos(\vartheta_0) - x\sin(\vartheta_0)] + mk_{ac}x. \tag{51}$$

$E_m(z)$ represents the amplitude of the mth diffracted light with circular frequency $\omega_m = \omega_{op} + m\omega_{ac}$.
Substituting (46) and (48) in (43) and neglecting second-order terms, we obtain the following difference-differential equation derived by Raman and Nath [31]

$$\frac{dE_m(z,t)}{dz} + \frac{\xi}{2L}(E_{m+1} - E_{m-1}) + j\frac{mk_{ac}}{\cos(\vartheta_o)}[\sin(\vartheta_0) - m\sin(\vartheta_B)]E_m = 0 \tag{52}$$

where

$$\xi = -\frac{k_f \Delta n L}{\cos(\vartheta_o)} \tag{53}$$

is a parameter related to the acoustic pressure and $2\sin\vartheta_B = \frac{k_{ac}}{k_{op}}$ (Snell's law). Here k_f is the optical wave number in free space and ϑ_B is the Bragg angle in the medium. The Bragg angle definition (52) implies a momentum-conservation relation in which the frequency shift of the diffracted light beam is neglected.

Solutions of (52) are obtained using exponential Fourier transform theory and numerical methods [9, 44, 47].

An approximate solution with a boundary condition $E_0(0,t) = 1$ and $E_m(0,t) = 0$ $(n \neq 0)$ when $\omega_a << \omega_o$, hence $m \sin\vartheta_B \approx 0$, is

$$E_m(z) = \exp\left(-j\frac{1}{2}mk_{ac}z\tan\vartheta_o\right)J_m\left(\xi\frac{\sin\left(k_{ac}z\tan\frac{\vartheta_o}{2}\right)}{k_{ac}L\tan\frac{\vartheta_o}{2}}\right) \quad (54)$$

where J_m is the Bessel function of order m [46]. The normalized intensity (to the incident beam) of the mth diffracted light at $z = L$ is given by

$$I_m(z)\,|_{z=L} = \left[E_m(z)E_m^*(z)\right]_{x=L} = J_m^2\left(\xi\frac{\sin(\gamma)}{\gamma}\right) \quad (55)$$

where $E_m^*(z)$ is the complex conjugate of $E_m(z)$ and $\gamma = k_{ac}L\tan\frac{\vartheta_o}{2} \approx Q\frac{k_{op}}{k_{ac}}\frac{\sin\theta_0}{2}$. The last approximation is acceptable because common applications require $\theta_0 << 1$.

The parameter Q, defined by Klein and Cook [18,19], is used in order to establish a criterion of diffraction feature. It is given by

$$Q = \frac{k_{ac}^2}{k_{op}}\frac{L}{\cos\vartheta_0} = 2k_{ac}L\frac{\sin\vartheta_B}{\sin\vartheta_0} \approx \frac{k_{ac}^2}{k_f n_0}L. \quad (56)$$

Again the last approximation is acceptable if $\theta_0 << 1$.

The Q value does not define severe limits for the working region, but in practice it is used as [19]

- $Q < \approx 0.3$ for the Raman-Nath region
- $Q \approx 1$ for the transition region
- $Q \geq 4\pi$ for the Bragg region

Figure 11 shows the zeroth and first order intensity levels as functions of Q at Bragg incidence with $\xi = \pi$.

In order to develop an acousto-optic device and to choose the diffraction region of operation (Raman-Nath or Bragg) one has to consider

- The optical path length L
- The optical angle of incidence ϑ_o
- The acoustic frequency f_a

So, for a short path $\frac{L}{\lambda} < 10$ and low frequency (usually $f < 10$ MHz), independent of the incidence angle, we will have Raman-Nath diffraction, and for high frequency $f > 100$ MHz, long path $\frac{L}{\lambda} > 10$ and incident angle $\vartheta_o \approx \pm\vartheta_B$ we will have Bragg diffraction [11].

5.1.3 Bragg diffraction

It is found that the maximum intensity of diffracted light occurs when the incident light beam angle ϑ_o is $\vartheta_o \approx \pm\vartheta_B$. In this situation, called Bragg reflection, only the zeroth and first order diffraction are predominant while higher orders are neglected.

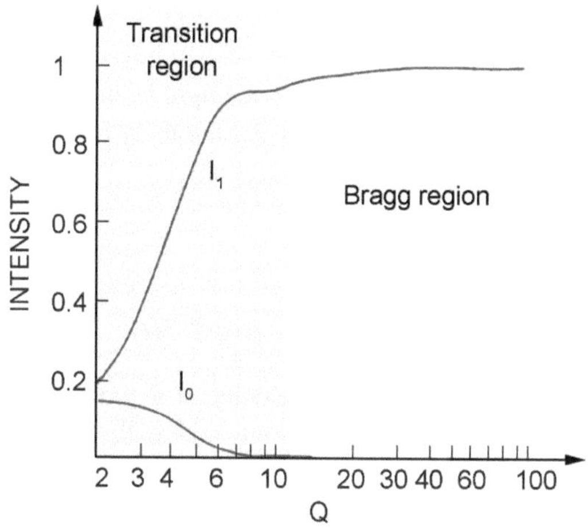

Fig. 11 Intensity of the zeroth and first order vs. Q at Bragg incidence $\xi = \pi$.

Thus for $\vartheta_0 \approx +\vartheta_B$ (52) is reduced to

$$\frac{dE_0(z,t)}{dx} + \frac{\xi}{2L} E_1(z,t) = 0 \tag{57}$$

and

$$\frac{dE_1(z,t)}{dx} - \frac{\xi}{2L} E_0(z,t) + \frac{\zeta}{L} E_1(z,t) = 0 \tag{58}$$

where

$$\zeta = \frac{k_{ac}L}{\cos(\vartheta_o)}[\sin(\vartheta_0) - m\sin(\vartheta_B)]. \tag{59}$$

Solutions for E_0 and E_1 using $E_0(0) = 1$ and $E_1(0) = 0$ are given by [36]. The normalized intensities (to the incident light I_i) I_0 and I_1 at $x = L$ are

$$I_0 = |E_0|^2 = 1 - I_1 \tag{60}$$

$$I_1 = \left(\frac{\xi}{2\gamma'}\right)^2 \sin^2 \gamma' \tag{61}$$

where

$$\gamma' = \zeta^2 + \left(\frac{\xi}{2}\right)^2. \tag{62}$$

If $\vartheta_0 = +\vartheta_B$ and hence $\zeta = 0$ then (61) is reduced to

$$I_1 = \sin^2\left(\frac{\xi}{2}\right) \quad and \quad I_0 = \cos^2\left(\frac{\xi}{2}\right) \tag{63}$$

Furthermore if ξ is small then $I_1 = \left(\frac{\xi}{2\zeta}\right)^2 \sin^2 \zeta$.

5.1.4 Diffraction efficiency

In order to relate the intensity I_1 of the refracted light beam to the power P_a $[W]$ and the figure of merit M we introduce the acoustic intensity I_{ac} $\left[Wm^{-2}\right]$ given by

$$I_{ac} = \frac{1}{2}\rho v^3 SS^*. \tag{64}$$

Thus

$$P_{as} = I_{ac}HL = \frac{1}{2}\rho v_{ac}^3 SS^* HL \tag{65}$$

where ρ is the density, v is the acoustic velocity, H is the thickness of the acoustic beam, and S^* is the complex conjugate of strain S. Here the suffix notation is omitted for simplicity.

Using (32), the variation of refractive index can be written as

$$\Delta n = -\frac{1}{2}n_0{}^3 pS = -\frac{1}{2}n_0^3 p\sqrt{\frac{2P_{ac}}{\rho v_{ac}^3 HL}}. \tag{66}$$

Substituting (66) in (53)

$$\xi = \frac{k_f \frac{1}{2}\frac{n_0^3 p}{\sqrt{\rho v_{ac}^3}}\sqrt{\frac{2P_{ac}L}{H}}}{\cos(\vartheta_o)}. \tag{67}$$

Thus the normalized intensity of the first order diffracted light at the Bragg incidence is (using 63)

$$I_1 = \sin^2\left(\frac{\pi}{\lambda_{op}\cos\vartheta_B}\sqrt{\frac{1}{2}M_2 P_{ac}\frac{L}{H}}\right) \tag{68}$$

where

$$M_2 = \frac{n^6 p^2}{\rho v_{ac}^3}. \tag{69}$$

The efficiency η is given by the ratio of the intensity of the diffracted light I_1 over the incident beam I_{inc}

$$\eta = \frac{I_1}{I_{inc}} = \sin^2\left(\frac{\pi}{\lambda_{op}}\sqrt{\frac{1}{2}M_2 P_{ac}\frac{L}{H}}\right). \tag{70}$$

Using the Taylor approximation,

$$\eta \approx \frac{\pi^2}{2\lambda_{op}^2}M_2 P_{ac}\frac{L}{H}. \tag{71}$$

So the efficiency of the diffraction (or deflection) is proportional to the acoustic power P_{ac}, the material figure of merit M_2 and the geometric factors L/H while it is inversely proportional to the square of the optical wavelength. From (71) a good acousto-optic material should have a high figure of merit in addition to good optical and acoustical characteristics such as low attenuation. The figure of merit M_2 is not the only one, but others have been defined and used according to the application. The most common figures of merit are [5]

$$M_1 = \frac{n^7 p^2}{\rho v}$$

$$M_2 = \frac{n^6 p^2}{\rho v^3}$$

$$M_3 = \frac{n^6 p^2}{\rho v^2}$$

$$M_3 = \frac{n^8 p^2}{\rho v^{-1}}$$

in which n is the optical index of refraction, p the appropriate component of the photoelastic tensor, ρ the mass density, and v the acoustic phase velocity.

It is common practice to use M_2 as a reference and to define M_1, M_3 and M_4 in relation to M_2, n and v.

In fact $M_1 = M_2 n v^2$. It is used to optimize the efficiency bandwidth $\Delta f \propto n v^2$ [13]. Thus, the efficiency can be written as

$$\eta \approx 9 M_1 \frac{P_{ac}}{\lambda_{op}^3 f_{op} \Delta f H}. \tag{72}$$

M_2 is used when the diffraction efficiency is directly related to the acoustic power P_{ac} and the medium geometry L and H

$$\eta \approx \frac{\pi^2}{2} M_2 \frac{P_{ac} L}{\lambda_{op}^2 H}. \tag{73}$$

$M_3 = M_2 n v$ is used to design a reflector where the thickness H is as large as the optical beam size [8], and hence the relative diffraction efficiency will be

$$\eta \approx 9 M_3 \frac{P_{ac}}{\lambda_{op}^3 f_{op}}. \tag{74}$$

$M_4 = M_2 \left(n v^2\right)^2$ is applicable in the design of wideband deflectors or modulators where power density is the limiting factor. Thus, the efficiency will be given by

$$\eta \approx \frac{16 M_4}{\lambda_{op}^4 f_{op}^2 \Delta f^2} \frac{P_{ac}}{LH}. \tag{75}$$

5.2 Acousto-optic devices applications

Acousto-optic devices have long been used in a variety of laser intracavity applications. These applications can be divided into two categories: zeroth beam order applications and diffracted beam applications. Diffracted beam applications are the most common (such as modulator, deflector, tunable filter, frequency shifter) while one of the zeroth order beam applications is A-O Q-switching.

5.2.1 Modulator

The acousto-optic interaction is also used to modulate light both in amplitude and in frequency. Usually this type of device operates in the Bragg region where only one diffracted order is predominant. For proper modulator operation, the divergence of the optical beam ϕ_{op} should be approximately equal to that of the acoustic beam ϕ_{ac} [5].

For a Gaussian beam the divergence can be written as

$$\phi_{op} = \frac{4\lambda_{op}}{\pi n d} \tag{76}$$

and for an acoustic wave generated by a flat transducer of width L'

$$\phi_{ac} = \frac{\lambda_{ac}}{L'}. \tag{77}$$

Thus the divergence ratio is given by

$$a = \frac{\phi_{op}}{\phi_{ac}}. \tag{78}$$

At low values of a the maximum modulation frequency f_m approaches its limit (see Figure 12a)

$$f_m \approx \frac{0.75}{\tau} \tag{79}$$

where $\tau = \frac{d}{v_{ac}}$ is the acoustical transient time across the optical beam. Increasing the a value, it also increases the intensity value of the diffracted light, as it is shown in Figure 12b.

The maximum product frequency bandwidth – peak intensity is given for a values between 1.5 and 2. That corresponds to

$$f_m \approx \frac{0.65}{\tau}. \tag{80}$$

In general, to characterize the modulator the rise time τ_R is used that is proportional to the acoustic traveling time τ through the laser beam. It is given by

$$\tau_R = \beta\tau \tag{81}$$

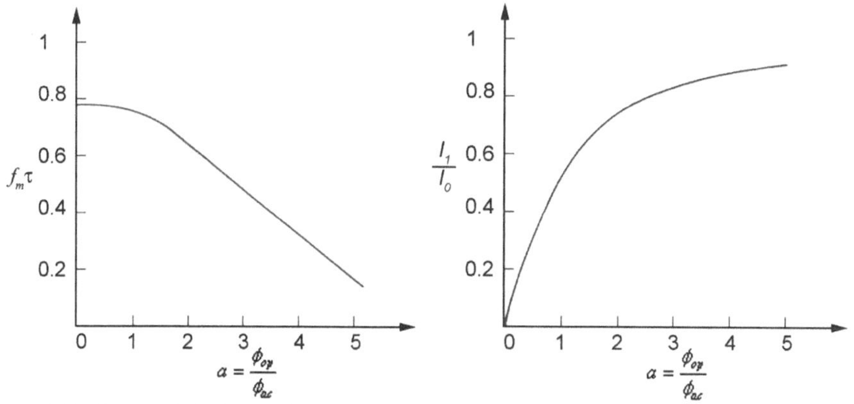

(a) Product bandwidth modulation vs. *a*. (b) Normalized maximum peak diffracted light vs. *a*.

Fig. 12 Modulator behavior according to the ratio of the optical and acoustic divergence.

where β is a constant depending on the laser beam profile. For example, for the TEM_{00} beam it is equal to 0.66.

Another requirement for the modulator device is that the diffracted and the undiffracted beam must be well separated. This implies that the Bragg angle ϑ_B should be at least equal to the optical divergence ϕ_{op}

$$\vartheta_B = \arcsin\left(\frac{\lambda_{op}}{2\lambda_{ac}}\right) \approx \frac{\lambda_{op}}{2\lambda_{ac}} = \frac{\lambda_{op} f_{ac}}{2v_{ac}} = \phi_{op} = \frac{4\lambda_{op}}{\pi d}, \tag{82}$$

hence the minimum center frequency is

$$f_{ac,\min} = \frac{8}{\pi\tau}. \tag{83}$$

Combining (83) and (80) we find the relation between center frequency and modulation bandwidth

$$f_{ac,\min} \approx 4 f_m. \tag{84}$$

The acousto-optic modulator has a nonlinear transfer function MTF (Figure 13b) defined by

$$MTF = \exp\left(-\left(\frac{f_m}{1.2 f_0}\right)^2\right) \qquad f_0 = \frac{0.35}{\tau_R}. \tag{85}$$

In order to measure the separation level between the light intensity of the zeroth order and that of the first order the contrast ratio CR [29], defined as

$$CR = \frac{I_1}{I_0}, \tag{86}$$

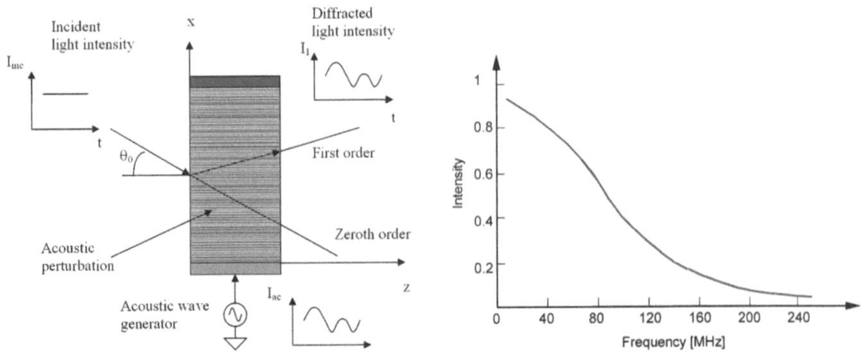

(a) Modulator: the incident light intensity is constant while the acoustical signal is variable in frequency and intensity.

(b) qualitative plot of the modulator transfer function according to the acoustic frequency.

Fig. 13 Modulator working principle.

is used, where I_1, is the light intensity of the first-order diffracted beam and I_0 is the light leakage of the incident beam in the direction of the first-order beam when the *AO* modulator is not energized. The CR value is defined for both pulse modulation mode conditions (dynamic CR) and static conditions (static CR). The dynamic CR has great importance in laser communication systems in order to measure the cross talk between zeroth and first-order channels. In general the CR value is limited by crystal imperfection and by the scattered light.

As shown by (79) the time taken for the acoustic wave to travel across the diameter of the light beam limits the modulation bandwidth. So to increase the bandwidth the diameter of the light beam must be as small as possible.

Modulators are used:

- In infrared communication applications due to the existence of several materials working in that spectral region [1, 3, 48]
- As multiplexer and demultiplexer in optical pulse code modulation using low acoustic power and standing acoustic waves
- Inside a laser cavity as Q-switching, Mode Locking and cavity dumping

5.2.2 Deflector

An acoustic-optic deflector changes the angle of the deflecting beam proportionally to the driver acoustic frequency, so that the higher the frequency, the larger the diffracted angle (Figure 14). It can be used to modulate the incident beam by shifting the position of the reflected beam on the output collimator.

The angle between the undiffracted beam and first-order diffracted beam is equal to two times ϑ_B:

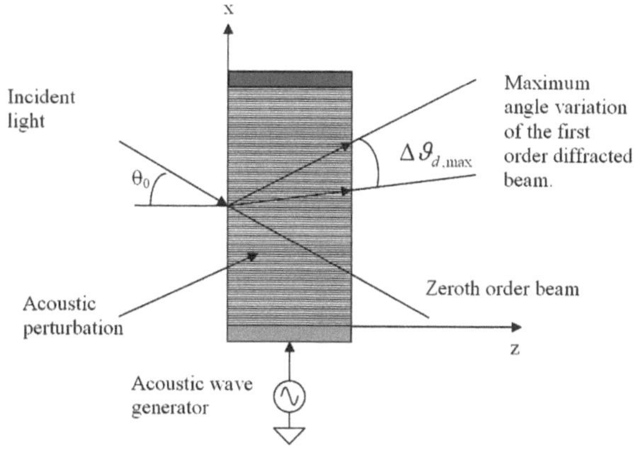

Fig. 14 Deflector working principle.

$$2\vartheta_B = \arcsin\left(\frac{\lambda_{op}}{2\lambda_{ac}}\right) = \arcsin\left(\frac{\lambda_{op} f_{ac}}{2v_{ac}}\right). \tag{87}$$

Thus the total angle of deflection $\Delta\vartheta_d$ for a frequency change Δf_{ac} is

$$\Delta\vartheta_d = \frac{\lambda_{op}}{v_a}\frac{\Delta f_{ac}}{\cos\vartheta_0} \approx \frac{\lambda_{op}}{v_a}\Delta f_{ac} \tag{88}$$

if ϑ_0 is neglected.

In a deflection system, there are two important performance parameters: resolution (maximum number of resolvable angular positions) and speed. The resolution N is defined as the range of deflection angles $\Delta\vartheta_{d,\max}$ divided by the angular spread of the diffracted beam ϕ_d(divergence)

$$N = \frac{\Delta\vartheta_{d,\max}}{\phi_d}. \tag{89}$$

The divergence of the diffracted beam ϕ_d will be equal to that of the incidence beam ϕ_{op} if a in (78) satisfies a $<<$ 1

$$\phi_{op} = \xi\frac{\lambda_{op}}{d}, \tag{90}$$

where ξ is a multiplication factor (near unity) that depends on the amplitude distribution of the optical beam and the criterion used for resolvability [39], and d is the diameter of the deflected beam (or of the incident beam if they are equal). Combining (89), (88) and (90)

$$N = \frac{\tau\Delta f_{ac}}{\xi} \tag{91}$$

where

$$\tau = \frac{d}{v_{ac} \cos \vartheta_0}. \quad (92)$$

As with the modulator, the performance of the deflector is related and limited to the transient time τ across the optical beam. But here, to the contrary, in order to increase the number of resolvable spots N, τ should be increased (directly proportional). Another limitation is due to the length L of the optical path. The spread angle in which the deflector works properly depends on whether a wide enough spectrum of plane waves is available in the radiation pattern of the transducer to satisfy the Bragg angle condition at all frequencies

$$\frac{\lambda_{ac,nom}}{L} > \frac{1}{2} \frac{\lambda}{v_{ac}} \Delta f_{ac} \quad (93)$$

or

$$L < \frac{1}{2} \frac{\lambda_{ac,nom}^2}{\lambda_{op}} \frac{f_{ac,nom}}{\Delta f_{ac}} \quad (94)$$

where $\lambda_{ac,nom}$ and $f_{ac,nom}$ are the nominal sound wavelength and frequency respectively. Equation (94) indicates that for large Δf the path length L must be decreased. This reduces the diffraction efficiency (73), and also increases the strength of additional orders by moving out of the Bragg region (56). Such difficulties may be overcome by using a phased array transducer.

Deflectors are used as:

- Scanner
- Switches
- Spectrum analyzer
- Hologram, printing, photolithography
- Display driver
- Cavity dumper

5.2.3 Tunable filter

The AOTF (Acousto-optic Tunable Filter) working principle is based on the anisotropic collinear/non collinear wave interaction or isotropic collinear interaction [37]. A collinear interaction is present when the light propagates in the same direction as the acoustic wave (the wavevectors k_{ac} and k_{op} are parallel). In this case, the grating fringes are perpendicular to the direction of light propagation and the grating behaves as a pure reflection grating. On the contrary if k_{ac} and k_{op} are not parallel then a non collinear interaction will occur. In a bulk acousto-optic device the polarization of light can be rotated 90 degrees by way of the photoelastic effect produced by the acoustic strain wave. The filtered light will be separate from the unfiltered if the interaction strength L is strong enough to allow polarization to flip, for example from TE to TM. This process is resonant and narrow in spectral width because the two polarization states propagate at different velocities. Coupling can be achieved only when the phase-matching condition is met, i.e., when the sound wave momentum just compensates the TE and TM momentum mismatch [43].

The conservation of energy and momentum conditions with the phase matching condition of an acousto-optic interaction of the collinear type give the following relation [10]

$$k_o = k_e + k_{ac} or \frac{\omega_o n_0(\omega_0)}{c} = \frac{\omega_e n_e(\vartheta_e, \omega_e)}{c} + \frac{\omega_{ac}}{v_{ac}} \tag{95}$$

$$\omega_\mathbf{o} = \omega_\mathbf{e} + \omega_\mathbf{ac} \tag{96}$$

$$n_e(\vartheta_e, \omega_e) = \sqrt{\frac{1}{\frac{n_o^2}{\cos^2 \vartheta_e} + \frac{n_e^2}{\sin^2 \vartheta_e}}} \tag{97}$$

where k_e and k_o are the incident and diffracted optical momentum at the propagation angles ϑ_e and ϑ_o respectively, $n_0(\omega_0)$ and $n_e(\vartheta_e, \omega_e)$ are the refractive indexes of the ordinary and extraordinary polarizations, and w_e and w_o are the frequencies of the incident and diffracted light.

Combining (95) and (96)

$$\frac{(\omega_e \pm \omega_{ac})\, n_o}{c} = \frac{\omega_e n_e}{c} \pm \frac{\omega_{ac}}{v_{ac}}. \tag{98}$$

Reassembling (98) yields

$$\pm \frac{\omega_{ac}}{v_{ac}} \pm \frac{\omega_{ac} n_o}{c} = (n_e - n_o)\frac{\omega_e}{c}. \tag{99}$$

That can be approximated by neglecting the term $\frac{\omega_{ac} n_o}{c}$

$$\omega_{ac} \approx |n_e - n_o| \frac{v_{ac}}{\lambda_{op}}. \tag{100}$$

As the doped surface refractive index, such as of the waveguide surface, is more than one percent of that of the body, and hence the difference between the modal index and the substrate refractive index is less than one percent, it is possible to write (100) as [35]

$$\omega_{ac} \approx |n_{TE} - n_{TM}| \frac{v_{ac}}{\lambda_{op}}. \tag{101}$$

5.2.4 Frequency shifter

This component is used to change the frequency of the diffracted light, but also it can be a modulator or e deflector. It uses the principle of energy momentum conservation so that the scattered circular frequency is given by

$$\omega_\mathbf{s} = \omega_\mathbf{op} + \omega_\mathbf{ac}. \tag{102}$$

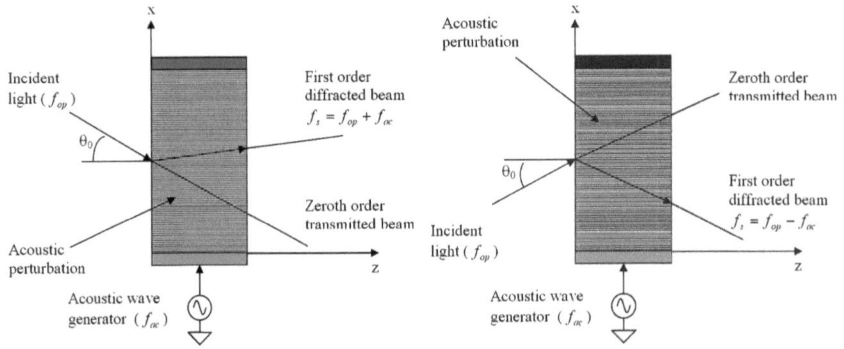

(a) Frequency up shifter. (b) Frequency down shifter.

Fig. 15 Schematic of frequency shifter.

This phenomenon is called the Doppler shift. If the incident acoustic wave is introduced in the direction of the incident optical wave, the scattered optical frequency is given by (Figure 15a).

$$f_s = f_{op} + f_{ac}. \tag{103}$$

If the incident acoustic wave is introduced in the opposite direction of the incident optical wave, the scattered optical frequency will be decreased (Figure 15b).

5.2.5 Q-switch

A Q-switch is a device that, inserted into a laser cavity, allows the production of a short high energy light pulse [4, 6, 33] (Figure 16). The term Q means quality factor of a laser cavity. It is defined as the ratio of the energy stored in the laser cavity to the energy loss per cycle. Changing the cavity loss allows a change in the Q factor. When a Q-switch is turned on, the cavity loss is large enough (low Q) to inhibit laser radiation, despite continual pumping of the gain medium. When the Q-switch is turned off, the cavity loss is reduced to its minimum (high Q) and all of the energy stored in the gain medium is released in a single high-power laser pulse. By repeating this process, a sequence of laser pulses is emitted. In the Q-switching operation, the repetition rate has been limited by the time to repump the population inversion [5].

The acousto-optic Q-switch acts as a fast optical shutter that changes its polarization. The low-Q state is achieved by applying an acoustic wave to the Q-switch such that the polarization is rotated by 90 degrees. In this way, the optical feedback is lacking and the cavity can't resonate. The high-Q state is achieved by turning off the acoustic wave so that polarized laser light can move through the optical path of the cavity with minimum loss.

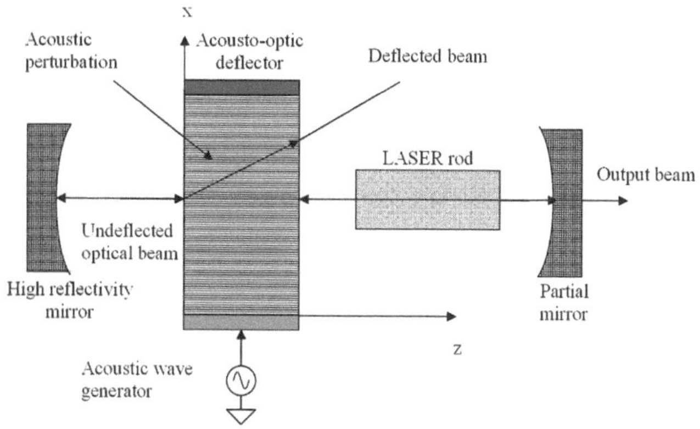

Fig. 16 Q switching LASER working principle.

The Q-switch laser can be used for: materials processing (marking, cutting, welding, drilling), medical (ophthalmology and dermatology), military (sensing, range finding, target illumination) etc.

5.2.6 Mode lockers

Normally a laser can oscillate in many longitudinal modes, with frequencies that are equally separated by the intermodal spacing f_m given by [16, 40]

$$f_m = \frac{c}{2L_m}, \tag{104}$$

where L_m is the cavity length and c the light velocity. Furthermore these modes oscillate independently in free-running modes, each with its own phase with respect to the other. The mode locker (Figure 17a) forces the phase of each mode to remain equal with respect to the other (all modes are locked in phase). In order to achieve its objective the mode locker laser uses a q-switcher inside the resonator working at frequency $f_{ac} = f_m$.

Figure 17b shows some parameters that characterize a mode locker laser. These parameters are:

Temporal period:

$$\tau_m = \frac{1}{f_m}, \tag{105}$$

Pulse width:

$$\tau_{pulse} = \frac{\tau_m}{N_m} \tag{106}$$

where N_m is the number of modes of oscillation,

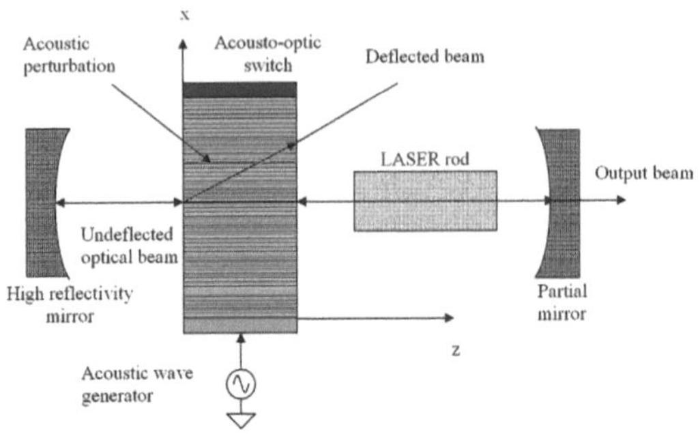

(a) Mode locker working principle.

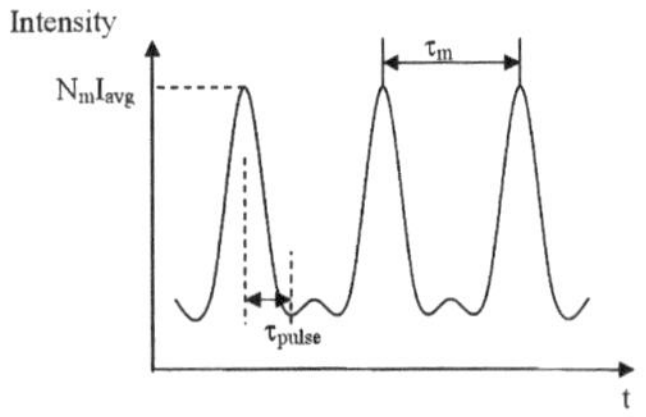

(b) Mode locker output pulses.

Fig. 17 Mode locker working principle.

Mean intensity:

$$I_{avg} = N_m I_0, \tag{107}$$

where I_0 is the intensity of each mode,

Peak intensity:

$$I_{peak} = (N_m)^2 I_0 = N_m I_{avg}, \tag{108}$$

Spatial period:

$$c\tau_m = 2L_m, \tag{109}$$

Pulse length:

$$c\tau_{pulse} = \frac{2L_m}{N_m}. \tag{110}$$

In order to have stable mode locking of a laser, temperature control of the laser's environment as well as temperature control of the mode locker modulator's crystal is necessary, as any change in cavity length will result in unstable mode locking [49]. Stable mode locking of a laser also requires a very clean cavity. Any dust or contaminant inside the laser cavity will influence the laser modes.

5.2.7 Cavity dumping

The Cavity Dumping technique is used to obtain a single high-intensity pulse. An acousto-optic device placed into the laser cavity allows the production of a high power pulse (Figure 18). The difference with Q-switching is that here a photon, rather than a population difference, is stored in the resonator during off-times and releasing during on-times [5]. This acousto-optic device replaces one mirror in the absence of the acoustic grating, and diffracts energy out of the cavity when an acoustic wave propagates into the mirror [17,28].

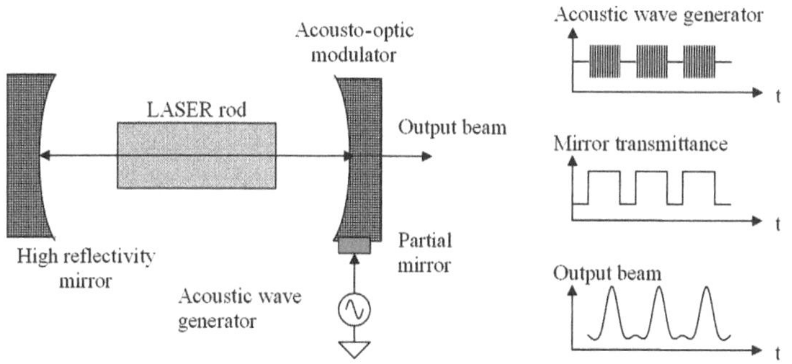

Fig. 18 Cavity dumping working principle.

6 Conclusions

This chapter gives an overview of work done over the years in the field of IR power generation and detection. In the first part we described the classical background, while in the second part acousto-optic effects are reviewed in order to give an idea of the complexity of the matter. The availability of stable IR sources in the $1 \div 14$ μm has opened the possibilities of detecting volatile compounds, even aggressive ones which manifest adsorption power, particularly in the $8 \div 14$ μm. Requirements remain for a significant miniaturization of the acousto-optic apparatus, which as of now are too expensive, large and heavy.

References

1. Abrams R. L. and Pinnow A., Efficient Acousto-Optic Modulator at 3.39 and 10.6 Microseconds in Crystalline Germanium, IEEE J. Quantum Electron. (Corresp.), Vol. QE-7, Mar. 1971, pp. 135–136.
2. Born M. and Wolf E., Principles of Optics, Third Edition, Pergamon Press, New York, 1965, ch. 12.
3. Chang C. and Moradian G., Frequency Modulated Acousto-Optic modulators for 10.6 pm laser Communications, Electro-Optics System Design Conf., San Francisco, Nov. 1974.
4. Chang Chih-Kang, Chang Jih-Yuan and Kuo Yen-Kuang, "Optical Performance of Cr:YSO Q-switched Cr:LiCAF and Cr:LiSAF Lasers," Proc. SPIE, Vol. 4914 2002, pp. 498–509.
5. Chang I.C., Acoustooptic Devices and Applications, IEEE Trans. Son. Ultrason. Vol n. 23, No. 1, Jan. 1976.
6. Delgado M., Zalvidea Pinar D., Dez A., Prez-Milln P. and Andrs M. V., Q-switching of an All-Fiber Laser by Acousto-Optic Modulation of a Fiber Bragg Grating, Opt. Express, Vol. 14, No. 3, 6 Feb 2006, pp. 1106–1112.
7. Dixon R. W., Acoustic Diffraction of Light in Anisotropic Coherent Light, Proc. ZEEE Vol. 54, Oct. 1966, pp. 1429–1437.
8. Dixon R. W., Photoelastic Properties of Selected Materials and Their Relevance for Applications to Acoustic Light Modulators and Scanners, J. Appl. Phys., Vol. 38, No. 13, 1967, pp. 5149–5153.

9. Dunn Derrek Butler, McNeill Mark, Jiangang Xia and Ting-Chung Poon, Three-Dimensional Analytical and Numerical Solutions for Acousto-Optic Interaction, IEEE 1997.
10. Gao Lu, Herriot Sandrine I. and Wagner Kelvin H., Novel Approach to RF Photonic Signal Processing Using an Ultrafast Laser Comb Modulated by Traveling-Wave Tunable Filters, IEEE J. Sel. Top. Quant. Electron., Vol. 12, No. 2, March/April 2006.
11. Gaylord T. K. and Moharam M.G., Analysis and Applications of Optical Diffraction by Gratings - Proc. IEEE, Vol. 73, No. 5, May 1985.
12. Gordon E. I., A Review of Acousto-Optical Deflection and Modulation Devices, Proc. IEEE, Vol. 54, Oct. 1966, pp.1391–1401.
13. Gordon, E., Figure of Merit for Acousto-Optical Deflection and Modulation Devices, IEEE J. Quan. Electron. Vol 2, No. 5, May 1966, pp. 104–105.
14. Henderson R., Wavelength Considerations. Instituts fr Umform- und Hochleistungs. Retrieved on 2007-10-18.
15. IPAC Staff, Near, Mid and Far-Infrared. NASA ipac. Retrieved on 2007-04-04.
16. Jhon Young Min and Kong Hong Jin, Self Q Switching of a CW Mode-Locked Nd : YLF Laser by Cavity Length Detuning, - IEEE J. Quant. Electron. Vol. 29, No. 4, April 1993, pp. 1042–1045.
17. Johnson R. H., Characteristics of Acousto-Optic Cavity Dumping in a Mode-Locked Laser, IEEE J. Quant. Electron., Vol. QE-19, Feb. 1973, pp. 255–257.
18. Klein W. R., Cook B. D., and Mayer W. C., Light Diffraction by Ultrasonic Gratings, Acustica, Vol. 15, Jan. 1065, pp. 67–74.
19. Klein W. R. and Cook B. D., Unified Approach to Ultrasonic Light Diffraction, IEEE Trans. Son. Ultrason., Vol. SU-14, July 1967, pp. 123–134.
20. A. R. Korpel Atller, Desmares P., and Watson W. A Televislon Display Using Acoustic Deflection and Modulation of Coherent Light, Proc. IEEE Vol. 54, N. 10 Oct. 1966.
21. Korpel A., Acousto-Optics-A Review of Fundamentals, - Proc. IEEE, Vol. 69, No. 1, Jan. 1981.
22. Korpel A., Acousto-optics, in Applied Solid State Science, Vol. 3, R. Wolfe, Ed, Academic, New York, 1972, ch. 2, p. 73–179.
23. Korpel A., Acousto-Optics, Marcel-Dekker, New York, 1988.
24. Krishman R. S., Progress in Crystal Physics, Vol. I. Interscience, New York, 1958.
25. Kruse P. W., McGlauchlin L. D. and McQuistan R. B., Elements of Infrared Technology, Wiley, New York, 1962.
26. Maldonado Theresa A. and Gaylord Thomas K., Electrooptic Effect Calculations: Simplified Procedure for Arbitrary Cases, Appl. Opt. Vol. 27, No. 24, 15 December 1988, pp. 5051–5066.
27. Mason W P., Optical Properties and the Electro-Optic and Photo-Elastic Effects in Crystals Expressed in Tensor Form, Bell Sys. Tech. J., Vol. 29, 1950, p. 161.
28. Maydan D., Fast Modulator for Extraction of Internal Laser power, Appl. Phys., Vol. 41, March 15, 1970, pp. 1552–1559.
29. Men A. and Rosenberg M., Acoustooptical Device with Extremely High Contrast Ratio Appl. Opt. Vol. 22, No. 6, 15 March 1983, pp 873–875.
30. Miller J. L., Principles of Infrared Technology (Van Nostrand Reinhold, 1992), and Miller and Friedman, Photonic Rules of Thumb, 2004.
31. Nath N. S. N., The Diffraction of Light by High Frequency Sound Waves: Generalized Theory, Proceedings of the Indian Academy of Science, Sect. A Vol. 4, No. 2, 1936. pp. 222–242.
32. Nelsen D. F. and Lax M., Theory of the Photoelastic Interaction, Physical Review B, Vol. 3 No. 8, (April) 1971 pp. 2778–2794.
33. Ngoi B. K. A., Venkatakrishnan K., Tan B., Stanley P. and Lim L. E. N., Angular Dispersion Compensation for Acoustooptic Devices Used for Ultrashort-Pulsed Laser Micromachining, Optics Express, Vol. 9, Iss. 4, (August) 2001, pp. 200–206.
34. Nye J. F., Physical Properties of Crystals, Oxford, New York, 1960.
35. Ohmachi Yoshiro and Noda Juichi, LiNb03 TE-TM Mode Converter Using Collinear Acoustooptic Interaction, IEEE J. Quant. Electron., Vol. QE-13, No. 2, Feb. 1977.

36. Phariseau P., On the Diffraction of Light by Progressive Supersonic, Indian Acad. Sci., Vol. 28A, 1948, pp. 54–62.
37. Quate C. F., Wilkinson C. D. W. and K. Winslow, Interaction of Light and Microwave Sound, Proc. IEEE, Vol. 53, Oct. 1965, pp. 1604–1623.
38. Raman C. V. and Nath N. S. N., The Diffraction of Light by High Frequency Sound Waves, Proc. Indian Acad. Sci. Pt. I, Vol. 2A, pp. 406–412, 1935; Pt. II, Vol. 2A, pp. 413–420, 1935; Pt. III, Vol. 3.4 pp. 459–465, 1936.
39. Randolph J. and Morrison J., Modulation Transfer Characteristics of an Acousto-Optic Deflector, Appl. Opt., Vol. 10, 1971, pp. 1383–1385. "Rayleigh-Equivalent Resolution of Acousto-Optic Deflection Cells" Appl. Optics, Vol. 10, 1971, pp. 1453–1454.
40. Saleh Bahaa E. A. and Teich Malvin Carl, Fundamentals of Photonics, Wiley, New York, 1991, pp. 826–827.
41. Sapriel J., Acousto-Optics, Wiley, New York, 1979.
42. Schmid M., Weber R., Graf Thomas, Roos M. and Weber Heinz P., Numerical Simulation and Analytical Description of Thermally Induced Birefringence in Laser Rods, IEEE J. Quant. Electron. Vol. 36, No. 5, May 2000.
43. Smith David A., Chakravarthy Rohini S., Bao Zhuoyu, Baran Jane E., Jackel Janet L. d'Alessandro Antonio, Fritz Daniel J., Huang S. H., Zou X. Y., Hwang S.-M., Willner Alan E. and Li Kathryn D., Evolution of the Acousto-Optic Wavelength Routing Switch, J. Lightwave Technol., Vol. 14, No. 6, June 1996.
44. Tarn Chen-Wen and Banerjee Partha P., A Fourier Transform Approach to Acoustooptic Interaction of Curved, CW Ultrasonic Waves and Optical Beams with Arbitrary Profiles, Proc. IEEE 1992, pp. 566–568.
45. Tsai Chen S., Guided-Wave Acoustooptic Bragg Modulators for Wide-Band Integrated Optic Communications and Signal Processing, IEEE Trans. Circ. Syst., Vol. CAS-26, No. 12, Dec. 1979.
46. Uchida Naoya and Niizeki Nobukazu, Acoustooptic Deflection Materials and Techniques, Proc. IEEE, Vol. 61, No. 8, Aug. 1973.
47. Venzke Clark, Korpel Adrian and Mehrl David, Improved Space-Marching Algorithm for Strong Acousto-Optic Interaction of Arbitrary Fields, Appl. Opt. Vol. 31, No. 5, 10 Feb. 1992.
48. Warner A. W. and Pinnow D. A., Miniature Acousto-Optic Modulators for Optical Communications, J. Quant. Electron.,Vol. QE-9. Dic 1973, p 1155–1157.
49. Wisoff P. J. K. and Young J. F., Active Mode Locking of a Microwave-Pumped XeCl Laser, IEEE J. Quant. Electron., Vol. QE-20, No. 3, Mar. 1984, pp. 195–197.

Knowledge Based Diversity Processing

Christopher John Baker[1] and Hugh Duncan Griffiths[2]

Abstract In the past, radar sensing has tended to consist of relatively monolithic, single entity systems that present their output (often in the form of detections on a PPI display) as reports to an operator. The function of the operator is to interpret these reports and subsequently either provide information via a command chain for a decision on how to react, or to make such a decision themselves. In this way the human operator has been the source of intelligence in the sensing system, often aided and abetted by training and experience that has allowed a remarkably wide set of tasks to be performed with a high level of ability. However, the advent of electronic scanning coupled with advances in digital signal processing leads to a class of radar known as 'Multi-Function' and these are now challenging traditional methods by placing demands on the radar itself to make well informed, reliable decisions as to how a mission should be conducted. This is leading to the concept of intelligent or cognitive sensing. As a simple example an electronically scanned radar system is able to re-point its beam in timescales that are much faster than human reaction times. Where the beam should next be pointed therefore has to be a decision made by the radar itself. To understand and exploit its environment as fully as possible the system has the option of varying its parameters in a way that is tailored to the information it is seeking. This we term 'diversity processing'. A logical strategy is to do this in the light of prior experience, its own perception of the world and an appreciation of the task to be carried out. This we term 'knowledge based processing'. In this chapter we explore the early development of the concept of 'knowledge based diversity', drawing upon examples from both synthetic and natural echo locating systems to indicate how, eventually, true intelligence might be incorporated into future sensors.

Keywords: Multi-Function Radar, knowledge based processing, intelligent processing, diversity, resource management

[1]University College London

[2]Defence College of Management and Technology, Shrivenham, Cranfield University

J. Byrnes (ed.), *Unexploded Ordnance Detection and Mitigation,*

1 Introduction

We begin by examining what is meant by terms such as 'intelligence' and 'knowledge' to set the scene for the challenges faced by knowledge based diversity processing systems. Artificial intelligence has been a topic of intense research for many years and so called artificial intelligent methods of processing radar data, such as neural networks, have been exploited for some time. However, these have been implemented in a relatively simplistic way not really representative of true intelligence. In fact intelligence might be better used to describe a property of the mind (or in our case the radar signal processor) that encompasses many related abilities, such as the capacities to reason, to plan, to solve problems, to think abstractly, to comprehend ideas, to use language, and to learn. There are several ways to define intelligence. In some cases, intelligence may include traits such as creativity, personality, character, knowledge, or wisdom. These are characteristics that go substantially beyond any existing radar sensor system, as they imply a dynamic interaction with the world in which they exist whilst attempting to achieve a chosen goal or objective. Similarly, knowledge is defined in the Oxford English Dictionary variously as (i) *expertise, and skills acquired by a person through experience or education; the theoretical or practical understanding of a subject,* (ii) *what is known in a particular field or in total; facts and information or* (iii) *awareness or familiarity gained by experience of a fact or situation.* Philosophical debates in general start with Plato's formulation of knowledge as 'justified true belief'. There is, however, no single agreed definition of knowledge presently, nor any prospect of one, and there remain numerous competing theories. Knowledge acquisition involves complex cognitive processes: perception, learning, communication, association and reasoning. The term 'knowledge' is also used to mean the confident understanding of a subject with the ability to use it for a specific purpose if appropriate.

Although imprecise, these descriptions convey a sense of known information that can be exploited to create a desired effect in the context of a perceived understanding of the local and globally significant operating environments. Together they will form the basis of future truly intelligent sensing systems and are providing the impetus for much current research as demonstrated by the US 'KASSPER' [1] and 'Sensors as Robots' [2] programmes. In this chapter we explore the underlying concepts that provide the keys to greater and greater levels of cognition in sensor systems. In particular we draw upon the lessons that are being learnt from nature and especially from echo-locating mammals which imply that a much more holistic approach to systems operation may be necessary. We begin by reviewing the nature of information that can be inferred from radar echoes, and then go on to explore how this is enhanced by exploiting the full parametric diversity available to maximize the quality of sensed information upon which radar systems can make robust and reliable decisions, and subsequently how this might be further enhanced by exploiting prior knowledge.

2 Acquiring Information from Radar

It is intuitive that the information acquired from a radar sensor is dependent on the parameters used to design the system. The more we are able to adaptively adjust these parameters, the more it is possible to maximize the quality and relevance of the information acquired and hence to maximize the likelihood of 'mission' success. Changes to the radar of this type we term diversity and the range of diversity possible is embraced by the following:

- ***Bandwidth and or frequency*:** this might be wide bandwidths for high spatial resolution or narrow bandwidth, long duration signals for high Doppler resolution.
- ***Orientation*:** this provides for spatial diversity. A simple example is SAR or ISAR which use angular diversity to reduce speckle in multi-looked SAR or ISAR imagery.
- ***Waveform design*:** where parameters such as pulse length, modulation and PRF can be adjusted dynamically.
- ***Signal strength*:** it is clear that a low echo strength of received signals will be corrupted by the effects of noise, and hence the signal-to-noise ratio must be sufficiently high to avoid this.
- ***Time*:** the time evolution of behavior of an object can give important clues as to its nature and intent.
- ***Phase centre*:** the use of multiple and adjustable phase centres can provide important location information.
- ***Polarization*:** man-made targets are often made up of polarization sensitive structures that allow them to be more easily recognized against a background comprising natural scatterers.
- ***Knowledge diversity*:** this can take almost any form of priors, memory mapping and their associations.

In the following sections we examine aspects of these forms of diversity, showing how some are already used in radar sensors but how performance could be improved if the concepts of true intelligence could be appropriately adopted.

3 Information in Current Radar Systems

We begin this section by examining how resolution in the radial-range and cross-range dimensions provides useful improvements in information content by creating localized zones of scattering. Frequency or bandwidth span is conventionally used to provide spatial resolution to isolate targets from one another. Wide-bandwidths can result in improvements in down range resolution which further increase information content by isolating scattering into separate discrete areas. This is an example of using coherence between the transmitted and received signals. Synthetic aperture techniques such as SAR and ISAR, e.g. [3], combine angle and bandwidth data coherently to provide high resolution in two dimensions. We also note that frequency diversity employing a multiplicity of illuminating frequencies of non-overlapping bandwidths can be used non-coherently to reduce the effects of speckle as in multi-looked SAR or MIMO radar [4]. In this case the multi-looking is attempting to provide an improved estimate of the underlying radar cross-section and hence improved information. Here we will examine in more detail how coherent use of frequency diversity provides better spatial information.

Radar is a relatively simple sensor able to provide information about the position of an object in 3-D space as a function of time. In essence this is facilitated by the measurement of radial range from the radar to a target and the rate of radial change of position of the target, both as a function of azimuth and elevation angles as determined by the azimuth and elevation beamwidths. In this way targets can be detected and tracked over time. By transmitting a short or modulated pulse a radar is able to resolve between multiple targets with a resolution given by:

$$\triangle r = \frac{c}{2B} \tag{1}$$

where

$\triangle r$ = range resolution (m)
c = velocity of light (ms^{-1})
B = bandwidth of radar signal (Hz)

Thus wideband signals are necessary to achieve high spatial resolution. If $\triangle r$ is a fraction of the target dimension as presented radially to the radar then it is possible to begin to measure important target characteristics such as length. Indeed, it is potentially possible to use difference in echo strengths from different parts of the target to uniquely discriminate the target from other possible candidates. Such an echo is termed a High Range Resolution Profile (HRRP) and Figure 1 shows schematically the type of response that might be observed.

However, there are limits on how wide a bandwidth it is realistically possible to transmit and receive within a single pulse. This limit can be overcome by

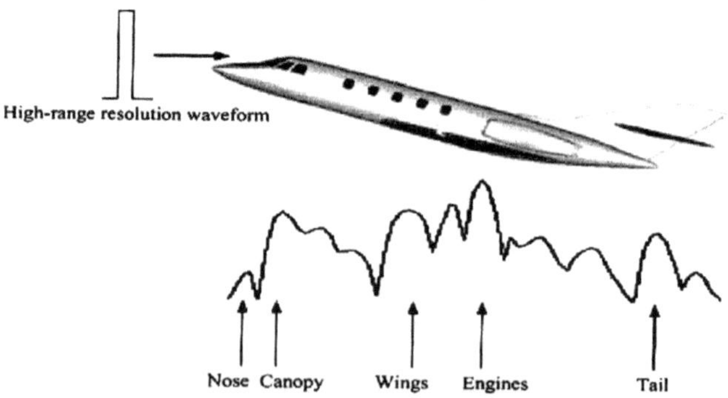

Fig. 1 The HRRP generated by a wide band pulse for an aircraft target.

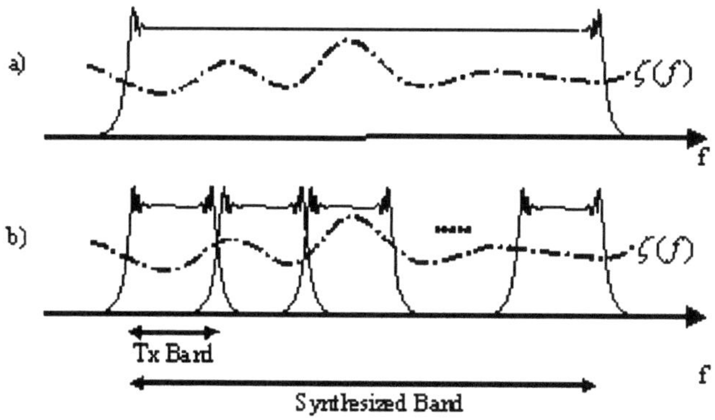

Fig. 2 Spectrum reconstruction of the target reflectivity function by one single chirp pulse **(a)** and by stepped-frequency coherent addition of sub-spectra **(b)**.

transmitting and receiving a series of pulses where the centre frequency of each is stepped incrementally such that the total bandwidth spanned is much greater that of any single pulse. Each echo is digitised and recorded so that the full bandwidth span can be re-constructed and significantly higher resolution achieved. Figure 2 shows this schematically for a series of pulses.

In this example the separation of pulses in the frequency domain is equal to their bandwidth. More typically they will be overlapped by as much as 50%. This reduces the achieved bandwidth span and degrades resolution but avoids potentially awkward bandwidth reconstruction at the band edges. Figure 3 shows an example of range profiles of a Land Rover vehicle plotted as a series of intensity modulations covering a total azimuth extent of 360. Discrete areas of high echo strength are clearly visible as is their angular span. The scattering characteristics appear quite varied, nevertheless this represents the base information for classification. However, this method only provides high resolution in the radial direction and hence any scatters lying at the same range will be measured as a single echo response. This means that small changes in the viewing geometry will cause large and rapid fluctuations in the range profile as scatters will interfere with one another constructively and destructively. One method to militate against this is to provide resolution in the cross range dimension using synthetic aperture techniques.

The real beamwidth of any radar antenna is determined approximately by the ratio of the wavelength to the physical size of the antenna. In simple form this is expressed as:

$$\triangle\Phi = \frac{\lambda}{D} \tag{2}$$

where

$\triangle\Phi$ = angular beamwidth (rads)
λ = wavelength (m)
D = diameter of antenna

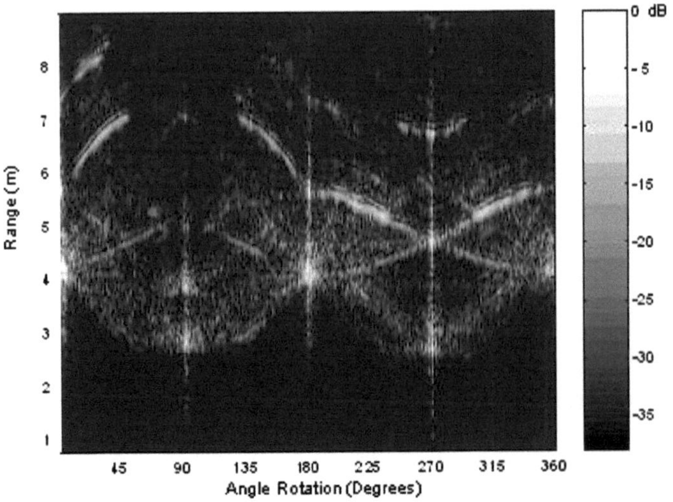

Fig. 3 History of HRR range profiles (8 cm of range resolution) from a series of X-Band stepped frequency chirps illuminating a ground vehicle as it rotates over 360 degrees. At zero degrees, the target (a Land Rover) is broadside oriented, while at 90 degrees has its end-view towards the radar.

If Equation (2) is multiplied by the range we get the width of the beam at that range. Thus at a range of 10 km the beam size of a 1 m dish antenna operating at a wavelength of 3 cm is 300 m which is much larger than, for example, vehicles and most aircraft. To overcome this limitation large apertures can be synthesised (i.e. effectively increasing D in Equation (2)) using the Synthetic Aperture Radar (SAR) or Inverse Synthetic Aperture Radar (ISAR) techniques. These techniques create large apertures (albeit on signal reception only) by collecting data over as a function of viewing angle, thus effectively mapping out a much larger aperture than that of the physical antenna. Figure 4 shows an example of ISAR imagery generated from the HRRPs displayed in Figure 3.

The image in Figure 4 has the expected rectangular outline typical of a Land Rover vehicle and the bonnet, cab and truck areas are clearly discernable. The scatterers observable as a function of angle in Figure 3 are now 'focussed' into discrete zones showing that in fact they emanate from the same physical part of the truck. In forming a 2-D image in this way we have now exploited both range resolution (frequency diversity) and angle resolution (angular diversity) and hence this is an example of exploiting diversity (i.e. combining range and angle) to improve the information obtained by the radar to make better decisions. Indeed, it seems much less of a leap of faith to make the assumption that the data shown in Figure 4 is in a form to support human based classification. However, automating this process has proved to be an extremely challenging problem in all but a few restrictive cases. The reasons for this are chiefly to do with the fact that often 'resolution cells' still contain multiple scatterers and hence they may again, scintillate rapidly with small changes in viewing angle. Additionally, there is often a multipath component which adds to

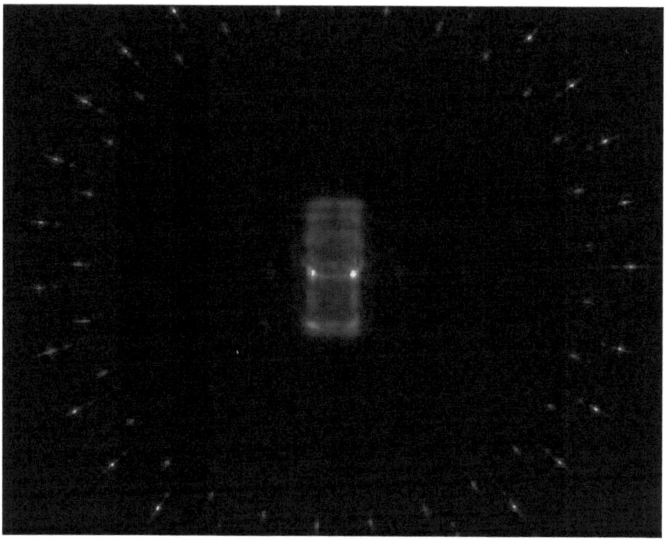

Fig. 4 Multi look ISAR image of the Land Rover target.

this scintillation. It is also possible for some resolution cells to contain part target and part clutter which further exacerbates these effects. In addition not all objects are comprised of discrete scatterers alone and other forms of re-radiation can take place. Overall the form of backscatter from extended targets is extremely complex and subject to great variability hence making consistent and reliable interpretation by automatic means a substantial challenge. However, tasks such as navigation, collision avoidance and object classification are executed with seeming ease by echo locating mammals. In the next section we examine this in more detail with a view to learning lessons that can be valuably employed in synthetic sensors.

Firstly, we examine one more component of diversity as provided by polarization. It is well known that polarimetric information may be useful for classification since it completes the information which can be obtained from the target returns. Radar targets have different responses to different polarization signals. By emitting a mixture of polarizations and using the receiving antenna with a specific polarization, several different signals can be collected and used for recognition. For this purpose it is necessary to illuminate the target with two signals having different polarization vectors, for instance when linear polarization is used the polarizations are vertical and horizontal. The polarization properties of the target can be completely represented by its scattering matrix [5]. The scattering matrix is defined as:

$$\begin{bmatrix} S_{11} & S_{12} \\ S_{21} & S_{22} \end{bmatrix} \tag{3}$$

where S represents the state of polarisation. The components of the matrix are called scattering components and are derived from the following relation:

Table 1 Scattering matrix for some common geometrical shapes.

Shape	Linear polarization	Circular polarization	Aspect
Sphere 	$S=\begin{bmatrix}1 & 0\\0 & 1\end{bmatrix}$	$S=\begin{bmatrix}0 & 1\\1 & 0\end{bmatrix}$	Every aspect
Flat Plate 	$S=\begin{bmatrix}1 & 0\\0 & 1\end{bmatrix}$	$S=\begin{bmatrix}0 & 1\\1 & 0\end{bmatrix}$	Radar signal perpendicular to the plate
Dihedral comer reflector 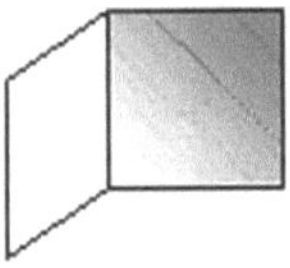	$S=\begin{bmatrix}1 & 0\\0 & 1\end{bmatrix}$	$S=\begin{bmatrix}0 & 1\\1 & 0\end{bmatrix}$	Radar signal perpendicular to the comer's axis
Trihedral comer reflector 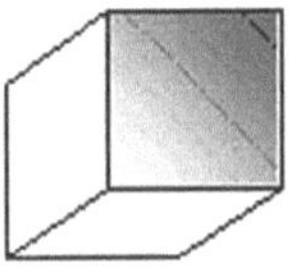	$S=\begin{bmatrix}1 & 0\\0 & 1\end{bmatrix}$	$S=\begin{bmatrix}0 & 1\\1 & 0\end{bmatrix}$	Radar signal perpendicular to the comer's axis

$$\begin{cases} E_H^R = S_{11}E_H^T + S_{12}E_V^T \\ E_V^R = S_{21}E_H^T + S_{22}E_V^T \end{cases} \tag{4}$$

where E is the electric field, the subscript means horizontal polarization, H, or vertical polarization, V, and the superscript means reflected, R, or transmitted, T. S_{11} and S_{22} are the co-polar coefficients and S_{12} and S_{21} are the cross-polar coefficients. Simple geometrical shapes present well-know polarization matrices; some examples are show in Table 1.

In HRRPs the scattering matrix obtained in each range bin is the result of the combination of the matrices of all the scattering points in the same range bin. Therefore, the shape extracted from this matrix is not necessarily reliable for recognition. In ISAR images, instead, if the resolution is high it is possible to begin to isolate the scattering matrix of the main scattering point and consequently to estimate the geometrical shapes of these points. The estimation of the shapes of scattering points can be subsequently used as features for target recognition [6].

The polarimetric information content for both HRRPs and Images are examined here qualitatively. Figure 5 shows the linear polarimetric range profiles as a function of rotation angle again for the Land Rover target. The information is largely the

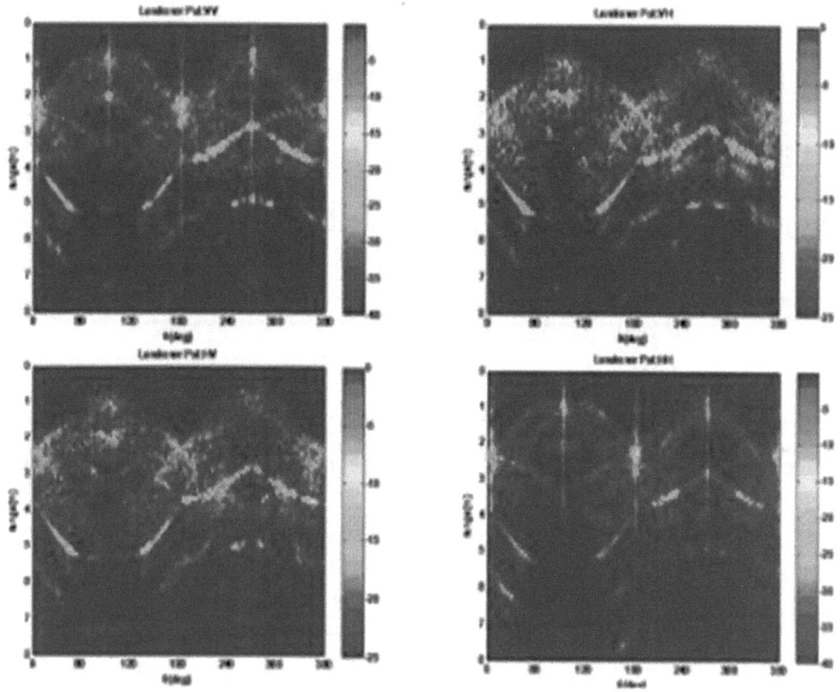

Fig. 5 Polarimetric range profiles of the Land Rover target.

same for features such as length and width but there are discernable differences in detail with some scatters appearing in one polarisation and not another as well as differences in echo strength and their angular persistence. Figure 6 shows the same base data but this time processed into 2-D imagery. Again the main characteristics are similar and there are small differences in detail but, perhaps, even less obvious. The overall advantage of polarimetric data in terms of improved classification performance is still to be proven.

4 Diversity in Biologically Sensing System

Reliable and robust navigation, collision avoidance and object classification are carried out with great success by echo locating mammals such as bats that are able to detect, select and attack prey even in a dense clutter and hostile countermeasures environment. Although the frequencies and waveform parameters used by synthetic sensors and by echo locating mammals are not the same there remain close parallels that suggest lessons can be usefully learnt. Here we examine methods that are exploited by mammals such as bats and dolphins and to see how they may be applied, particularly concentrating on how diversity is utilized.

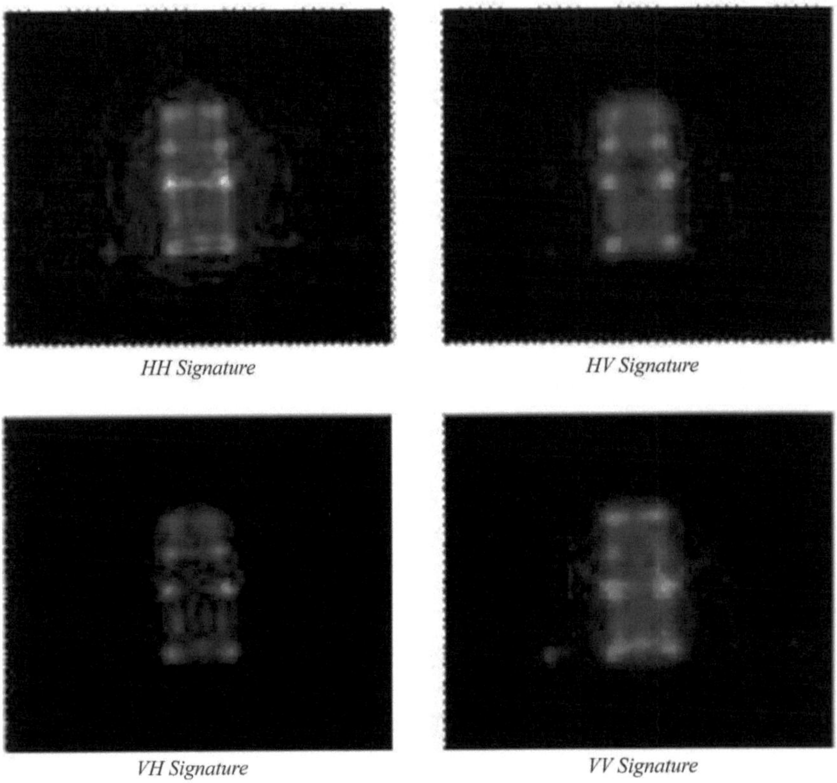

Fig. 6 Polarimetric images of the Land Rover target.

Bats have evolved echo location as a means of detecting, selecting and attacking prey over a period of more than 50 million years into a highly sophisticated capability on which they depend for their survival. It therefore seems self evident that there is potentially a great deal that can be learnt from understanding how they use this capability and applying such an understanding to synthetic sensing systems such as radar and sonar. Bats are able to modify their PRF, transmitted power, waveform type, and bandwidth. They also use multiple perspectives as part of their hunt strategies. These are all examples of exploiting diversity and could be implemented in modern radar systems if we are able to understand how to do so appropriately.

Bats use two very different types of waveform for classification. The first has the form of an un-modulated pulse that contains a number of harmonics. This is well suited to the classification of moving targets exhibiting a micro-Doppler component and there is evidence from biological studies that supports this hypothesis. The second waveform consists of a near hyperbolic chirp, also with two or three harmonics typically present. As the bats approaches the target the pulse length decreases and the degree of hyperbolic curvature increases. In this way a greater bandwidth is generated leading to finer and finer resolution as the bat gets closer and closer to the target. Again there is evidence from biological studies that this type of waveform is used when engaging static targets such as flowers. There also seems to be a strong relationship between the orientation of the bat, the position of the target, the precise nature of the transmitted waveform and the clutter and reverberation environment.

Here, we examine both forms of real bats calls recorded during the classification phase when attempting to identify potential prey supported by full scale measurements of real targets. A data base of real bat calls has been used to extract digitised versions that seem best suited to classification tasks. These include calls of horseshoe bats (Rhinolophidae) that use long constant frequency components that encode acoustic glints in echoes from fluttering targets and the hyperbolic modulation of an Eptesicus Nilssoni used when extracting nectar from a flower head.

Figure 7 shows the spectrogram of the calls of a Pearson's horseshoe bat Rhinolophus pearsonii when attempting target classification. The waveform is characterized by a relatively large constant frequency component with a short wider bandwidth modulation at either end. There are a series of harmonic replicas spanning quite a considerable overall bandwidth. The method of classification being used is based on exploiting the micro-Doppler signature which is best achieved with an extended constant frequency component. The short wider band part of the sign is probably used for ranging. The response to a simple model for the beating wing of an insect is shown in Figure 8. The Doppler signal generated is a function of the harmonic frequency and shows different degrees of sensitivity to the different constant frequency transmissions. Thus the bat is able to select the best signal for classification as well as using multiple copies for further confirmation.

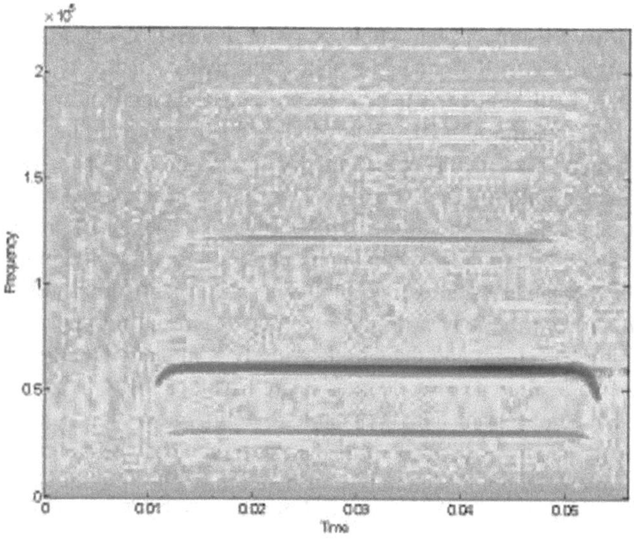

Fig. 7 Spectrogram of the call made by a bat when attempting target classification.

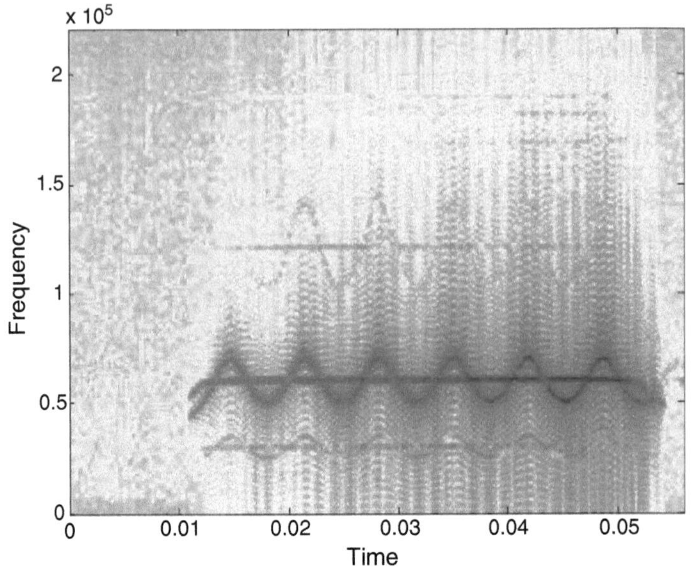

Fig. 8 Change in the received signal when modified by a simulation of a wing beat at an orientation broadside to the target.

An additional possibility is that the bat also uses this harmonic type of waveform to generate high resolution in range by effectively mimicking a frequency stepping system (albeit in a single pulse). It does, potentially, have very attractive properties that make it worthy of further investigation for a variety of applications. For example, Passive Bistatic Radar often uses illuminators of opportunity that emit simultaneously on several frequencies that collectively span a much wider bandwidth than that of any single frequency. By utilizing the total bandwidth transmitted a relatively high resolution mode could be developed. However, the signals will be under-sampled and there could be potentially significant sidelobes. This is another example of the exploitation of diversity.

A further factor that is generally very important in the attack of prey is the trajectory that is used. For example, when approaching a target located in a background of leaf clutter the bats uses a low line of attack, presumably, to minimise backscatter (i.e. clutter) from the leaves. Figure 9 shows the resulting interaction between the illuminating waveform and the simple wing beat model when the relative angle between the two is altered to 45 degrees. Here a reduction in the amplitude of the modulation can be observed. The bat is able to deliberately alter its orientation and transmission frequency with respect to the prey such that it maximises its sensitivity to the micro-Doppler signature and hence has the best information set for classification. Also, this provides important information as to the orientation of the prey and that micro-Doppler aspect angle dependent information may be gleaned that is used in the classification process. Clearly, there is a degree of speculation in these hypotheses and further research is required to develop a more complete understanding.

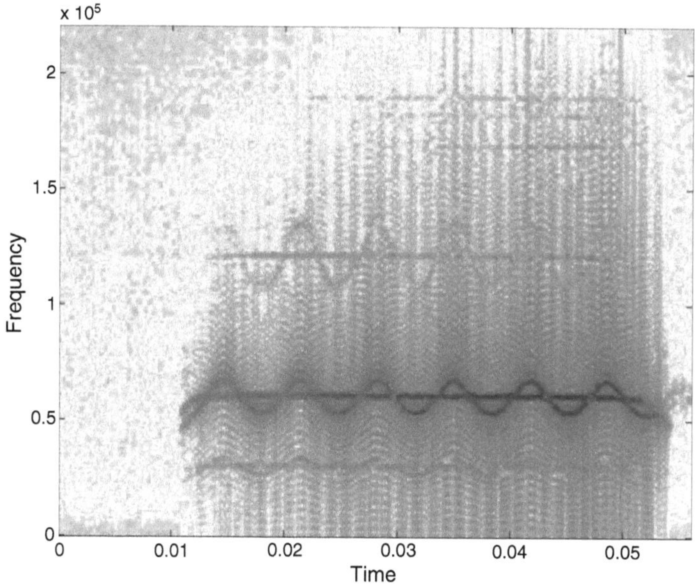

Fig. 9 Change in the received signal when modified by a simulation of a wing beat at an orientation angle of 45 degrees to that of Figure 6.

Figure 10 shows a sequence of waveforms that an Eptesicus Nilssoni bat used when attacking a stationary target. The bat was constantly changing its orientation such that it views the target over an angle range of approximately 270 degrees and is gradually getting closer and closer to the target. In other words the bat is benefitting by using a combination of angle and waveform diversity.

For each pulse the waveform approximates well to a hyperbolic function. As successive pulse are emitted and received and the bat closes in on the target the pulse length is reduced to conserve energy and avoid ambiguity. The degree of hyperbolic curvature increases to improve resolution and tolerance to any differential Doppler. Therefore, just prior to attack the bat operates with the highest resolution (of the order of 1 mm) and has gained multi perspectives for classification. Indeed, the use of the two receivers (ears) combined with very high resolution is suggestive of the selection of a particular part of the target. Figure 11 shows one of the final pulses and its ambiguity diagram in more detail. The high degree of Doppler curvature and subsequent high range resolution and Doppler tolerance may be observed. The waveform and its dynamic adjustment as a function of the previously received echoes are quite typical of nectar feeding bats. Thus we see that the waveform is being altered in a number of ways as the viewing geometry is also changed. This suggests diversity in a wide sense may be a key ingredient along the path to employment of truly intelligent synthetic sensing systems. For synthetic sensor systems the equivalent dynamical adjustment of sensor parameters is well within the scope of current technology. The more demanding question is 'what are the appropriate adjustments that need to be made to maximise the success of a mission? This requires considerable further research.

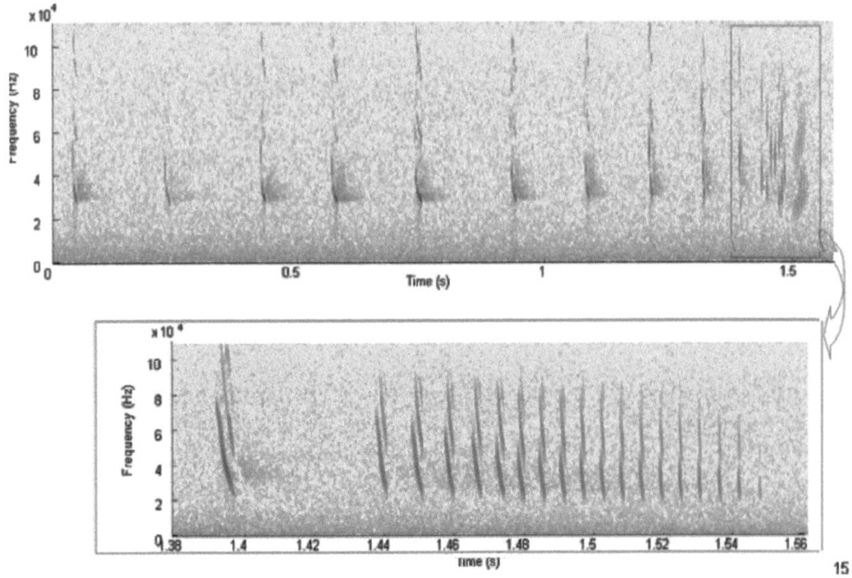

Fig. 10 Sequence of waveforms that an Eptesicus Nilssoni bat used when attacking a stationary target.

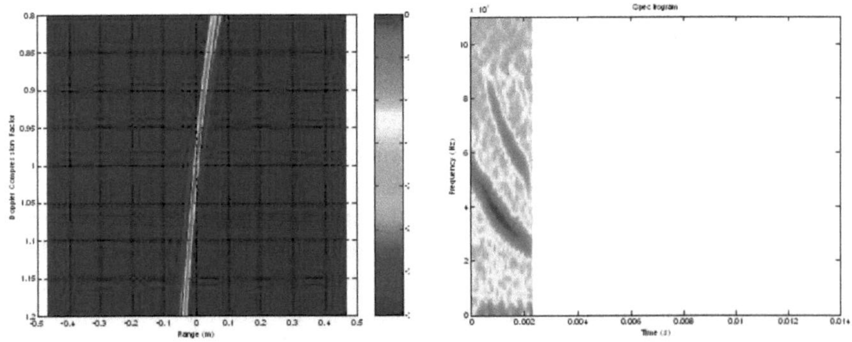

Fig. 11 Example ambiguity diagram and waveform for an Eptesicus Nilssoni used in the attack phase.

We now examine how the bat is able to recognize nectar providing plants against a complex background as classification is a key component in intelligent sensing. To investigate static target classification in more detail, high resolution range profile data was acquired from experiments carried out at Bristol University [7]. The experiments generate a wideband waveform that provides a resolution of approximately 1mm. This is used to illuminate a number of flower head targets mounted on a narrow pin and recording the received echo as a function of orientation angle in the ground plane. Figure 12 shows this experimental layout in schematic form.

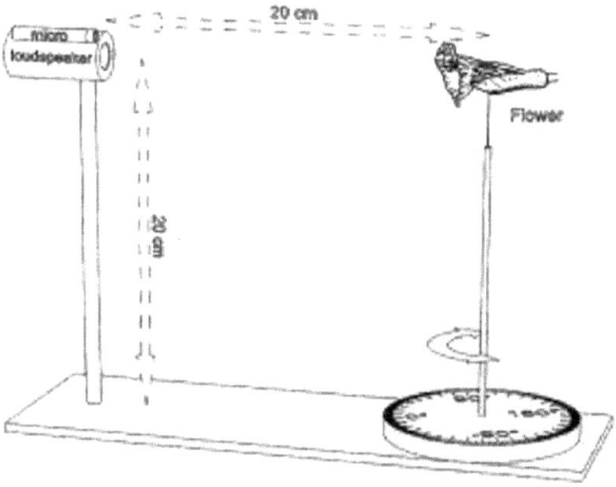

Fig. 12 Experimental geometry used to gather high range resolution profiles of flower heads as a function of orientation angle (Courtesy of G Jones and M Holderoid, University of Bristol.)

This particular configuration has been chosen as bats and plants have co-evolved to enable classification to take place reliably. In other words it is in the interests of the bat to be able to recognise the flower as it is a source of food and for the flower to be recognised by the bat as the bat is a means of pollination and thus the ingredients for successful classification are in-built.

One strategy used by plants is to expose their flowers e.g. by suspending them on modified long leafless braches into the sub-canopy. This helps bats to approach the flower and the uncommon presentation stands out against other vegetation. Indeed it might be that the local scattering environment offers many clues (prior knowledge) that there are likely to be flowers present and is this an important classification aid. However, many flowers are also presented closer to the plant or grow on stems and branches (cauliflory). In such cases several echo acoustic cues might render the flower unique such as:

– Echo strength can be higher than in leaves of the same size, because the bell-shaped corolla of many chiropterophilous flowers collects and focuses sound back towards the bat.
– Floral echoes last longer than echoes of leaves and branches, because sound is reflected within the corolla.
– Floral echo fields are often omni-directional, which means that most sound energy is always projected back into the direction of sound incidence. This contrasts with other plant structures, such as leaves, which produce echoes of maximum amplitude only when ensonified perpendicularly.
– Because flowers are complex targets and consist of many different reflectors at different distances, interference generates specific peaks and notches in the echo spectra, giving them a 'coloured' spectral appearance. This not usually occurs in convex structures such as branches or leaves.

Fig. 13 Example flower head (Crescentia cujete) and its corresponding high range resolution profile together with a spectral representation (both as a function of orientation angle.

Figure 13 shows an example flower head (Crescentia cujete) and its corresponding high range resolution profile together with a spectral representation (both as a function of orientation angle).

The range profiles show clear structure that is present at all angles and exhibits both similarities and differences to the high resolution radar profiles of the Land Rover target. One difference that seems to be apparent from a visual inspection of the profiles is that more features appear to persist over larger angular ambits. In the radar profiles there appears much more variability and indeed, it is the angular dependence that is responsible for improved classification performance as additional perspectives are included. Figure 14 shows the same form of plot as 13 but for a Viresa gladiflora. Again, similar behaviour may be observed although the detail of the form and structure is significantly different.

In order to examine classification using such data in more detail a multi perspective classifier was applied. Three flower targets are used to make the case quite demanding as in reality it is unlikely that the bat has to differentiate between two flower heads in order to feed. Figure 15 shows the results of applying a Forward feed neural network multi-perspective classifier. The plot also includes increasing levels of noise.

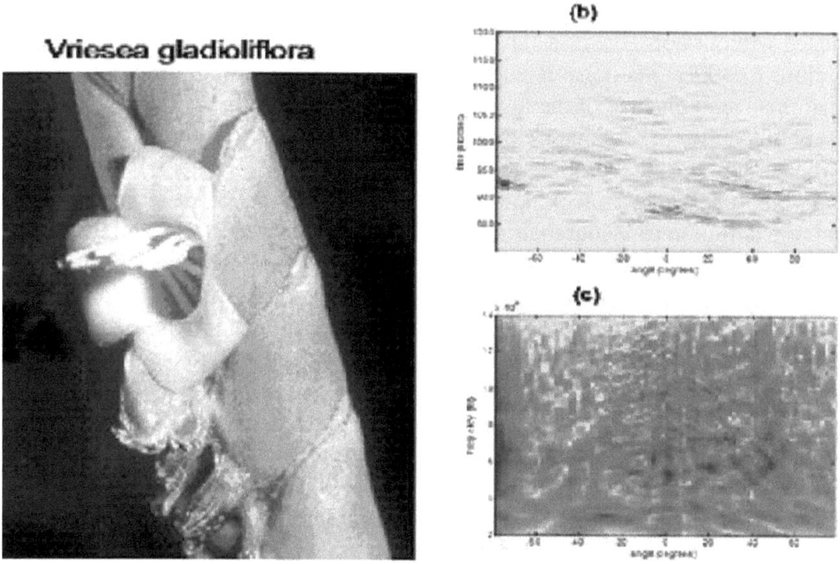

Fig. 14 Example flower head (Vires gladiflora) and its corresponding high range resolution profile together with a spectral representation (both as a function of orientation angle.

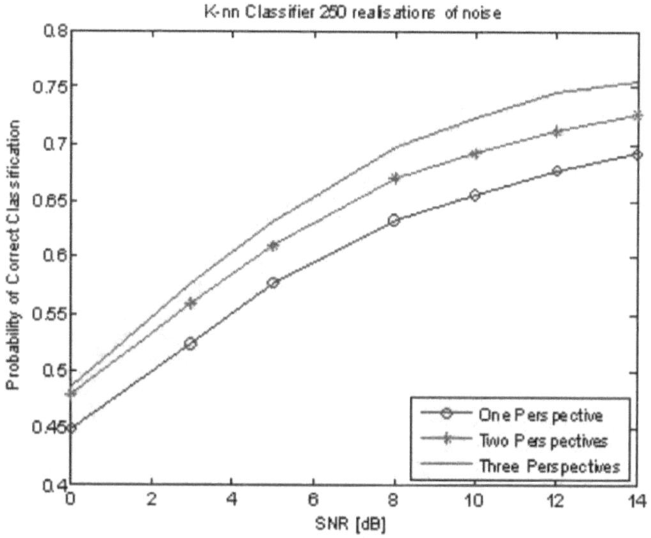

Fig. 15 Multi perspective classification performance of the three flowers versus signal to noise ratio.

There are two main conclusions that may be drawn. The first is that there is a significant increase in classification performance in going from one to two to three perspectives i.e. exploiting angle diversity. Secondly, as the signal to noise ratio increases the classification performance, as might be expected, also increases. Indeed without noise the classification performance, even with a single perspective is close to 100%. As noise is added eventually there is a more rapid drop off in performance, indicative of the loss of key information, probably embedded in scatterers of smaller echo value? This therefore highlights the importance of having a sufficiently high signal to noise ratio when echo locating.

In this section it was seen how the bats use a combination of waveform design and orientation to carry out their mission. In other words they are exhibiting all the requirements of a truly intelligent system (and with a processor the size of a 'pea') and are utilizing diversity in doing so. The bat gathers information about its local environment and then adjusts the parameters of the next call as well as changing its position to ensure that it extracts the right or best information to continue and complete its task as successfully as possible. Indeed, this simple examination of how diversity is exploited by bats highlights that this natural sensing system is continually adapting its entire diversity range, seemingly to ever improve the information the bat needs to fulfill its mission. In the next section we examine how angle diversity via multiple viewing angles can be used to improve classification performance in much the same way it might be used by bats.

5 Exploiting Diversity in Synthetic Sensors

In this section we look at the role of trajectory in the recognition process as invoked in a synthetic radar sensor [8]. The same form of radar data as displayed in Figure 3 is utilized as an input to a classifier. As in the previous section we compare classification performance as a function of the number of perspectives used. In this example three classifiers are employed to ensure that the results are not biased by the classifier itself. The three classifiers are (i) A Bayesian classifier, (ii) a neural network and (iii) a K nearest neighbor classifier. Range profiles at approximately every eighteen degrees are used to train the classifiers and then removed from the data set to be classified. The results are displayed in Figure 16.

Figure 16 shows that the classification performance increases with increasing number of perspectives no matter which type of classifier is used consistent with that seen for the case of the bats data. This is also consistent with our own experiences. Generally, if we wish to positively identify something that we expect to recognize we will move our viewing position either to get a 'better look' or to re-enforce an expectation of the objects identity. We might further conjecture that there will be preferential look directions to exploit determined by the first look direction. For example, objects that exhibit symmetry would not yield large additional amounts of information if the viewing angle were altered by 180 degrees. This is further illustrated in Figure 17 which shows the multiperspective classification performance for a Land Rover target as a function of viewing angle for a three perspective classifier.

Figure 17 shows that at certain angle combinations the classification performance falls. The performance at position 0,0 is the monostatic case and is the worst of all.

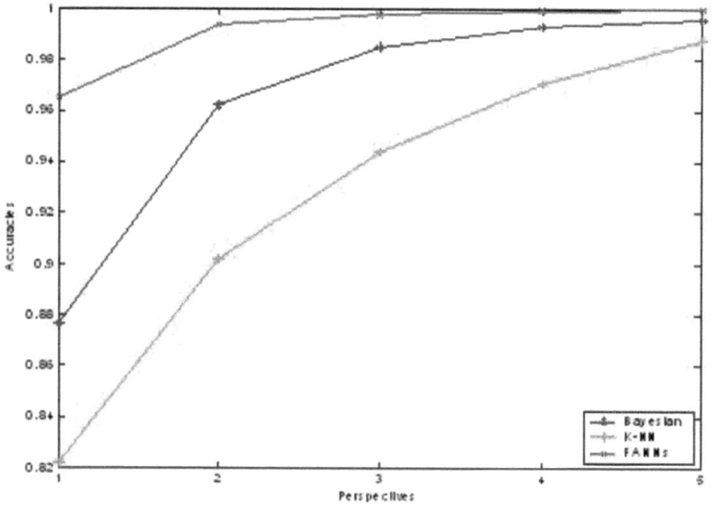

Fig. 16 Multi perspective classification performance of four vehicle targets versus number of perspectives using high resolution range profiles as input data.

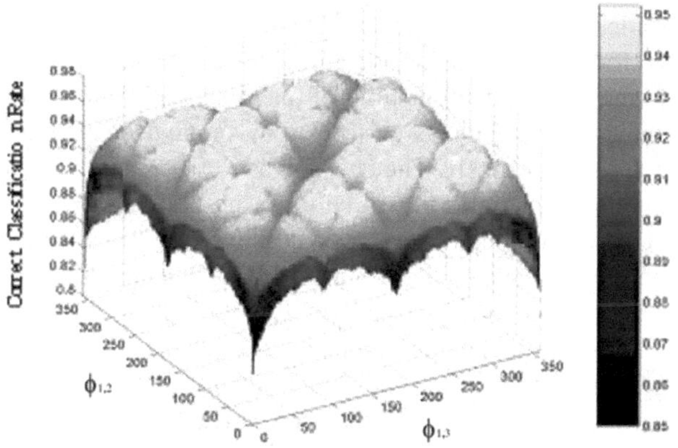

Fig. 17 3-P correct classification rates versus the angular displacements and for the three-class problem.

The diagonal represent the two perspective case and again performance is inferior to the peak. The other troughs tend to occur at separation angles such as 180 degrees, where the symmetry of the target means that there is little additional information to be gleaned that can benefit the classifier. If the orientation of the target can be determined then this prior information can be used to arrange a second look to be one that compliments the first adding maximal new information thus optimising overall classification performance. This would be an example of a simple but genuinely intelligent radar sensor i.e. one that uses previous information to improve subsequent information and hence maximizes its chances of carrying out the desired task successfully. In the next section we examine other prior information methods to improve mission performance.

6 Prior Knowledge

As we have begun to explore in the previous sections, knowledge based radar systems might be thought of as the precursor to future systems that will employ artificial intelligence [9]. Indeed, it is logical that information already known from other sources about the target area or task to be carried out can be used to direct the operation of the radar or to interpret the radar's findings more usefully. As we saw in the section above the possibility also exists for data gathered in real time (or 'on the fly') to be used for generation of relevant information or for better cuing of further information. Another example could include an MTI radar that is also able to operate in a SAR mode which can be used to provide coarse image information as an aid to more accurate detection of moving targets. Historical data of the same type (i.e. MTI data) for a given scene should also theoretically provide a useful 'memory' if a way can be found to exploit it when performing new MTI detection in real time.

The electronically scanned radar system is another example of a sensor that brings the need for intelligent radar operation into sharp focus. Instantaneous, adaptive beam pointing enables combinations of functions such as tracking, surveillance, and weapons guidance (previously performed by single dedicated radar systems) to be implemented simultaneously. The decision as to where and when to re-point the radar beam can and sometimes has to be taken in the interpulse period, far faster than the rate at which a human operator could intervene. Thus the radar itself must make these decision based upon a combination of what it has already observed, prior knowledge and the mission objectives which is embracing the concept of intelligence. One area where this might occur is that of sector prioritization. Military understanding is used as the prior knowledge to allocate the sectors and determine high level metrics for assigning priority. We consider such an example that also utilizes fuzzy logic to give both a more humanistic decision making logic and to preserve where possible radar resources for further allocation of tasks.

The attribution of priority to regions and targets of interest may be done in a variety of ways [10]. For example the decision tree structure shown in Figure 18 could be used. The information required to take the decision is supplied by radar data operating in tracking and high resolution modes. The different variables provide differing types of information used to set priority. These are threat, hostility, weapons system capabilities, track quality and the target position. The selection of these variables has been by expert judgment based upon operational experience. They may require further refinement with further experience. The logical relationships between these variables then determine the setting of priorities. Fuzzy logic can be used to provide a softer way of making decision by allowing variables to take

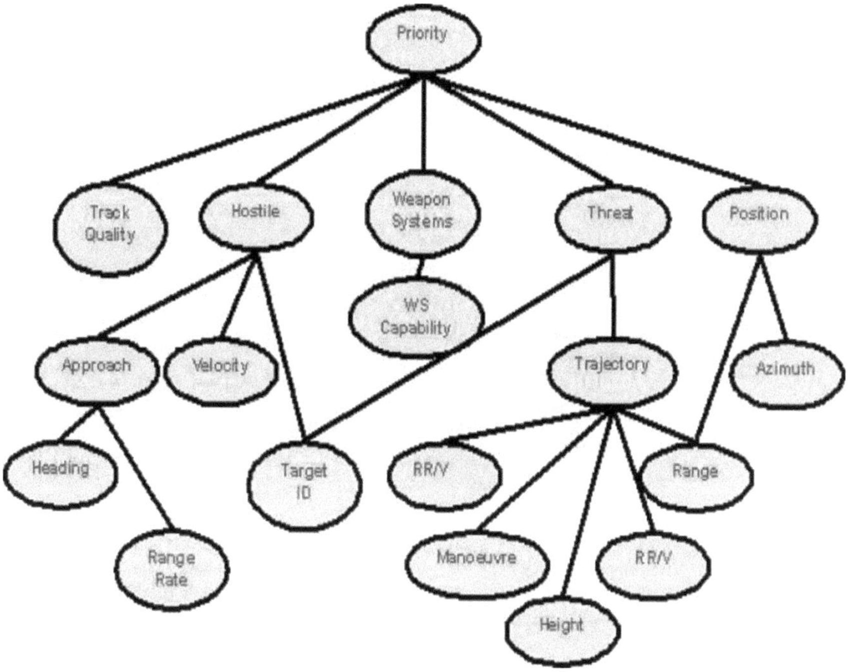

Fig. 18 Decision tree for sectors for target priority assessment.

values in between an either 'on' or 'off' state. The nature of the inferential rules linking the fuzzy variables can be written again using expert judgment and then tuned using simulation and real experience. In fact the actual number of rules used in the inference system may be less than the number initially set. This is because some combinations of rules are unlikely to be found in real systems. The reduced number of rules does not reduce system performance as associations used to determine the truth of an assertion is largely determined by the dominating term.

The evaluation of the fuzzy rules must follow the sequence proposed in the decision tree. Thus the system inputs are fuzzified and successively used to assess other fuzzy variables in the cascade to the point where the final priority is evaluated. Graphic representations are invaluable in helping to assess how the fuzzy rules are operating. These may be generated by fixing all the variables except the two being assessed. Figure 19 shows an example of this where it is assumed that three variables (track quality, position and weapons capabilities) are maintained at a fuzzy value of 0.5 and both the threat and the hostility are varied over their entire ranges. This configuration might represent a situation in which the target is located at a medium range and has medium importance with respect to the weapons system of the radar platform.

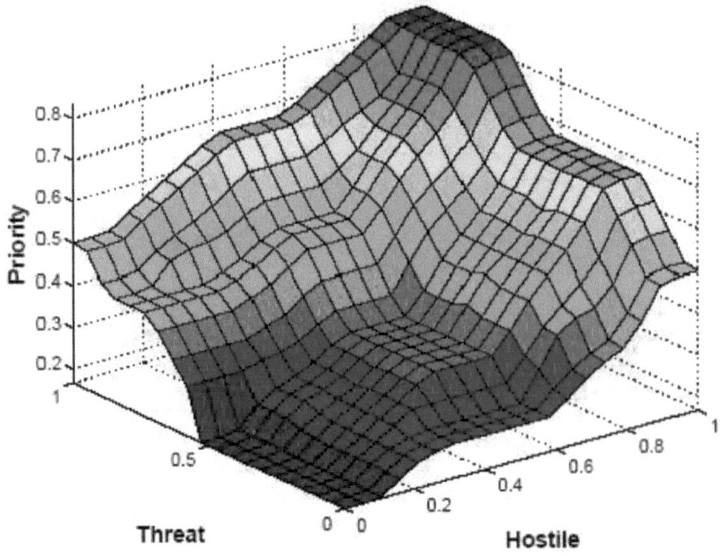

Fig. 19 Graphic representation of the fuzzy rules with the position, track quality and weapons systems fixed.

It is observed that, as might be expected, the priority increases as a consequence of increases in the degree of threat and hostility of a target. Conversely, low degrees of threat and hostility place the priority at a low level. Two other areas may also be identified on this surface. The first is related to the degree of hostility varying between 0.5 and 1 (medium to very high) and the degree of threat varying between 0 and 0.5 (very low to medium). The resulting priority increases as a result of rises in the degree of threat or hostility. However, the sensitivity to rise in the degree of hostility is greater than that of the threat. This behavior is explained by examining situation in which targets with medium and high probability of being the enemy are moving away from the radar platform. The hostility of the target is determined by its probability of being an enemy but the threat is determined more by its trajectory and position. Thus the situation is dominated by the identity of the target. The second area corresponds to degrees of threat varying between 0.5 and 1 and low levels of hostility. The resulting priority increases are a consequence of rises in the degree of threat or hostility. However, the behavior is different to the previous area. Thus the sensitivity to increases in the degree of threat is greater than the sensitivity to increases in the degree of hostility. This is explained by considering situations where, having low probabilities of being the enemy, targets move on threatening trajectories towards the radar platform. In this case, the way the target is approaching the radar platform has a greater effect in determining the final priority than its identity. Of course the manner in which these relationships are formulated is itself a variable and is one in which the expert judgment plays a key role. Thus there is a learning process during which these rules and relationships will be refined in the light of experience.

Having defined and tuned the fuzzy if-then rules the method for prioritising the relative importance of tracked targets can be validated against test trajectories. In the example presented here the scenario consists of targets with different identities and velocities. The analysis shows that by knowing the identity of the targets their priorities may vary. This provides valuable information to be accounted for when deciding how to allocate radar resources in overload situations. Two cases are presented for targets moving towards the radar platform on constant-velocity straight line trajectories. These have been chosen as they represent situations of a high degree of threat where targets may be moving towards the radar platform in order to start an attack. In addition, they represent the behaviour of the method when a variable such as approach is fixed. This helps simplify the analysis and the evaluation of the reasons for the results of the prioritisation. The system can also be examined in more complex scenarios where all variables involved in the prioritisation are changing over the simulation.

The left hand side of Figure 20 shows the first test trajectory where a target moves towards the radar platform on a straight line, having a constant velocity of 300 m/s. The red dot indicates the origin of the trajectory. Three targets are assumed in the analysis. They have the same dynamics and flight height; however, their probabilities of being enemy are different as follows: 1 (enemy), 0.5 (unknown) and 0.1 (friendly), corresponding to the red, blue and green curves respectively. The evolution of the resulting priorities is seen in right hand-side figure shows that, in general, all priorities increase as the targets move towards the radar platform; and the greater the probability of being enemy, the greater the resulting priority. Figure 18 also suggests that priorities of targets which have unknown identity present a similar behaviour to friendly targets in the early stages of the trajectory. This may be explained by the fact that during that period, the range of the targets is longer than the tactical range of the platform weapon systems. This happens until around 80 s. From that instant, as the target is moving close to the boundaries of this weapon systems

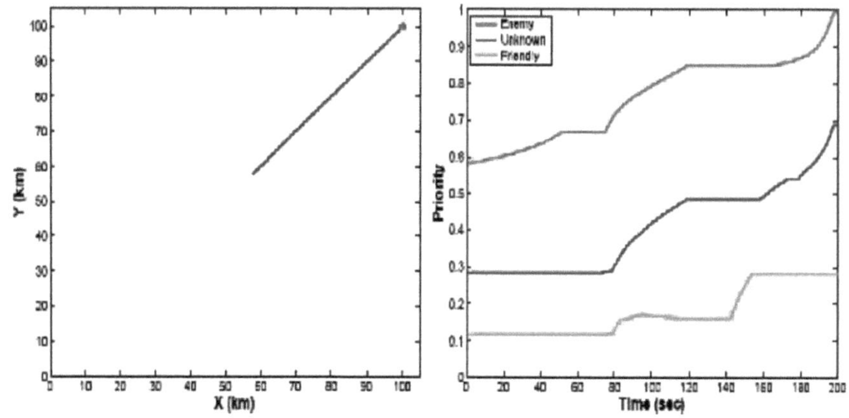

Fig. 20 Resulting priorities for three targets with different probabilities of being enemy, moving on the same trajectory.

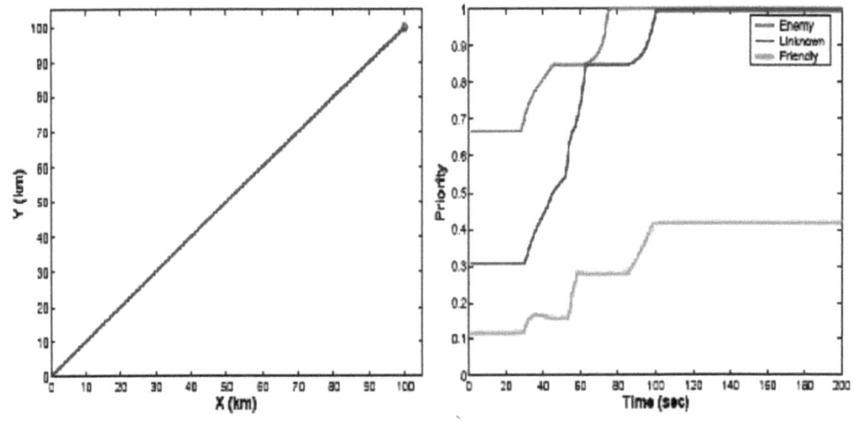

Fig. 21 Resulting priorities for three targets with different probabilities of being enemy, moving on the same trajectory. Target velocity: 800 m/s.

tactical range, the degree of threat of the unknown target is likely to increase. Thus, its priority evolution has the similar behaviour to the priority evolution of the enemy target. The closer the unknown target is, the higher and the closer to the enemy target its priority will be. At short ranges, if the identity of the target is still unknown, the target is assumed to be enemy, and its resulting priority is assessed as that.

Figure 21 presents the results of a simulation where targets are assumed to move on a straight line trajectory but this time with a velocity of 800 m/s. The same probabilities of being enemy as in the previous case are considered. Due to the high velocity and short ranges, the evolution of the priorities is now rather different. During the first few seconds of simulation, both unknown and enemy targets have slightly higher priorities than in the first example. This may be explained by their high velocities.

All target priorities remain fixed until about 30 s, when the target position is getting close to the weapon systems operational range. Before 30 s, all targets have the maximum priority possible for the set of characteristics of their dynamics, identity and the capabilities of the weapon systems. Thereafter, the priorities are increased in order to allow the radar platform to face the threat. It is observed that, from around 30 s to 60 s of simulation time, the priority of the unknown target presents a high rate of increase. The analysis indicates that more importance is progressively given to this target which is gradually assumed to be like an enemy target, because its velocity is very high, the target is approaching the radar fast, and its identity is unknown over this period. From around 60 s to 85 s, the unknown target has the highest priority possible for the combination of input variables which determine its importance. From 85 s, its priority increases again, reaching its highest at around 100 s, when the target position is within operational range of the platform weapon systems and as a consequence both enemy and unknown targets have the same priority. Such an unknown approaching target is considered to be of highest importance because of its potential degree of danger, represented by its velocity, the way it is approaching the radar platform. Like the unknown target, the priorities of both enemy and friendly targets increase from around 30 s, as they are getting close to the weapon system operational range. These priorities continue to increase reaching their maximum values not later than 100 s of simulation, when the position in within the operational range with a degree of membership of 100%.

The results of the situations examined here suggest that the fuzzy logic approach is an intelligent and valid means for evaluating the priority of targets. By imitating the human decision-making process, and by combining dynamic characteristics about radar tracking and military aspects, such as the ability of the weapon systems of the radar platform to face potential threats, the fuzzy approach may represent an effective and intelligent support for decisions regarding radar resource management. It also demonstrates the range of behaviours that such an approach can cope with and that it does so in a manner consistent with our definition of an intelligent system.

Prior knowledge can also be introduced in a variety of other ways, but has only been addressed by a few researchers [1, 2, 10]. For example, whilst there has been much work carried out on DPCA and STAP for GMTI there has been much less examining the role of prior information. This may well be a reflection of the fact that the processing needed for useful simulation and implementation of such Knowledge Based (KB) systems has only become available in recent years. However, it seems intuitive that there is a lot to gain from using external information to improve the performance of the MTI process. Some initial studies have examined the use of Geographic Information Systems (GIS), which have information relating to aspects such as land use, building disposition, terrain data etc and hence may be rewarding sources of information.

A major question to be answered is 'how to integrate these disparate data types with the MTI process and hence what is the likely maximum improvement in MTI accuracy that can be expected'? As yet there appears to be no emergent convincing answers and methodologies. It is complicated by the fact that there are different qualities of terrain data, GIS data etc and so one would expect there to be a relationship between the quality of data and the benefit it can bring to the MTI process. For example, there are data of higher resolution than others but there has been no work to confirm that improvements in resolution of the data used by KB techniques are proportional to the performance improvement in MTI radar. Indeed, the key to good KB data may lie in another property such as the existence and relative location of roads and buildings. Similarly, should there be a lower limit on the quality of KB data needed to 'add value' to the MTI processor and is there a level of MTI fidelity that is required for KB to work? The issue of data fusion is clearly important in KB research. Additionally, the amount of data available will continue to increase and the ability of human analysis of it all will become an unrealistic prospect, if this is not already the case. Yet refraining from any analysis of this data will almost certainly mean missing out on useful information that could be the difference between detecting a moving target of interest on the ground or not.

There are two strands for information fusion in this context: fusing data of the same type and fusing data of different types. A KB MTI system can be thought of as a fusion of different types of data. Fusion of the same type of data is of equal relevance: obviously the cumulative information from two maps at different scales will be higher than the information from just one map, so how should the two maps be fed into an MTI process? Should the maps be fused beforehand, or should the fusion be one of a fusion of different types within the MTI system as described above? The fusion of similar-type data may be thought of as a separate area. For example, how should two maps or SAR images be fused for performance improvement as part of a KB MTI system? There is significant work on the fusion of SAR images from different radars, but not on using the result to provide knowledge for GMTI. Using SAR data itself for MTI processing has been attempted with promising results, and this type of approach would involve the fusion of two different types of MTI data, in which there has been little exploration. Furthermore maps and terrain data can be considered historical compared to real-time MTI data. Could therefore historical MTI data be used to provide extra knowledge? The literature on KB techniques for MTI radar appears to show only that there is good potential for increased MTI performance. The main focus of KB techniques in MTI radar has been to intelligently select training data for STAP. There are many more areas that are not yet investigated. Finally, it should be noted that, in published literature both on STAP/DPCA and KB techniques, is that the application to real radar systems is lacking. Real data provides the ultimate test for new systems because it brings with it real world errors that are often difficult to simulate or even unexpected.

A summary of information sources that could be obtained before live GMTI gathering but used for knowledge based processing with live data is shown below. Multiple sources of the same time, but gathered at different times, could be used with data fusion methods outlined above. Some sources are evidently more readily available than others.

- GIS data from digital maps, encompassing terrain and land elevation information, ground cover and transport routes
- Airborne, look-down optical imagery from reconnaissance aircraft or satellites
- GMTI data from historical runs over the area of interest
- Known clutter information of the scene (clutter maps for a radar with given parameters)

In a STAP system GIS data could be used to select more accurate training data for generation of the interference covariance matrix. Optical imagery or high resolution SAR imagery from the scene of interest could also be used to either select suitable training data or to identify targets of interest. Image processing techniques would be required in this case. Clearly this is an area of development with much further research necessary before KB diversity systems become common place.

7 Conclusions and Summary

In this chapter we have examined and demonstrated the value of exploiting sensor and platform diversity together with prior knowledge in improving system performance. Analogy with echo locating bats provides encouragement that such techniques will lead to future developments that embody genuine intelligence, potentially offering vastly superior capability over the systems of today. However, there remains much research to be done before this potential can be realized and there are still many questions remaining.

References

1. J.R. Guerci and E.J. Baranoski, 'Knowledge aided adaptive radar at DARPA', *IEEE-AES Magazine*, **23**(1), 41–50, Jan 2006.
2. M.C. Wicks, M. Rangaswamy, R. Adve and T.B. Hale, 'Space-Time adaptive processing', *IEEE-AES Magazine*, **23**(1), 51–65, Jan 2006.
3. W.G. Carrara, R.S. Goodman and R.M. Majewski, *Spotlight Synthetic Aperture Radar: Signal Processing Algorithms*, Artech House, Boston, MA, 1995.
4. P.F. Sammartino, C.J. Baker and H.D. Griffiths, 'Adaptive MIMO radar system in clutter', *IEEE Radar Conference 2007*, Waltham, MA, 276–281, 17–19 April 2007.
5. F.T. Ulaby and C. Elachi, *Radar Polarimetry for Geosciences Applications*, Artech House, Norwood, MA, 1990.
6. P. Tait, *An Introduction to Radar Target Recognition*, IET Publications, London, UK, 2005.
7. M.W. Holderied and O.V. Helversen, 'Binaural echo disparity as a potential indicator of object orientation and cue for object recognition in echolocating nectar-feeding bats', *Journal of Experimental Biology*, **209**, 3457–3468, 2006.
8. M. Vespe, C.J. Baker and H.D. Griffiths, 'Radar target classification using multiple perspectives', *IET Radar, Sonar and Navigation*, **1**(4), 300–307, 2007.
9. S. Haykin, 'Cognitive Radar', *IEEE-AES Magazine*, **23**(1), 30–40, Jan 2006.
10. S. Miranda, C.J. Baker, K. Woodbridge and H.D. Griffiths, 'Knowledge based resource management for Multifunction Radar', *IEEE-AES Magazine*, **23**(1), 66–76, Jan 2006.

Ground Penetrating Radar for Buried Landmine and IED Detection

David J. Daniels*

Abstract Detection of landmines using electromagnetic induction (EMI) techniques is well established and a range of metal detectors is commercially available. Recent developments using dual sensor technology combining EMI and ground penetrating radar (GPR) have enabled improved discrimination against small metal fragments to be demonstrated in live minefields. Reductions of up to 7:1 compared with the standard metal detector have been achieved in the field by hand held systems such as the UK-German MINEHOUND/VMR2 system and the US AN/PSS-14 (formerly HSTAMIDS: Handheld Standoff Mine Detection System).

Stand off vehicle based radar systems are now being trialled in realistic conditions. Airborne systems have also been trialled, but as yet have some way to go to deliver useful performance. These three distinct modes of operation pose fundamentally different challenges in terms of the physics of propagation and the radar system design and will be discussed.

End user expectations in terms of performance are challenging and at present only the hand held detectors are approaching these needs. This chapter reviews the high-level performance requirements from an OA perspective in order to set the performance envelopes of the radar designs. We also address the fundamental challenges in terms of propagation, proximity to the ground surface; ground topography and signal to noise and signal to clutter bandwidth of operation with reference to both close in and stand off landmine and IED detection. A review of the performance of GPR systems at the higher TRL levels is provided.

A key issue in comparing the published results of controlled trials relates to statistics of the depth of cover, the soil propagation characteristics, and the type of landmine, the sample size, the physical placement of the landmine as well as the characteristics of the clutter. This chapter will also highlight the future engineering challenges to achieve not only detection but recognition and identification using GPR.

Keywords: Landmine detection, radar, ground penetrating radar

*Chief Consultant Sensors, ERA Technology, UK, Cleeve Road, Leatherhead, Surrey, KT22 7SA, UK, e-mail: david.daniels@era.co.uk, www.era.co.uk

J. Byrnes (ed.), *Unexploded Ordnance Detection and Mitigation*,

1 Introduction

Landmine detection using electromagnetic induction (EMI) techniques (commonly termed metal detector (MD)) is well established and a range of these devices is commercially available. Recent developments using dual sensor technology combining EMI and ground penetrating radar (GPR) [1] have enabled improved discrimination against metal fragments to be demonstrated in live minefields and reductions of up to 7:1 compared with the standard metal detector have been achieved in the field by hand held systems such as MINEHOUND [3] and AN/PSS-14 [5]. These systems have reached the stage where they are being produced in large numbers.

Stand off vehicle based radar systems are now being trialled in realistic conditions. Airborne systems have also been trialled, but as yet have some way to go to deliver useful performance. These three distinct modes of operation pose fundamentally different challenges in terms of the physics of propagation and the radar system design and will be discussed in this chapter.

End user expectations in terms of performance are challenging and at present only the hand held detectors approach these expectations. This chapter will review the high-level performance requirements from an OA perspective in order to set the performance envelopes of the radar designs. We also address the fundamental challenges in terms of propagation, proximity to the ground surface; ground topography and signal to noise and signal to clutter bandwidth of operation with reference to both close in and stand off landmine and IED detection.

A review of the performance of GPR systems at the higher TRL levels will be provided as well as an introduction to the various algorithmic approaches to the classification of landmines. A key issue in comparing the published results of controlled trials relates to statistics of the depth of cover, the soil propagation characteristics, and the type of landmine, the sample size, the physical placement of the landmine as well as the characteristics of the clutter. We highlight the future engineering challenges to achieve not only detection but recognition and identification using GPR.

2 Background

Landmines can be either buried or laid on the surface of the ground or buried flush with the surface of the ground. They are emplaced by a variety of techniques, including being scattered on the surface by vehicles or helicopters. Thus landmines may be found in regular patterns, or in irregular distributions. Where environmental conditions result in soil erosion and movement caused by rain over several seasons the landmines may be lifted and moved to new locations and can be covered or exposed. Landmines are encountered in desert regions (i.e. Somalia, Kuwait), mountains (i.e. Afghanistan, El Salvador), jungles (i.e. Cambodia, Vietnam) as well as urban areas (i.e. Beirut, former Yugoslavia).

In general, most pressure sensitive landmines are not designed to operate when buried deeply. In these circumstances the overburden ground material acts as a mechanical bridge and inhibits triggering of the detonator mechanism and also reduces the force of the explosion. This fact is often taken into account in the specification of performance for a mine detector. For example a hand held mine detector should be able to detect AT landmines at depths up to 300 mm and AP Landmines at depths up to 100 mm with spacing between the detector head and ground surface of up to 100 mm. Users of vehicle based close-in landmine detectors prefer a greater ground clearance, although very successful operation of EMI arrays has been achieved with very close (proximal) ground clearance. Landmines can also be encountered at depths well beyond the range of most detection systems due to movement of the soil. Mine detection systems can be employed in several different roles: for close-in hand-held detection, for vehicle mounted standoff detection or as a remote sensor mounted on low flying fixed or rotary wing aircraft. These are mostly synthetic aperture radars (SAR).

The variety of environmental conditions in which landmines can be found is enormous. Minefields are not only neat ordered rows of landmines in flat deserts but can also be found among the debris of burnt out buildings and post-conflict urban and rural environments. Clearly, mine detection equipment has to be designed to work in a wide range of physical environments and the statement of operational requirements issued by end-users will reflect this need. Detection equipment must be able to be operated in climatic conditions, which range from arid desert, hillside scree to overgrown jungle. Ambient operating temperatures can range from below 20°C to 60°C. Rain, dust, humidity and solar insolation all must be considered in the design and operation of equipment. The transport conditions of equipment can be arduous and these as well as man-machine interface issues are vitally important to the design of detectors.

The large majority of civilian casualties are caused by anti-personnel landmines, which come in a wide variety of types. Many are designed only to maim. The blast type anti-personnel landmine will cause a traumatic amputation to a foot or leg, often injuring the other leg and genitals as well. Fragmentation land mines are far more deadly. Some models shoot hundreds of metal fragments in an arc that reaches out 50 m. Other types spring into the air when triggered and then explode at waist level. Anti-personnel mines can be buried in the ground or placed on the surface and can be set off by pressure, trip wire, remote control or sensors. They can be laid by hand, dropped from airplanes or spread by artillery. Many are made of plastic, which means they cannot be located by metal detectors during clean-up operations. Mine clearance has come a long way since the procedures adopted in the Second World War as shown in Figures 1 and 2.

Anti-vehicle mines are less numerous but more powerful. A mine that can disable a tank will destroy a civilian vehicle and kill its occupants. These mines usually cannot be detonated by a person's body weight alone, although when they are fitted with an anti-handling device they become anti-personnel weapons. Anti-vehicle mines are a particular threat to humanitarian aid workers who must travel roads before they have been systematically cleared. Examples of manual clearance methods are shown in Figures 1, 2 and 3 showing the historical improvements in methods and safety procedures.

Fig. 1 UK Army landmine detection 1945. (Photo: IWM.)

Fig. 2 Sappers learning mine detecting and clearance methods at the Royal Engineers School of Mine Warfare, Middle East 1942. (Photo: IWM.)

Fig. 3 A woman deminer working for MAG excavating anti-personnel mine in Battambang province in Cambodia. (Photo: MAG.)

3 Types of Landmines

In terms of detection techniques, landmines can be classified into several groups. These are metallic landmines, minimum metal landmines and non-metallic landmines. The latter type is in a minority and cannot be detected with the metal detector, in contrast to the metallic and minimum metal landmines. In addition to conventionally manufactured landmines there are numerous examples of other versions, which fall in the category of improvised explosive device (IED). These are often not buried and hence are not landmines. The Geneva International Centre for Humanitarian Demining (GICHD) provides a useful introduction to types of landmine [7]. Other sources of information on landmines can be found on the US Department of Defense CD Minefacts © which contains details of over 675 landmines as well as the US Department of Defense, Naval Explosive Ordnance, CD Ordata © which is a guide to UXO identification or many of the websites of Mine Action Centres and Non-Governmental Organisations NGO's. Some examples of typical landmines are shown in Figures 4 and 5.

These landmines are generally detectable with standard metal detectors but the completely non-metallic mine, though rarely encountered, can only be detected using a radar based detector. The French 1947 AT shown in Figure 6 has been found in Southern Lebanon and is constructed from bakelite and uses a glass based chemical detonator. Mines that are flush buried are a major problem as can be seen in

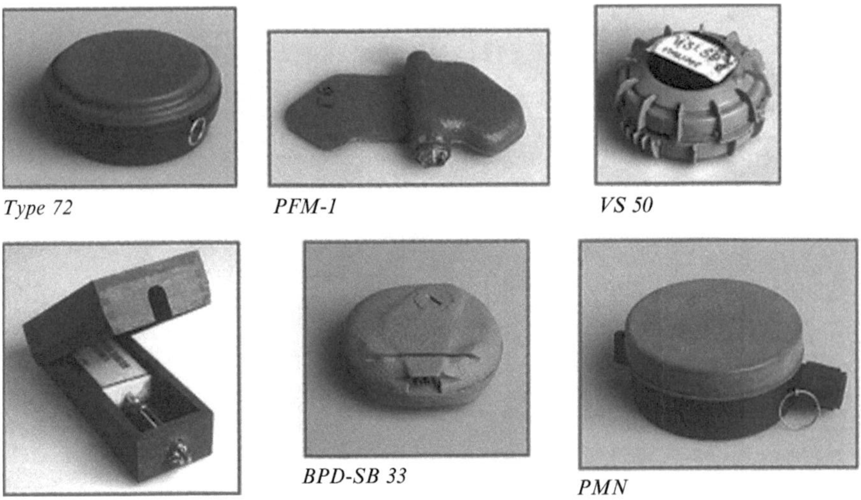

Type 72 *PFM-1* *VS 50*

BPD-SB 33 *PMN*

PMD-6

Fig. 4 Various anti-personnel blast landmines. (Photo: GICHD.)

Fig. 5 TM-57 metallic landmine and TM-62 P2 minimum metal antitank landmines. (GICHD.)

Fig. 6 Examples of the French 1947 AT landmine in Southern Lebanon. (Photo: Bactec.)

Fig. 7 PMA2 AP landmine in Bosnia (photo: D J Daniels) and internal construction (ORDATA).

Figure 7, which shows the detonator just above the surface of the ground. The search techniques must allow for this situation, as too close an approach is inadvisable.

It is very important to understand the physical construction of landmines as this has a major influence on their radar cross-section (RCS). Some minimum metal landmines are substantially solid explosive, but others have significant air gaps and these enhance the radar scattering cross-section of the landmine. Landmines such as the PMD-6 and PFM-1 are asymmetric and this affects the polarisation characteristic of the RCS as well as causing differences between the centres of detection of radar and a MD.

3.1 Performance requirements

The key performance factors of the specification of a landmine are its probability of detection (PD) and its probability of false alarm (PFA). For a hand held system the requirement is to achieve a PD = 1 and PFA = 0. The threshold between the populations of true/false reports can be plotted as a sensitivity/specificity graph and generates a receiver operating characteristic (ROC) curve. A typical example is shown in Figure 8 which plots true positives against true negatives in a sample population.

It will be noted that for a true positive or PD value of 1 incurs a true negative or PFA of 0.4. The closer the ROC curve is to a step function the lower the PFA for a PD equal to 1. The ROC curve is used as a means of evaluating detector performance or aspects of the performance of a detector. It should be noted that where human decision-making is involved the ROC curve could also be used to assess both equipment and human performance. Given a certain spatial density of landmines the chance of survival can also be determined. In the case of vehicle-based systems that provide route clearance this is an important parameter. For example, if

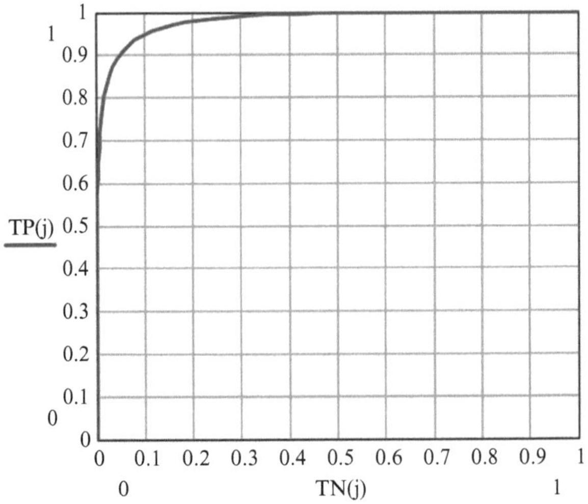

Fig. 8 Receiver operating characteristic.

a mine density of 1 per km along a route 4 m wide is assumed, then for a sensor PD of 0.9, the land mine detection system has a 60% probability of encountering a landmine explosion within 10 km, on the basis of a probability of explosion of 0.5 for each of the mines encountered. The implications of this are that the attrition rate of such systems will be high and the vehicle protection and cost and replace-ability of the sensors and vehicle drive train are important system parameters.

The density of mines, PD and PFA of the sensor system fundamentally determine the rate of advance of vehicle-based systems in the following ways. The density of mines clearly affects the number of potential encounters, the PD affects the chances of the vehicle being damaged by a landmine as well as the time spent in neutralising the landmine and the PFA affects the time spent in clearing false alarms. Typical example values are that the rate of advance is limited to a maximum of 11 kph for 1 mine per kilometer of a 4 m wide swathe, assuming a sensor probability of detection of 0.9. The effect of clutter is to reduce this speed even further. For a situation with 0.1 mines per kilometer and 100 items of clutter per kilometer, a system probability of detection = 0.9 and PFA of 0.01, the maximum speed that a vehicle could advance would be 10 kph. Only an improvement in the probability of false alarm to 0.001 would enable the vehicle to increase significantly its rate of advance.

4 GPR for Landmine Detection

Ground penetrating radar (GPR) is an electromagnetic technique which is used to measure the range and position of landmines buried within the ground or dielectric material. The energy radiated by a GPR system occupies a frequency band of a decade or more from several hundred megahertz up to several gigahertz with commensurate wavelengths of 1 m down to 10 cm in air but appropriately reduced by the dielectric constant of the ground. The wavelengths are therefore the same order of magnitude as the dimensions of the landmine and are very different from conventional radar systems where the landmine dimensions are much larger than the wavelength of the incident radiation. The typical average radiated power, integrated over the band of interest, may be on the order of a few tens of milliwatts, but the power per hertz may be as low as picowatts. For landmine detection it is important that the radiated power is lower than that required to initiate some types of fuse. The loss of the soil is often measured as a propagation loss in decibels per meter and is dependent on the conductivity of the soil and the frequency of operation. At 1 GHz it is possible to encounter attenuation losses of many tens of decibels per meter. Some GPR systems are operated so that the landmine, which is within a lossy dielectric, may be only a few wavelengths from the aperture of the antenna. The total path losses within a few wavelengths may be as much as 100 dB depending on the material. As GPR systems do not have a total loop gain much in excess of 120 dB the designer has a major challenge to detect landmines signatures within very short ranges of typically 20 ns.

Additionally GPR can be operated so that the antenna is very close to the ground surface and landmine such that the energy transfer is predominantly either induction or quasi-stationary (the near field), or can be operated such that the energy transfer is in the far field region. GPR encounters extremely high levels of clutter at short ranges and this as well as signal/noise ratio is its major technical challenge. All these aspects pose special design problems for GPR, which is described in detail by Daniels and Curtis [3]. The landmine is surrounded by soil, which is a lossy dielectric whose relative dielectric constant depends mainly upon the water content. Typically the relative dielectric constant of the soil varies from 3 in dry sand to greater than 16 in wet and waterlogged soils.

The explosive used in landmines is typically nitrogen based with a relative dielectric constant between 2.7 and 3.5, ammonium nitrate being the exception as shown in Table 1. Landmines can also be found in fresh water, which has a relative dielectric constant of approximately 80, but a very low loss tangent, hence it is quite feasible to detect landmines in fresh water or soils saturated in fresh water, which also has the benefit of increasing impedance contrast. Salt water on the other hand completely attenuates radar signals. It should be noted that the ground and surface are quite likely to be inhomogeneous and contain inclusions of other rocks of various size as well as man-made debris. Thus the signal to clutter performance of the radar is likely to be an important performance factor. Clutter may be regarded as any radar return that is not associated with the wanted landmine and needs to be defined with respect to a particular application.

Scattering of electromagnetic energy from a landmine results from the impedance differences of the landmine compared with the host material. Canonical targets such as cylinders, which are similar to landmines, have well understood free space scattering characteristics that will be modified by the dielectric of the soil. The mine may have a number of scattering centres, each with their own angular

Table 1 Relative dielectric constants of explosives.

Substance	Name	Relative dielectric constant
TNT	2,4,6-Trinitrotoluene	2.70
Detasheet	PETN	2.72
PETN	Pentaerythritol tetranitrate	2.72
Comp B	RDX TNT	2.90
Octol	HMX TNT	2.90
Tetryl	2,4,6-Trinitrophenyl-N-methylnitramine	2.90
Semtex-H	RDX-PETN	3.00
HMX	Cyclotetramethylene-tetranitramine	3.08
Comp C-4	RDX	3.14
RDX	RDX Hexahydro-1,3,5-trinitro-1,3,5-triazine	3.14
AN	Ammonium nitrate	7.10
NG	Nitroglycerin	19.00

radiation pattern and, in the case of plastic landmines, the internal structure of the mine may generate additional scatterers. Most minimum metal landmines may be considered as multiple layered dielectric cylinders, each interface causing a reflection, the impact of the small internal metallic fuse being minimal. A simple transmission line model representing the case where the angle of incidence is equal to the angle of reflection can simulate the time domain signature of the latter.

GPR system design can be classified into two classes. Systems that transmit an impulse and receive and process the reflected signal from the landmine using a sampling receiver can be considered to operate in the time domain. Systems that transmit individual frequencies in a sequential manner or as a swept frequency and receive the reflected signal from the landmine using a frequency conversion receiver can be considered to operate in the frequency domain. Handheld GPR systems use separate, man-portable, transmit and receive antennas, which are placed just above the surface of the ground and moved in a known pattern over the surface of the ground under investigation. This generates, in real time, data or an image. By systematically surveying the area in a regular pattern, a radar image of the ground can be built up. Alternatively, the GPR may be designed to provide an audible warning of landmine presence while the antenna is moved. Vehicle based or airborne systems use much larger arrays of antennas to illuminate a swathe of the ground surface ahead of the platform and rely on the movement of the vehicle to create the data, which may be processed using SAR techniques.

The GPR image of a landmine is very different from its optical image because the wavelengths of the illuminating radiation are similar in dimension to the landmine. This results in a much lower definition in the GPR image and one that is highly dependent on the propagation characteristics of the ground. The beam pattern of the antenna is widely spread in the dielectric and this degrades the spatial resolution of the image, unless corrected. Refraction and anisotropic characteristics of the ground may also distort the image. For some longer-range systems, synthetic aperture processing techniques are used to optimise the resolution of the image.

Unprocessed GPR images often show "bright spots" caused by multiple internal reflections within the landmine as well as a distortion of the aspect ratio of the image of the landmine caused by variations in the velocity of propagation. Symmetrical targets, such as spheres, cause migration of the reflected energy to a hyperbolic pattern. Radar images can be processed to compensate for these effects and this is usually carried out off-line. A radar can be designed to detect specific landmines by means of polarised radiation. This chapter considers the practical limitations of radar for detecting buried landmines. The types of radar considered are those in which the antenna is very close to the ground surface (proximal operation), radar systems whereby the antenna is operated a few wavelengths from the surface of the ground and finally radar systems whereby the antenna is many wavelengths from the surface of the ground (stand-off operation).

There is an extensive literature on radar methods for landmine detection and a variety of sophisticated modelling and processing methods have been applied to the problem. However the ill-posed nature of operation in real soils has meant that few of these techniques have proved robust when moved from the laboratory to the field and simpler methods have often proved more reliable.

4.1 Attenuation

Electromagnetic waves propagating through soil incur an attenuation loss given by

$$L_a = 8.686 \cdot 2 \cdot R \cdot 2\pi f \sqrt{\left(\frac{\mu_0 \mu_r \varepsilon_0 \varepsilon_r}{2}\left(\sqrt{(1+\tan^2\delta)}\right) - 1\right)}$$

where

f = frequency in Hz

$\tan d$ = loss tangent of material

ε_r = relative permittivity of material

ε_0 = absolute permittivity of free space

μ_r = relative magnetic susceptibility of material

μ_0 = absolute magnetic susceptibility of free space

R = range in metres

The graph in Figure 9 shows the two-way attenuation loss in decibels per meter plotted against frequency for a material with a relative dielectric constant of 9 and loss tangents of 0.1 to 0.9 in steps of 0.3 respectively. As the frequency is increased from 1 GHz to 5 GHz, the attenuation loss for a soil with a loss tangent of 0.3 increases from 20 to 100 dB.

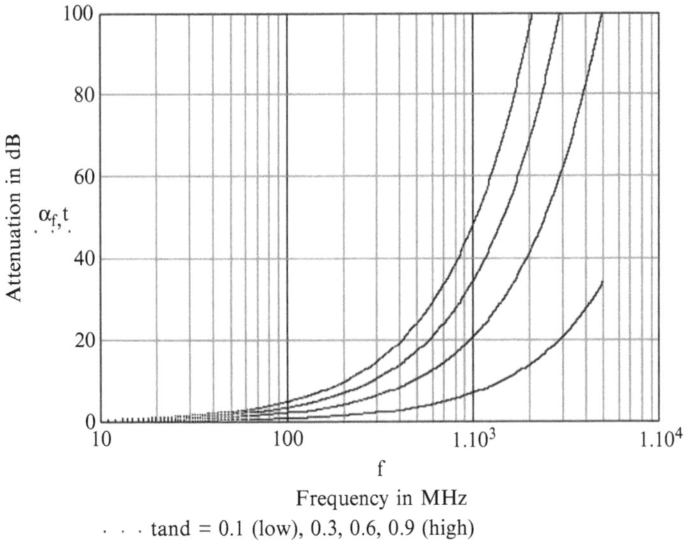

Fig. 9 Material losses in dBm^{-1} plotted against frequency in Hz for values of tan$d$ of 0.1 to 0.9.

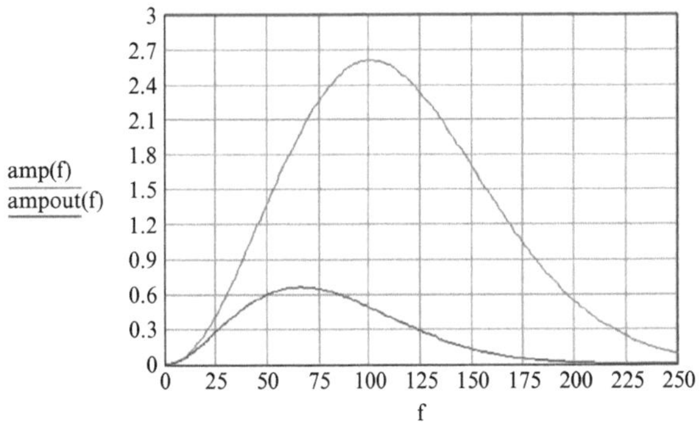

Fig. 10 Spectrum of transmitted and received signals after passing through lossy ground.

The effect on the spectrum of typical radar is shown in Figure 10 which shows the peak of the spectrum shifted to lower frequencies and the higher frequencies considerably reduced.

4.2 Coupling energy into the ground

Buried mines pose a difficult detection problem for radars and their performance is strongly influenced by the ground conditions. For close-in operation the efficiency of the coupling process is high but this is not the case for standoff radar systems since, where lossy materials are involved, complex angles of refraction may occur. With vertical polarisation at incidence angles less than the Brewster angle, transmission losses at the air/ground interface are relatively small but at larger incidence angles than the Brewster angle the losses increase more rapidly. Hence to maximise the operating range the radar should be mounted as high off the ground as is possible. Thus for a given height, the performance of the radar will be set by the relative dielectric constant of the ground. In addition to the problem of coupling energy into the ground the effective cross section of all landmines decreases when they are buried. Measurements and modelling suggest that under conditions of negligible attenuation losses, as are expected in very arid ground or for shallow burial depths, metal landmine to clutter ratios are expected to be degraded on burial by approximately 10 dB. Under the same conditions the cross section of plastic mines is reduced by a larger factor because of reduced dielectric contrast between the mine material and the surrounding soil, so that, in wet sandy soils, plastic mines are more readily detected than in dry conditions. However plastic mines are subject to substantially smaller burial losses in dry sand when they contain air voids. This is beneficial for detection as plastic mines generally contain such voids to allow movement behind the pressure plate. The radar system must have at least a 20 dB signal to clutter ratio to detect buried landmines in all weather conditions. Thus in order to detect buried plastic landmines with air voids the corresponding signal to clutter ratio for surface-laid metal landmines must be better than 12 dB for dry conditions and 18 dB for wet conditions.

4.3 Depth resolution

For traditional radar systems it is accepted that two identical targets can be separated in range if they are 0.8 of a pulse width apart. Essentially range resolution is defined by the bandwidth of the received signal and in this context it is the bandwidth of the received signal which is important, rather than that of the transmitted signal. The earth material acts as a low pass filter, which modifies the received spectrum in accordance with the electrical properties of the propagating medium. A receiver bandwidth in excess of 500 MHz and typically 1 GHz is required to provide a typical resolution of between 5 and 20 cm, depending on the relative permittivity of the material. Where interfaces are spaced more closely than one half wavelength the reflected signal from one interface will become difficult to resolve with that from another. It should be noted that the normal radar criteria for range resolution is less appropriate for the case of a weak target adjacent to a strong target and there is no accepted definition of resolution for the case of unequal size targets.

4.4 Plan resolution

The plan resolution is defined by the characteristics of the antenna and the signal processing employed. In general radar systems (apart from SAR) require a high gain antenna to achieve an acceptable plan resolution. This necessitates a sufficiently large aperture at the lowest frequency to be transmitted. To achieve small antenna dimensions and high gain therefore requires the use of a high carrier frequency, which may not penetrate the material to sufficient depth. When selecting equipment for a particular application it is necessary to compromise between plan resolution, size of antenna, the scope for signal processing and the ability to penetrate the material. Plan resolution improves as attenuation increases, provided that there is sufficient signal to discriminate under the prevailing clutter conditions. In low attenuation media the resolution obtained by the horizontal scanning technique is degraded, but only under these conditions do synthetic aperture techniques increase the plan resolution. Essentially the ground attenuation has the effect of placing a "window" across the SAR aperture and the higher the attenuation the more severe the window. Hence in high attenuation soils SAR techniques may not provide any useful improvement to radar systems. SAR techniques have been applied to GPR, but very often in dry soils with low attenuation.

A key feature of non-contacting ground antennas is their illuminating footprint. As a landmine radar image is effectively the convolution of the antenna footprint with the landmine radar spatial cross section, the landmine image becomes blurred. This effect increases with antenna to ground spacing and eventually results in landmines with small radar cross-section (AP mines) becoming vanishingly small.

Plan resolution actually improves as attenuation increases, assuming that there is sufficient signal to discriminate under the prevailing clutter conditions. In low attenuation media the resolution obtained by the horizontal scanning technique is degraded, but under these conditions the use of advanced signal processing techniques becomes feasible. These techniques typically require measurements made using transmitter and receiver pairs at a number of antenna positions to generate a synthetic aperture or focus the image. Unlike conventional radars, which generally use a single antenna, most GPR systems use separate transmit and receive antennas in what has been termed a bistatic mode. SAR techniques typically require measurements made using transmitter and receiver pairs at a number of antenna positions to generate a synthetic aperture or focus the image. Unlike conventional radars, which generally use a single antenna, most landmine radar systems use separate transmit and receive antennas to provide receiver isolation. The GPR community refer to this as a bistatic mode, although actually the antenna system is closely spaced and mobile. This is different from the traditional radar community that associates the term bistatic with large separations.

4.5 Frequency of operation

The most basic model for assessment of signal level is derived from the far field radar range equation, which does however have limitations with respect to correct representation of the actual operation of very short-range system. However, it does enable a first order assessment of expected signal levels. In the absence of

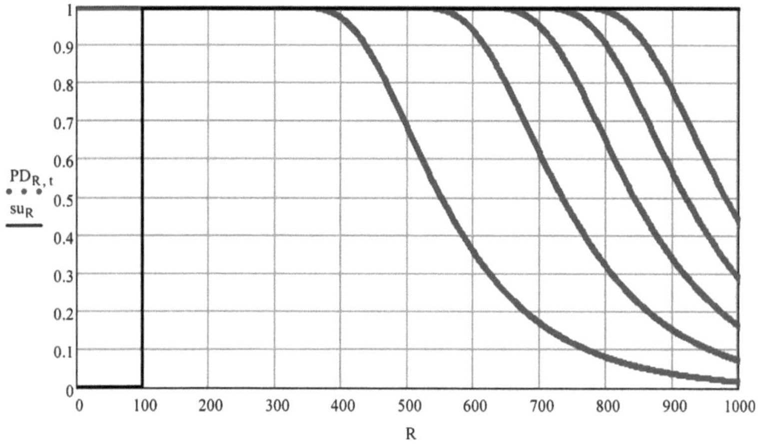

Fig. 11 Probability of detection of dielectric cylinders 10–50 cm at 1 GHz.

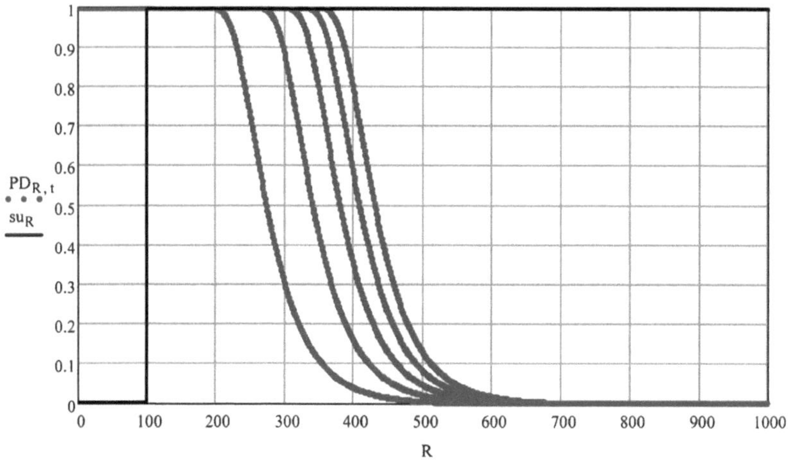

Fig. 12 Probability of detection of dielectric cylinders 10–50 cm at 3 GHz.

any clutter whatsoever in the ground and assuming a complete removal of the front surface reflection it is possible to calculate the probability of detection as a function of landmine range and landmine size. This is shown in Figure 11. The family of curves represents the probability of detection versus range for dielectric cylinder of diameters 10–50 cm in increments of 10 cm, working from left to right.

The signal to noise ratio (SNR) of the radar receiver is 14.6 dB and the mine signal is 6 dB greater than the SNR. A frequency of 1 GHz was used with a landmine $e_r = 2.2$, a soil relative permittivity e_r of 9 and ground attenuation of 27 dB m^{-1} at 1 GHz. The antenna to ground spacing is 10 cm. The smallest cylinder can be detected at a depth of cover of 25 cm. At 3 GHz the radar performance as a function of range is considerably reduced as the attenuation has increased to 82 dB m^{-1} as shown in Fig. 12.

4.6 Landmine scattering characteristics

Scattering of electromagnetic energy results from impedance differences in the landmine compared with the host material. Canonical landmines such as cylinders have well understood radiation characteristics as described by Skolnik [10], that can be modified for the dielectric of the soil. The mine may have a number of scattering centres, each with their own angular radiation pattern and in the case of plastic landmines the internal structure of the mine may generate additional scatterers. Most plastic landmines can be considered as multiple layered dielectric cylinders, of which each interface causes a reflection. A simple transmission line model representing the case where the angle of incidence is equal to the angle of reflection can simulate the time domain signature of the latter as shown in Fig. 13. The first reflection is due to the ground surface and the subsequent reflections are due to the landmine air void and explosive. The depth of cover of the mine is 10 cm and it is 10 cm in thickness.

For comparison the time domain signatures of various landmines buried at 5 cm are shown in Fig. 14 and it can be seen that the simulation most closely resembles the VS50 in shape. On the horizontal scale ten samples equals 0.25 ns and the vertical scale represents relative amplitude.

If the metallic landmine is at an angle to the plane of the surface the peak response may well be to one side of the actual physical position of the landmine. This is particularly critical for hand held radar systems. Other aspects of the radar cross section of landmines are concerned with the relative contributions of specular reflection, diffraction off discontinuities, travelling waves including direct illumination running wave, creeping wave on metal, trapped guided wave on dielectric as well as the contribution due to resonant scatterers, which are a combination of discontinuities that allow the echo to build up. Much effort has been applied to accurate modelling, as described by Streich and Van der Kruk [11].

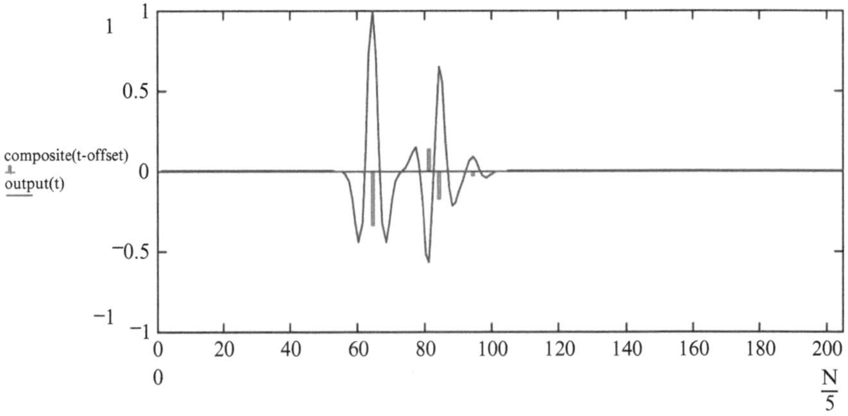

Fig. 13 Simulated time domain signature of a buried landmine simulation.

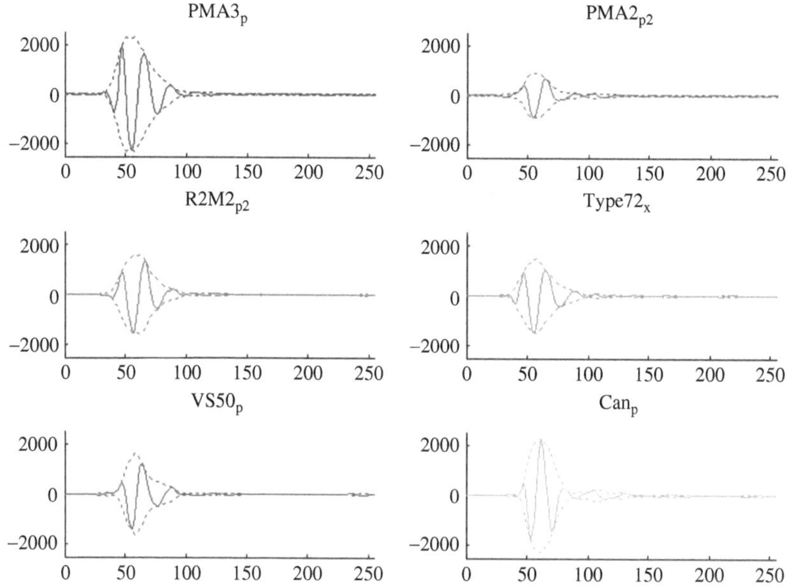

Fig. 14 Time domain signatures of buried mines taken with radar transmitting 1 ns duration impulses.

4.7 Clutter

A major difficulty for operation of GPR systems is the presence of clutter within or on the surface of the material or in the side and back lobes of the antenna and sources of surface clutter. These has been modelled by Firoozabadi et al. [6]. Clutter is defined as sources of unwanted reflections that occur within the effective bandwidth and search window of the radar and are present as spatially coherent reflectors. Animal burrows and cracks in the ground are examples of features that will cause reflections. Careful definition and understanding are critically important in selecting and operating the best system and processing algorithms. Clutter can completely obscure the buried landmine and a proper understanding of its source and impact on the radar is essential. A key issue is the effect on the radar of variations in the topography of the ground surface caused by potholes or ruts. Methods of processing the radar signals that adjust the delay time to the front surface to "flatten it" will actually distort the radar signature of buried landmines. Abrupt discontinuities can also cause multiple reflections, which become superimposed on later arriving reflected energy. Such "interference" will be extremely difficult to remove. Radar systems should not provide indications on the following small sources (small being defined as not exceeding a surface area of 1.5 cm^2):

- Small metal fragments
- Shrapnel
- Spent bullet and cartridge cases

- Ground topographical variations less than 3 cm in any dimension
- Puddles of water up to 15 cm diameter
- Tufts of grass up to 5 cm in diameter and 5 cm high
- Rocks, stones less that 5 cm in maximum dimension
- Animal burrows less than 5 cm diameter
- Cracks and fissures in ground less than 1.5 cm in width

5 Vehicle Based Radar Systems

Vehicle based systems have been developed that use arrays of antennas and generate 3-D data, which is then processed to provide a rolling map of detections. The signal and image processing options for vehicle based landmine detection are more extensive because the radar and its platform generate 3-D data. In general vehicle based systems concentrate on anti-tank landmines because it is difficult to achieve adequate cross range resolution at realistic budgets. Options for signal and image processing include image inversion and synthetic aperture techniques for image enhancement principal component analysis (PCA) and independent component analysis (ICA) techniques and hidden Markov models. ERA Technology developed a 4 m wide antenna close-coupled GPR system for the UK Minder CAP programme as shown in Figure 15. Against minimum metal mines buried up to 17.5 cm it achieved a PD of 0.77. It should be noted that during these trials 80% of

Fig. 15 UK Minder CAP programme countermine system.

the on-road AT mines were buried low metal (TMA4, Type72) and the maximum depth of burial was 17.5 cm. The most common depth of burial was 6" (15 cm) and approximately 65% of the mines were buried with a depth of cover greater than 10 cm. Using the trial results we get an extended estimate of GPR performance for off ground radars based on an average PD = 0.8 against a depth of cover of 10 cm and for proximal GPR systems with an average PD = 0.8 against a depth of 17 cm. It can be seen that the depth performance of the proximal GPR is greater because of the improved coupling and reduced range-spreading losses. Although a number of developmental vehicle-based GPR systems have been trialled and reported on over the last 5 years, even the most extensively reported NIITEK radar system has yet to move into production.

6 Handheld Radar Systems

Recent developments using hand held dual sensor technology combining electromagnetic induction EMI and ground penetrating radar (GPR) have enabled improved discrimination against metal fragments to be demonstrated in live minefields. Reductions of up to 7:1 compared with the standard metal detector have been achieved in the field by hand held systems such as MINEHOUND [3] shown in Figure 17 and AN/PSS-14 [5]. Handheld landmine systems are more limited in the signal processing algorithms that can be applied because they usually only have a single transmit-receive antenna pair and with only a few exceptions do not form an image. Research into landmine discrimination based on the analysis of A-scans by means of complex resonances, wavelets, time-frequency characteristics, neural networks, fuzzy sets, Gaussian mixture models, order statistics and template matching, has been carried out. Methods based on time-frequency characteristics are reported by Wong et al. [13], Lopera et al. [8], as well as Daniels et al. [4] who showed the feasibility of discriminating between AP landmines and typical false landmines on a small data set.

The AN/PSS-14 hand held detector trialled in Angola [5] reports the following results for probability of detection using experienced operators at a 90% confidence level (CL) range with a false alarm rate of 0.23–0.28 m^{-2}. The trials carried out in Bosnia, Cambodia and Angola using the MINEHOUND detector [3] reported a ROC curve as shown in Figure 16. The ROC curve relates to over 1,069 encounters in live minefields of which seven were actually mines. It should be noted that the trials did not test the MINEHOUND in a blind test, but compared the MINEHOUND with the CEIA MIL D1 metal detector, which was used first. The MINEHOUND does not currently incorporate mine classification algorithms.

6.1 Assessment of radar performance

There is an potentially overwhelming body of literature on GPR for landmine detection and an approach for its assessment is to consider the comparability of the data,

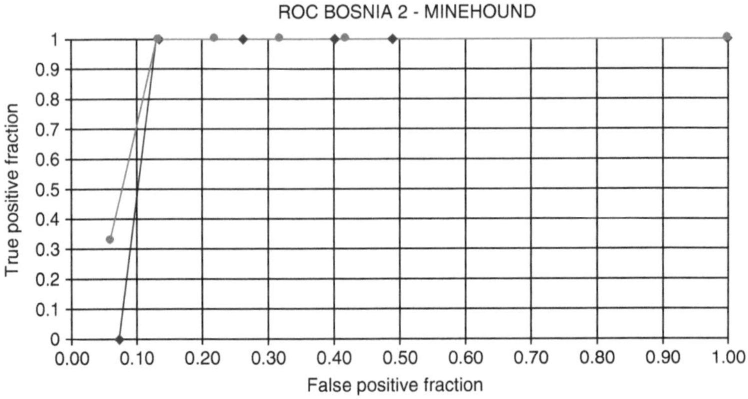

Fig. 16 ROC curve for MINEHOUND for trial in Bosnia.

Fig. 17 MINEHOUND dual sensor landmine detector.

the maturity of the technology, the feasibility of implementing the proposed algorithm and, most importantly, the performance in terms of the probability of detection and the probability of false alarm.

There are a number of fundamental issues that govern the probability of detection. Both propagation parameters as well as the radar cross-section define the fundamental system performance as discussed by Daniels [2].

In comparing published results relating to controlled trials, it is critical to know the statistics of the depth of cover, soil propagation characteristics, type of landmine, the physical placement of the landmine as well as the characteristics of the clutter. In the case of field trials in, or close to, live minefields it is more difficult to gather such information; however, a statistically based approach may be a realistic alternative. In reviewing published receiver operating characteristic (ROC) curves the statistics of the sample should be known in order to understand the confidence that can be placed in any data. Simonsen [9] provides a useful treatment of sampling statistics as applied to landmine data. Clearly any assessment of the performance of algorithms should also state the confidence limits that apply to ROC curves. However if the sample size is known, it is relatively straightforward to determine bounds. Simonson notes that 39 or more mines are needed to ensure at least an 80% chance of detecting a difference when the two systems have detection probabilities of 0.90 and 0.60, respectively.

Voles [12] considered this issue and showed that based on a Poisson distribution, even if no mines were missed by a sensor in a test of 100 then at the 95% confidence limit the highest value of probability of detection that can be claimed is 97%. Voles also showed that to achieve a 99.6% probability of detection at a confidence level of 95% would require a test of 750 mines and none should be missed.

6.2 *Future development of radar*

The main challenge for hand held radar is the further reduction in the rate of false alarm. At present the EMI detector encounters around 200 false targets to every AP mine. The current generation of dual sensor detectors reduces the ratio to around 5:1. If robust classification techniques can be developed that reduce the ratio down to around 30:1 the efficiency improvement in humanitarian operations will be even greater.

Vehicle based radar has to achieve orders of magnitude performance improvement to enable route clearing military operation to proceed at speed. A total system performance of a probability of detection better than 0.99, with a probability of false alarm less than 10^{-4}, is called for if route clearance at convoy speeds is to be achieved. Humanitarian clearance may tolerate speed reduction but still requires high detection rates. This applies to both stand off and close in GPR systems.

Airborne radar is an enormous technical challenge. However a new generation of unmanned airborne vehicles may provide suitable platforms for the close in GPR systems if ground skimming can be achieved. This would allow reconnaissance vehicles to run ahead of convoys and would reduce the need to mine protect vehicles.

7 Summary

- GPR systems for landmines have a loop gain on the order of 120 dB, which sets their order of magnitude performance.
- The radiated power is limited by licence restrictions and EMC considerations as well as the need to avoid detonation of certain types of fuses.

- Most path losses are such that penetration is limited to 50 cm depth of cover for most GPR systems.
- The propagation losses decrease as the fourth power of range to landmine for far field conditions.
- The propagation losses may decrease at lower rates depending on the landmine dimensions for near field boundary conditions.
- The received signal may be augmented by induction and quasi-stationary contributions for landmines within the near field.
- The attenuation losses in materials rapidly increase with frequency, which means that most systems will receive frequencies in the range 300 MHz to 1.5 GHz. The use of transmitted frequencies above 2 GHz is unlikely to provide useful performance in real world conditions and will severely limit depth performance.
- The attenuation losses in materials will reduce the effectiveness of multi-look antenna arrays by effectively putting a window taper across the array.
- At 1 GHz the total losses in typical soils mean that, in ideal conditions, detection ranges of 20–30 cm are feasible.
- In dry soils the dielectric contrast between the soil and mine reduces and this can make the detection of mines with minimal air voids more difficult.
- Most GPR systems will achieve optimum performance in terms of range when the antennas are operated in close proximity to the ground. As the antenna to ground spacing increases, the antenna radiation pattern results in reduction of the received signal from small landmines and increased vulnerability to clutter from free space sources.
- Rough surfaces, ruts, potholes etc. degrade the signal to clutter ratio and reduce the system performance.
- The angular response of mines that are tilted relative to the ground surface may not be co-incident with their physical position and this should be considered when neutralising.
- Stand off SAR radar systems have fundamental limits to performance at shallow grazing angles, which constrains their forward look range to between 10 and 20 m.

References

1. Daniels D J, Ground Penetrating Radar, ISBN 0863413609, IEE (Radar, Sonar and Navigation) 2004
2. Daniels D J, An Assessment of the fundamental performance of GPR against buried landmines, SPIE Detection and Remediation Technologies for Mines and Minelike Targets XII, Paper 6553-16, SPIE 2007, 13 April, 2006, Orlando, FL
3. Daniels D J, Curtis P, MINEHOUND trials in Bosnia, Angola and Cambodia, Proceedings of the SPIE Defense and Security Conference 2006, 17–23 April, 2006, Orlando, FL
4. Daniels D J, Curtis P, Lockwood O, Classification of landmines using GPR, IEEE RadarCon 2008, 26–30 May, 2008, Rome, Italy
5. Doheny R C et al., Handheld Standoff Mine Detection System (HSTAMIDS) field evaluation in Namibia, Proceedings of the SPIE Defense and Security Conference, 16–21 April 2006, Orlando, FL

6. Firoozabadi R, Miller E L, Rappaport, C M, Morgenthaler A W, Sub-surface sensing of buried objects under a randomly rough surface using scattered electromagnetic field data, IEEE Trans on Geoscience and Remote Sensing, Jan 2007, Volume 45, No 1, pp 93–104
7. Guidebook on Detection Technologies and Systems for Humanitarian Demining, Geneva International Centre for Humanitarian Demining (GICHD)
8. Lopera O, Milisavljevie N, Daniels D, Macq B, Time-frequency domain signature analysis of GPR data for landmine identification, Advanced Ground Penetrating Radar, 2007 4th International Workshop 27–29 June 2007, pp 159–162
9. Simonsen K, Statistical considerations in designing tests of mine detection systems 1 Measures related to the probability of detection, Sandia Report SAN98-1769/1
10. Skolnik M, Radar Handbook, 2nd Edition, ISBN 007057913X, McGraw-Hill, New York, Chap 11, Radar Cross Section of Targets, 1990
11. Streich R, Van der Kruk J, Accurate imaging of multicomponent GPR data based on exact radiation patterns, IEEE Trans on Geoscience and Remote Sensing, Jan 2007, Volume 45 No 1 pp 93–104
12. Voles R, Confidence in trials of landmine detection systems, Mathematics Today April 2000
13. Wong D, Nguyen L, Gaunaurd D, Radar classification of landmines by time-frequency analysis, Proc. SPIE 6566, 65660F, 2007

Overview of Statistical Tests for Unexploded Ordnance Detection

Hakan Deliç*

Abstract In this chapter, we outline the statistical procedures that can be employed for the detection of unexploded ordnance (UXO). Phenomenological modeling is first developed to relate the collected data to a sensor's feature parameters, which in turn allow for physics-based signal processing. Starting with the Bayesian framework, we introduce minimax and robust detection that do not require prior probabilities and distributional information on the measurement uncertainty, respectively. Nonparametric tests that perform well for broad classes of distributions are also presented. Finally, the generalized likelihood ratio test is described as a joint estimation-detection method which first estimates the feature parameters and then tests for the presence-absence of the UXO.

Keywords: Detection, Gaussian distribution, likelihood ratio test, modeling, minimax, Neyman-Pearson, nonparametric test, UXO

1 Introduction

Unexploded ordnance (UXO) refers to explosive devices that lie below ground or sea surface. Magnetometers, electromagnetic induction (EMI) and radar are typically used for sensing UXO. Excavation operations are risky and costly, and therefore false alarms should not exceed some acceptable level.

Detection of UXO involves several stages. Data are collected by sensors, preferably several distinct ones. Model parameters are extracted and refined in which information from one device may help constrain the parameter space of another sensor to minimize uncertainties. One example is the location estimate supplied by a magnetometer serving as a constraint when analyzing the EMI data. The next step

*Wireless Communications Laboratory, Department of Electrical and Electronics Engineering, Boğaziçi University, Bebek 34342 Istanbul, Turkey
e-mail: delic@boun.edu.tr

J. Byrnes (ed.), *Unexploded Ordnance Detection and Mitigation*,

is the detection of UXO; that is, distinguishing between UXO and non-UXO objects based on statistical tests performed on the measured parameter values. This may be followed by classification of the UXO type.

In this chapter, we concentrate on various detection techniques that primarily differ in the modeling assumptions. We characterize the problem in the form of two hypotheses described as

$$H_1: \text{UXO present},$$

$$H_0: \text{UXO absent, or non-UXO present.}$$

We will assume that the measurement uncertainties are represented by the multivariate Gaussian distribution. We adopt this distribution because of its maximum entropy property and the mathematical convenience it brings along, rather than being justified by empirical observation. However, we will also show methodologies that accommodate variations in the distributional form.

The simple, approximate magnetic-dipole model uses the magnetometer field measurements to determine the UXO depth below ground and the magnetic-dipole orientation [1, 9]. The EMI response can be modeled by generalizing the magnetometer model through a tensor that ties the excitation magnetic field and the magnetic dipole moment. Assigning unique magnetic dipoles to distinct components of the same UXO and providing more information, the EMI models work with more parameters such as the ordnance's center location constrained by the magnetometer data, UXO orientation (characterized by a unitary transformation matrix on the magnetization tensor), magnetization induced by ferrous elements, and the EMI resonant frequencies [15]. A multisensor towed array system of magnetometers and EMI sensors is shown to perform with a detection probability greater than 0.95 in [9]. The detection of deeply buried UXO by means of a magnetometer equipped with cone penetrometer technology is investigated in [14].

Ground-penetrating radar (GPR) in conjunction with synthetic aperture radar (SAR) processing can be used from airborne [3] or ground platforms for UXO location identification. In [4], a GPR in the 50–500 MHz range is deployed along with magnetometers. The SAR processing produces three-dimensional images of possible UXO locations. Further data analysis is needed to decide on the actual UXO presence [2]. For the application of a directional borehole radar, see [11].

2 Detection

Let $\mathbf{x}$ denote the feature vector whose elements are the magnetometer or EMI model parameters. The signal measurements received from the sensor is the n-dimensional vector $\mathbf{v}$, which is a function of $\mathbf{x}$. Extraction of $\mathbf{x}$ from $\mathbf{v}$ based on a phenomenological model is known as inversion. As an example, consider the EMI dipole model [1, 15], where $\mathbf{H}'$ denotes the excitation magnetic field and $\mathbf{M}$ is the magnetization tensor. The magnetic dipole moment is $\mathbf{m} = \mathbf{M} \cdot \mathbf{H}'$. Assuming that the rotationally symmetric UXO is aligned along the z-axis, $\mathbf{M}$ can be represented as the diagonal matrix

$$\mathbf{M}(\omega) = \mathbf{z}^T\mathbf{z}\left(m_z(0) + \sum_k \frac{\omega m_{zk}}{\omega - j\omega_{zk}}\right) + (\mathbf{x}^T\mathbf{x} + \mathbf{y}^T\mathbf{y})\left(m_p(0) + \sum_k \frac{\omega m_{pk}}{\omega - j\omega_{pk}}\right) \quad (1)$$

where $\mathbf{x}, \mathbf{y}, \mathbf{z}$ are orthonormal row vectors and $m_z(0), m_p(0)$ stand for the magnetization induced by ferrous objects. Keeping only the first terms in each summation in (1) is sufficient to have a physics-based signal model [1], where ω_{z1} and ω_{p1} can be used as features in the detection set-up because the imaginary resonant frequencies are functions of the UXO material properties and size.

2.1 Bayesian framework

Let $\pi_i, i = 0,1$, be the *a priori* probability of $H_i, i = 0,1$, with $\pi_0 + \pi_1 = 1$. The probability π_1 represents the prior knowledge, expectation or guess regarding the likelihood of encountering an actual UXO. Thus, the higher π_1, the greater the chance of running into an UXO at the area of exploration.

Suppose that we can devise a cost coefficient $c_{ij}, i, j = 0,1$ which represents the cost of deciding on H_i when H_j is true. Assuming that $c_{11} < c_{01}$, the expected Bayesian risk associated with the choice between the presence and absence of a UXO is minimized by the decision rule δ_B:

$$\delta_B = \begin{cases} 1 \text{ if } L(\mathbf{v}) \geq t, \\ 0 \text{ if } L(\mathbf{v}) < t \end{cases}$$

where the likelihood ratio (LR) and the threshold are respectively defined as

$$L(\mathbf{v}) = \frac{f_1(\mathbf{v})}{f_0(\mathbf{v})},$$

$$t = \frac{(c_{10} - c_{00})\pi_0}{(c_{01} - c_{11})\pi_1}$$

with $f_i(\mathbf{v}) = f(\mathbf{v}|H_i), i = 0,1$.

Determining the value of π_1 with good accuracy is critical for the detection performance when the Bayesian framework in (2.1) is employed. If the estimated π_1 value is overshot, the test will yield more false alarms than necessary. Similarly, an unrealistically low π_1 may result in excessive number of undetected UXOs with possibly catastrophic consequences. Therefore, it is desirable to work with detectors that do not require the priors, or that are insensitive to deviations from true values.

2.2 Minimax solution

Suppose that the prior probabilities (π_0, π_1) are unknown and they cannot be estimated with sufficient precision. A conservative approach to detector design would be to ensure good performance under the "least favorable" conditions, which is characterized by the priors that maximize the Bayesian cost. Such a worst-case design guarantees a minimum performance level in the event of parametric uncertainty: for any other (π_0, π_1) pair, the detector will do even better.

Let $R(\delta|H_i)$ denote the Bayesian risk associated with the decision rule δ given that hypothesis $H_i, i = 0, 1$, is true. For the unknown prior π_0, the expected risk is

$$R(\delta, \pi_0) = R(\delta|H_0)\pi_0 + R(\delta|H_1)(1 - \pi_0).$$

In accordance with the Bayesian paradigm, the goal now is to find a decision rule-least favorable prior pair (δ_M, π_{0M}) which solves the minimax problem:

$$(\delta_M, \pi_{0M}) = \arg\min_{\delta} \max_{\pi_{0M} \in (0,1)} R(\delta, \pi_0). \tag{2}$$

The formulation in (2) can be viewed as a competitive game between the engineer and nature. While the engineer attempts to minimize the cost by designing the best detector, nature tries to maximize the penalty involved by selecting the least favorable prior. From the engineer's perspective, nature wants to maximize the minimum cost induced by his/her δ decision. In contrast, the engineer's objective is to minimize the maximum cost that occurs from the (π_0, π_1) selection. The pair (δ_M, π_{0M}) exhibits a so-called saddle point behavior described by

$$R(\delta_M, \pi_0) \leq R(\delta_M, \pi_{0M}) \leq R(\delta, \pi_{0M})$$

for all δ and π_0.

The saddle point property stipulates that the following condition is satisfied for any $R(\delta, \pi_0)$.

$$R(\delta_M, \pi_{0M}) = \max_{\pi_0 \in (0,1)} \min_{\delta} R(\delta, \pi_0) = \min_{\delta} \max_{\pi_0 \in (0,1)} R(\delta, \pi_0). \tag{3}$$

The interpretation of (3) is simple and useful: If (δ_M, π_{0M}) is a saddle point, then it solves the minimax problem in (2), and vice versa. Moreover, the minimax solution is the same as the maximin solution, and one can opt for one or the other depending on their relative ease and complexity.

2.3 Neyman-Pearson framework

In addition to the lack of reliable information about the prior probabilities, it may not be possible to formulate meaningful cost coefficients as required by the Bayesian set-up. The two error types may have asymmetrical penalties involved. Specifically, let

$$e_{\mathrm{I}}(\delta) = P\{H_1 | H_0 \,\text{is true}\}$$

which is known as the false alarm probability, or type I error in statistics. Similarly, the miss probability, or type II error is defined as

$$e_{\mathrm{II}}(\delta) = P\{H_0 | H_1 \,\text{is true}\}.$$

It is clear that a false alarm event merely triggers a costly UXO removal operation whereas a miss leaves the UXO undetected. It is impossible to minimize e_{I} and e_{II} simultaneously. Recognizing that minimizing misses is far more important than avoiding false alarms leads to the following constraint optimization problem:

$$\text{Minimize } e_{\mathrm{II}}(\delta) \text{ subject to } e_{\mathrm{I}}(\delta) \leq \alpha. \tag{4}$$

Note that without the bound on the false alarm probability, one could achieve $e_{\mathrm{II}}(\delta) = 0$ by simply having $\delta = 1$ at all times. Unfortunately, this is an infeasible solution because it requires infinite time and resource budgets.

The solution of (4), which follows the construction of the appropriate Lagrangian and the application of Kuhn-Tucker conditions, is stated in the Neyman-Pearson Lemma.

$$\delta_{NP} = \begin{cases} 1 & \text{if } L(\mathbf{v}) > \lambda(\alpha), \\ r(\alpha) & \text{if } L(\mathbf{v}) = \lambda(\alpha), \\ 0 & \text{if } L(\mathbf{v}) < \lambda(\alpha), \end{cases}$$

where the threshold $\lambda(\alpha)$ and the randomization constant $r(\alpha) \in [0,1]$ are such that

$$\int_{\mathcal{V}_1} f_0(\mathbf{v}) d\mathbf{v} + r(\alpha) \int_{\mathcal{V}_2} f_0(\mathbf{v}) d\mathbf{v} = \lambda(\alpha),$$

and

$$\mathcal{V}_1 = \left\{ \mathbf{v} : \frac{f_1(\mathbf{v})}{f_0(\mathbf{v})} > \lambda(\alpha) \right\},$$

$$\mathcal{V}_2 = \left\{ \mathbf{v} : \frac{f_1(\mathbf{v})}{f_0(\mathbf{v})} = \lambda(\alpha) \right\}.$$

Once again, the optimal detector takes the form of a likelihood ratio test but in Neyman-Pearson set-up, the threshold is determined by the false alarm rate α, instead of priors or cost coefficients. If the density functions $f_i(\mathrm{v}), i = 0, 1$, are continuous everywhere, then randomization is not necessary and $r(\alpha)$ can be set to unity.

3 Gaussian Uncertainty

Suppose that the measurement uncertainties are represented by the multivariate Gaussian random variable so that

$$f_i(\mathbf{v}) = \frac{1}{(2\pi)^{n/2}|\mathbf{\Sigma}_i|^{1/2}} \exp\left\{-\frac{1}{2}(\mathbf{v}-\mathbf{m}_i)^T \mathbf{\Sigma}_i^{-1}(\mathbf{v}-\mathbf{m}_i)\right\}$$

where $\mathbf{m}_i$ and $\mathbf{\Sigma}_i, i=0,1$, are respectively the mean vector and the covariance matrix under $H_i, i=0,1$. The natural logarithm of the likelihood ratio takes the form

$$L'(\mathbf{v}) = \log_e L(\mathbf{v}) = (\mathbf{v}-\mathbf{m}_0)^T \mathbf{\Sigma}_0^{-1}(\mathbf{v}-\mathbf{m}_0) - (\mathbf{v}-\mathbf{m}_1)^T \mathbf{\Sigma}_1^{-1}(\mathbf{v}-\mathbf{m}_1). \quad (5)$$

The degree of correlation between successive measurements is hard to determine, and as an approximation and to keep the design simple, one can assume that $\mathbf{\Sigma}_i = \text{diag}\{\sigma_{i1}^2, \sigma_{i2}^2, \ldots, \sigma_{in}^2\}, i=0,1$. Letting $\mathbf{m}_i = [m_{i1} \cdots m_{in}]^T$ and $\mathbf{v}_i = [v_{i1} \cdots v_{in}]^T$, (5) becomes

$$L'(\mathbf{v}) = \sum_{k=1}^{n} \frac{(v_{0k}-m_{0k})^2}{\sigma_{0k}^2} - \sum_{k=1}^{n} \frac{(v_{1k}-m_{1k})^2}{\sigma_{1k}^2}$$

The optimal Bayesian test is

$$\delta_{B,G} = \begin{cases} 1 \text{ if } L'(\mathbf{v}) \geq t', \\ 0 \text{ if } L'(\mathbf{v}) < t' \end{cases}$$

where

$$t' = \log_e \frac{(c_{10}-c_{00})\pi_0}{(c_{01}-c_{11})\pi_1}.$$

If there is sufficient evidence that the underlying uncertainty cannot be adequately described by the Gaussian distribution, then it is possible to resort to robust formulations that ensure good performance even if there are deviations from the nominally assumed probability distribution. Let $\mathcal{F}_0$ and $\mathcal{F}_1$ respectively denote two disjoint classes of multivariate probability density functions (PDFs) that represent H_0 and H_1. Following a similar game as in the minimax construction, we then seek to find a pair $(\delta^*, f_1^*(\mathbf{v}))$, where δ^* is an admissible decision rule and $f_1^*(\mathbf{v}) \in \mathcal{F}_1$, such that

$$e_{\text{II}}(\delta, f_1^*) \leq e_{\text{II}}(\delta^*, f_1^*) \leq e_{\text{II}}(\delta^*, f_1), \forall \delta \in \mathcal{D}, \forall f_1 \in \mathcal{F}_1, \quad (6)$$

and

$$e_{\text{I}}(\delta^*, f_0) \leq \alpha, \forall f_0 \in \mathcal{F}_0, \quad (7)$$

where $\mathcal{D}$ is the class of admissible decision rules and α is a prespecified false alarm rate. If there exists a δ^* that satisfies (6) and (7), then it is referred to as a robust rule. The pair of density functions, f_0^* and f_1^* that satisfy (6) and (7) for the rule δ^* are called least favorable in $\mathcal{F}_0 \cup \mathcal{F}_1$. Moreover, δ^* is clearly a Neyman-Pearson test at (f_0^*, f_1^*) and α.

An interesting special case where $\mathcal{F}_0$ and $\mathcal{F}_1$ represent the following classes of stationary and memoryless processes has been extensively studied by Huber [6].

$$\mathcal{F}_0 = \{f(v) = (1-\varepsilon_0)f_0(v) + \varepsilon_0 h(v),\ v \in \mathcal{R}, h \in \mathcal{H}\}, \tag{8}$$

$$\mathcal{F}_1 = \{f(v) = (1-\varepsilon_1)f_1(v) + \varepsilon_1 h(v),\ v \in \mathcal{R}, h \in \mathcal{H}\} \tag{9}$$

where $\mathcal{R}$ is the real line, $\mathcal{H}$ is the class of all symmetric density functions, $\varepsilon_0, \varepsilon_1 \in (0,1)$ and v is some element of the feature vector $\mathbf{v}$ [7]. The robust detector under the classes of distributions defined in (8) and (9) is the likelihood ratio test designed for the corresponding least favorable f_0^*, f_1^*, which trims data that exceed certain threshold values, thereby eliminating the outliers.

4 Nonparametric Detection

The measurement and modeling uncertainty is specified by a relatively restricted family of distributions in (8) and (9). A detector that performs well for a broader class arises from the nonparametric procedures. Let $\mathcal{F}_{-\theta}$ represent the class of stationary and memoryless discrete-time processes with common mean at $-\theta$. Each member of the class is denoted by the first-order probability density function $f_{-\theta}$. The hypotheses H_1 and H_0 are described by $\mathcal{F}_\theta$ and $\mathcal{F}_{-\theta}$, respectively. A decision rule consists of the triplet $(T(\mathbf{v}), \lambda, r)$ where $T(\mathbf{v})$ is the corresponding test function, λ is the threshold and r is the randomization constant. The rule or test $(T(\mathbf{v}), \lambda, r)$ is nonparametric in $(\mathcal{F}_\theta, \mathcal{F}_{-\theta})$ if and only if it induces the same false alarm probability for all $f_{-\theta} \in \mathcal{F}_{-\theta}$ [7].

Let $\mathcal{F}$ denote the class of distributions obtained from either $\mathcal{F}_\theta$ or $\mathcal{F}_{-\theta}$ when the mean is set to zero for all members. For some $f \in \mathcal{F}$, suppose that f_θ is the PDF induced by f when its mean is changed from zero to θ. Let $n(\alpha, \beta, T_{f_\theta})$ be the number of data required by a Neyman-Pearson rule to attain the detection probability (also known as the power of the test) β while satisfying the false alarm constraint α when testing f_θ against $f_{-\theta}$. Likewise, let $n(\alpha, \beta, T, f_\theta, f_{-\theta})$ be the sample size needed by the nonparametric test of f_θ versus $f_{-\theta}$ to achieve the power β with false alarm α. The efficacy, EFF, and the asymptotic relative efficiency, ARE, are defined as

$$\mathrm{EFF} = \lim_{n\to\infty} \frac{\left(\frac{\partial}{\partial\theta} E[T(\mathbf{v})|f_\theta]\big|_{\theta=0}\right)^2}{n \cdot \mathrm{var}(T(\mathbf{v})|f_\theta)},$$

and

$$\mathrm{ARE} = \lim_{\theta\to 0} \frac{n(\alpha, \beta, T_{f_\theta})}{n(\alpha, \beta, T, f_\theta, f_{-\theta})}.$$

Efficacy indicates the asymptotic discrimination ability of the test when the hypotheses are close to each other. Asymptotic relative efficiency measures the additional sample size needed by the nonparametric test to yield the same power as the optimal Neyman-Pearson rule when the two hypotheses are asymptotically close to each other.

4.1 Sign test

The sign test was originally introduced as an ad hoc formalization but it also evolves as a limiting case of the robust test for the classes in (8) and (9). The associated test function is

$$T(\mathbf{v}) = \frac{1}{n}\sum_{i=1}^{n} \operatorname{sgn} v_i$$

where

$$\operatorname{sgn} v_i = \begin{cases} 1 \text{ if } v_i > 0, \\ 0 \text{ if } v_i \leq 0. \end{cases}$$

The corresponding decision rule is

$$\delta_S = \begin{cases} 1 \text{ if } T(\mathbf{v}) > \lambda, \\ 0 \text{ if } T(\mathbf{v}) \leq \lambda, \end{cases}$$

where the threshold λ is chosen such that the false alarm constraint is satisfied.

The sign test is nonparametric in $(\mathcal{F}_\theta, \mathcal{F}_{-\theta})$ for any n, and for f_θ and $f_{-\theta}$ generated by Gaussian PDF with variance σ^2, its efficacy and asymptotic relative efficiency are $\text{EFF} = 8/\pi\sigma^4$ and $\text{ARE} = 2/\pi$ [7]. Thus, the sign test requires about 57% more samples to reach the same performance level as the Gaussian-optimal Neyman-Pearson rule (as $\theta \to 0$), but the latter experiences performance degradation when the Gaussian distribution is not actually a valid uncertainty model.

4.2 Optimal rank test

The rank tests first order the measurements $\{v_1, \ldots, v_n\}$ from smallest to the largest and then take the signs of the ordered data. The new vector $\mathbf{z} = [z_1 \; \cdots \; z_n]^T$ of the signs, where

$$z_i = \begin{cases} 1 \text{ if the } i\text{th ranked datum in } \mathbf{v} \text{ has nonnegative sign}, \\ 0 \text{ if the } i\text{th ranked datum in } \mathbf{v} \text{ has negative sign}, \end{cases}$$

is called the rank vector.

Given $\theta > 0$ and some $f \in \mathcal{F}$, the optimal-at-f rank test is

$$\delta_O = \begin{cases} 1 \text{ if } \frac{K_{f_\theta}(\mathbf{z})}{K_{f_{-\theta}}(\mathbf{z})} > \lambda, \\ r \text{ if } \frac{K_{f_\theta}(\mathbf{z})}{K_{f_{-\theta}}(\mathbf{z})} = \lambda, \\ 0 \text{ if } \frac{K_{f_\theta}(\mathbf{z})}{K_{f_{-\theta}}(\mathbf{z})} < \lambda \end{cases}$$

where

$$K_{f_\theta}(\mathbf{z}) = n! \int \cdots \int \prod_{i=1}^{n} f(v_i - \theta z_i) d\mathbf{v},$$

$$K_{f_{-\theta}}(\mathbf{z}) = K_{f_\theta}(-\mathbf{z}),$$

and λ and r satisfy

$$P\left\{\frac{K_{f_\theta}(\mathbf{z})}{K_{f_{-\theta}}(\mathbf{z})} > \lambda | f_{-\theta}\right\} + r \cdot P\left\{\frac{K_{f_\theta}(\mathbf{z})}{K_{f_{-\theta}}(\mathbf{z})} = \lambda | f_{-\theta}\right\} = \alpha.$$

For the optimal-at-f rank test, ARE = 1, and the efficacy at f is the Fisher information, i.e.,

$$\text{EFF} = \int_{\mathcal{R}} \frac{[f'(x)]^2}{f(x)} dx$$

so long as $f \in \mathcal{F}$ possesses a Taylor series expansion [7].

4.3 Wilcoxon rank test

The Wilcoxon rank test [12] is as follows.

$$\delta_W = \begin{cases} 1 \text{ if } \sum_{i=1}^{n} i z_i > \lambda, \\ r \text{ if } \sum_{i=1}^{n} i z_i = \lambda, \\ 0 \text{ if } \sum_{i=1}^{n} i z_i < \lambda, \end{cases}$$

with λ and r such that

$$\sup_{f_{-\theta} \in \mathcal{F}_{-\theta}} \left(P\left\{ \sum_{i=1}^{n} i z_i > \lambda | f_{-\theta} \right\} + r \cdot P\left\{ \sum_{i=1}^{n} i z_i = \lambda | f_{-\theta} \right\} \right) = \alpha.$$

While the optimal rank test is the Neyman-Pearson test for the rank vector $\mathbf{z}$ extracted from a particular $f \in \mathcal{F}$, the Wilcoxon test is designed for the entire classes $(\mathcal{F}_\theta, \mathcal{F}_{-\theta})$. For Gaussian PDF with variance σ^2, the Wilcoxon rank test has ARE $= 3/\pi < 1$ and EFF $= 6/\pi\sigma^2$ [7].

The tests described in this section are nonparametric in both $(\mathcal{F}_\theta, \mathcal{F}_{-\theta})$ and $(\mathcal{F}_{2\theta}, \mathcal{F})$, which ensures broader applicability.

5 Generalized Likelihood Ratio Test

So far we have assumed that model parameters were known. In practice, these have to be estimated beforehand or simultaneously with the detection procedure. In the latter case, the generalized likelihood ratio test (GLRT) offers a joint estimation and detection methodology to solve the composite hypotheses that are represented by the corresponding density functions as below.

$$H_1: f_1(\mathbf{v}|\mathbf{x}), \mathbf{x} \in \mathcal{X}_1,$$

$$H_0: f_0(\mathbf{v}|\mathbf{x}), \mathbf{x} \in \mathcal{X}_0.$$

Suppose that $\hat{\mathbf{x}}_i$ is the maximum likelihood estimate of $\mathbf{x}$ under $H_i, i = 0, 1$, i.e., $\hat{\mathbf{x}}_i = \arg\max_{\mathbf{x}\in\mathcal{X}_i} f_i(\mathbf{v}|\mathbf{x})$. Then, the generalized likelihood ratio (GLR) is

$$L_{\text{GLR}}(\mathbf{v}) = \frac{f_1(\mathbf{v}|\hat{\mathbf{x}}_1)}{f_0(\mathbf{v}|\hat{\mathbf{x}}_0)}.$$

If the null hypothesis H_0 is the absence of UXO, then $\mathbf{x} = \mathbf{x}_0$ with no need for estimation. The GLRT is

$$\delta_{\text{GLRT}} = \begin{cases} 1 \text{ if } L_{\text{GLR}}(\mathbf{v}) \geq \lambda(\alpha), \\ 0 \text{ if } L_{\text{GLR}}(\mathbf{v}) < \lambda(\alpha) \end{cases}$$

where $\lambda(\alpha)$ is such that

$$\max_{\mathbf{x}\in\mathcal{X}_0} e_{\text{I}}(\delta_{\text{GLRT}}, \mathbf{x}) = \alpha$$

with the type I error now defined as

$$e_{\text{I}}(\delta_{\text{GLRT}}, \mathbf{x}) = P\{L_{\text{GLR}}(\mathbf{v}) > \lambda(\alpha)|\mathbf{x}, H_0 \text{ is true}\} = \alpha.$$

Model inversion to estimate $\mathbf{x}$ from the measurements is sensitive to errors in sensor positions, and Bayesian methods employing the GLRT that improve the detection performance are proposed in [10].

6 Conclusion

We have provided a brief summary of the statistical tools that are available for the detection of UXO. The application of the tests require accurate models that relate various UXO parameters to observed data, as well as a good distributional description for the uncertainties in measurements and modeling. Performance can be assessed by plotting e_{II} versus e_{I}, the so-called receiver operating characteristic (ROC). The ROC curve gives the detection probability for a given false alarm rate. To combine detection and ordnance classification, it is possible to set up a multiple hypothesis testing problem where each hypothesis $H_i, i \neq 0$, corresponds to a possible UXO type with the associated feature vector, and H_0 still represents the absence of UXO. The need for accurate representation of measurement uncertainties can be alleviated by using model-free approaches such as the support vector machine which relies on preprocessing with training data [15]. For multiple closely spaced UXO parts, blind source separation methods such as independent component analysis can precede the UXO detector [5]. Deploying multiple sensors with distinct distances from the buried UXO can deliver significant classification performance improvement as demonstrated in [13]. The reader is referred to the bibliography and the references therein for further exploration into UXO detection.

References

1. L. Carin, H. Yu, Y. Dalichaouch, A. R. Perry, P. V. Czipott and C. E. Baum, "On the wideband EMI response of a rotationally symmetric permeable and conducting target," *IEEE Transactions on Geoscience and Remote Sensing*, Vol. 39, No. 6, pp. 1206–1213, June 2001.
2. C. -C. Chen and L. Peters, "Buried unexploded ordnance identification via complex natural resonances," *IEEE Transactions on Antennas and Propagation*, Vol. 45, No. 11, pp. 1645–1654, November 1997.
3. A. Fijany, J. B. Collier and A. Citak, "Recent advances in unexploded ordnance (UXO) detection using airborne ground penetrating SAR," *Proceedings of the Aerospace Conference*, 1999, pp. 429–441. Snowmass, Colorado, USA, March 1999.
4. J. I. Halman, K. A. Shubert and G. T. Ruck, "SAR processing of ground-penetrating data for buried UXO detection: results from a surface-based system," *IEEE Transactions on Antennas and Propagation*, Vol. 46, No. 7, pp. 1023–1027, July 1998.
5. W. Hu, S. L. Tantum and L. M. Collins, "EMI-based classification of multiple closely spaced subsurface objects via independent component analysis," *IEEE Transactions on Geoscience and Remote Sensing*, Vol. 42, No. 11, pp. 2544–2554, September 2004.
6. P. J. Huber, *Robust Statistics*, New York: Wiley, 1981.
7. D. Kazakos and P. Papantoni-Kazakos, *Detection and Estimation*, New York: Computer Science Press, 1990.
8. C. V. Nelson, C. C. Cooperman, W. Schneider, D. S. Wenstrand and D. G. Smith, "Wide bandwidth time-domain electromagnetic sensor for metal target classification," *IEEE Transactions on Geoscience and Remote Sensing*, Vol. 39, No. 6, pp. 1129–1138, June 2001.
9. H. H. Nelson and J. R. MacDonald, "Multisensor towed array detection system for UXO detection," *IEEE Transactions on Geoscience and Remote Sensing*, Vol. 39, No. 6, pp. 1139–1145, June 2001.
10. S. L. Tantum, Y. Yu and L. M. Collins, "Bayesian mitigation of sensor position errors to improve unexploded ordnance detection," *IEEE Geoscience and Remote Sensing Letters*, Vol. 5, No. 1, pp. 103–107, January 2008.
11. R. van Waard, S. van der Baan and K. W. A. van Dongen, "Experimental data of a directional borehole radar system for UXO detection," *Proceedings of the Tenth International Conference on Ground Penetrating Radar*, Delft, The Netherlands, June 2004, pp. 225–228.
12. F. Wilcoxon, "Individual comparisons by ranking methods," *Biometrics*, Vol. 1, 1945, pp. 80–83.
13. D. Williams, C. Wang, X. Liao and L. Carin, "Classification of unexploded ordnance using incomplete multisensor multiresolution data," *IEEE Transactions on Geoscience and Remote Sensing*, Vol. 45, No. 7, pp. 2364–2373, July 2007.
14. Q. Zhang, W. Al-Muaimy and Y. Huang, "Detection of deeply buried UXO using CPT magnetometers," *IEEE Transactions on Geoscience and Remote Sensing*, Vol. 45, No. 2, pp. 410–417, February 2007.
15. Y. Zhang, L. Collins, H. Yu, C. E. Baum and L. Carin, "Sensing of unexploded ordnance with magnetometer and induction data: theory and signal processing," *IEEE Transactions on Geoscience and Remote Sensing*, Vol. 41, No. 5, pp. 1005–1015, May 2003.

Low Frequency Radar for Buried Target Detection

Hugh Griffiths[1] and Alastair McAslan[2]

Abstract The detection and mitigation of unexploded ordnance (UXO) is recognised to be a serious global issue. Many millions of landmines have been deployed in recent conflicts, with few records of what has been laid and where. As well as landmines, other types of UXO include unexploded shells, mortar bombs and missiles, scatterable mines fired from mortars or artillery or dropped from aircraft or helicopters, and cluster munitions. Not only do such weapons cause injury and death to innocent civilians, but also they deny the use of substantial areas of land for agricultural and other economic purposes, which may be critical in countries where the threshold of poverty is already low. Ground-penetrating radar (GPR) is one of a family of sensors that may be used to detect UXO. In addition, GPR may also be used to detect other classes of target such as Improvised Explosive Devices (IEDs), weapons caches, and tunnels; further applications of GPR include archaeology, forensics, and the detection of buried pipes and cables. The purpose of this chapter is to present an account of the principles of ground-penetrating radar and their use in detecting buried UXO.

Keywords: Landmines, unexploded ordnance, radar, impulse radar

1 Historical Background

The history of landmines goes back a long way [7]. The Emperor Caesar used pits, arrays of stakes, and devices called caltrops to impede the progress of the Gauls in the siege of Alesia in 52 BC [5]. Similar devices were used in the battle of

[1]Defence College of Management and Technology, Cranfield University, Shrivenham, UK, e-mail: h.griffiths@cranfield.ac.uk
[2]Defence College of Management and Technology, Cranfield University, Shrivenham, UK, e-mail: a.r.r.mcaslan@cranfield.ac.uk

J. Byrnes (ed.), *Unexploded Ordnance Detection and Mitigation*,

Table 1 Numbers of the German anti-tank Tellermine deployed in the World War II [7].

1939	108,100
1940	102,100
1941	220,900
1942	1,063,600
1943	3,414,000
1944	8,535,500

Bannockburn (1314) and the Wars of the Roses (1455–1485). After the discovery of gunpowder in the 13th century explosive charges were used in siege warfare.

This led to the development of the fougasse – essentially an underground cannon, placed forward of a defensive position to shower rocks and debris over a wide area.

The naval mine was developed and used during the American Civil War (1861). In both the American Civil War and the Boer War, electrically-operated fougasses and mines were laid, as well as pressure-operated landmines. In the First World War, British engineers tunnelled under the German trenches and laid huge explosive charges [4]. Anti-personnel mines were not much used, but with the introduction of the tank in September 1916, anti-tank mines were soon introduced, initially improvised from shells.

In the Second World War, both Anti-tank (AT) and Anti-personnel (AP) mines were extensively used, especially by the Germans. Considerable advances were made in mine technology, and in the technology of mine detection and mine clearance.

Table 1 above indicates that the Germans kept careful records of the number, and indeed the locations, of mines that they laid. However, whilst in post-World War II conflicts mines have been used extensively, armies have not necessarily been so careful in marking and recording the location of minefields.

Of the 48 countries in Africa, more than half are known to be mine-affected. There are minefields in North Africa that remain from World War II. In Zimbabwe (formerly Rhodesia) there are an estimated 1.5 million landmines, some of which have been laid at random and only 10% of which have been removed. Somalia, South Africa, Rwanda, Chad, Angola and Mozambique are also heavily affected.

Afghanistan and Cambodia are two of the most mine-infested countries of the world. In the Korean War (1951–1953) some ten different countries made use of anti-personnel mines. Some fields were so thick with AP landmines that they were a constant threat even to those that laid them. In the Vietnam War entire villages were surrounded by landmines, hand laid or dropped from the air, and no records were kept of the mines laid. In Cambodia, humanitarian groups have demined areas just to have them remined again. Cambodia has more amputees as a percentage of the population than any other country in the world.

In Bosnia–Herzegovina, an estimated 3–6 million mines still remain uncleared. Some maps were kept and have been turned over to the UN. In WWII landmines were not used extensively in Europe until the end of the war. Minefield clearance

Fig. 1 A minefield in the Falkland Islands.

is still being undertaken in countries such as Belgium, while in France land is still contaminated by landmines.

In El Salvador in 1980–1991, mining was done without any charting, so many of the original mine-layers were recruited for the demining operations. In a 1-month conflict in 1995, tens of thousands of landmines were laid down on the borders between Ecuador and Peru. Some efforts have been made to demine the area, but about 6,000 mines still remain. In the Falklands War (1982) extensive use of anti-personnel mines was made by the Argentine forces. Some clearance programmes were established, but were short-lived due to heavy casualties on demining units, so the minefields still remain (Figure 1).

Over 175 million landmines have been deployed since the end of World War II, including more than 65 million since 1980. Mines are seen by warring factions as attractive weapons as they are relatively cheap to acquire, easy to lay and invariably have a devastating effect on the target. They differ from most other weapons, however, by remaining active in the ground long after hostilities have ended. They lie in fields and woodlands, alongside roads and footpaths, and in villages creating a humanitarian problem – with social, economic and environmental dimensions. Anti-personnel landmines are designed to maim rather than to kill, since a wounded combatant is more trouble to an army than a dead one. Not only do such weapons take their toll on victims and families, but the presence of landmines in and around communities, on roads, in farmland, and near rivers and wells prevents the productive use of land, water and infrastructure for development.

The term 'minefield' conjures up an image of flat open countryside, in which rows of anti-tank and anti-personnel mines have been carefully laid, surveyed and recorded, and which are bounded by minefield fences marked with white tape and red warning triangles. In reality the situation is quite different. Minefields are often laid in a hurry by poorly trained and ill equipped armies; mines are rarely laid according to a pattern; booby traps may have been set up; and the area may be scattered with other forms of unexploded ordnance (UXO), from small items such as phosphorus grenades, to artillery shells and missiles containing a deadly cocktail of explosives and fuel.

In some situations the ground may be contaminated by scatterable mines fired from mortars and artillery, or dropped from helicopters and aircraft. It is estimated that two million tons of bomblets were dropped from US aircraft on Vietnam, Laos and Thailand in the 1970s, aimed at disrupting movement along the Ho Chi Min Trail. The bomblets were anti-personnel devices designed to explode on impact with the ground, although it is now assessed that 25% failed to explode and they remain an ongoing hazard to communities.

Of more recent concern is the use of cluster munitions. These are small weapons – often no larger than a small cola can – containing a powerful explosive charge. They are packed into containers and dropped from aircraft or fired from artillery systems. Cluster munitions have a high failure rate; more than 20% fail to detonate on reaching the ground and remain hazardous until they are cleared. Large numbers were dropped in the Balkans, Afghanistan, Iraq, and more recently in the Lebanon.

So after the guns fall silent, and when the mines and UXO no longer have a military purpose, the battlefield remains dangerous, and explosive remnants of war have a major impact on communities attempting to recover from years of conflict.

2 The Role of Technology

Over 1,000 km^2 of land have been cleared of mines and UXO since the start of modern humanitarian demining in the early 1990s. In its 2007 report, the international NGO Landmine Monitor estimates some 140 km^2 were cleared in 2006, as well as over 310 km^2 of UXO and other explosive remnants of war. This is a remarkable achievement, and is a significant improvement on clearance rates of a decade ago. But a massive challenge remains, and will continue for many more years – long after international interest and funding has moved on to address other issues and humanitarian concerns.

There is a pressing need to find smarter ways of clearing landmines and UXO. This can be achieved in three ways: first, by improving the quality of the information on the threat and its impact, and from this improved information to prioritise better the use of clearance teams; second, by developing new survey and clearance procedures; and third, by developing and deploying better equipment, including improved sensors.

Over the past 15 years there has been substantial interest in finding a technical 'silver bullet'. These ideas have included experimental prodders (with acoustic sensors to detect the presence of metals and plastics), improved handheld metal detectors, nuclear quadrupole detection, X-ray backscatter, vapour and chemical analysis detectors, laser detection, the use of animals and insects, infrared detectors and exploiting other parts of the electromagnetic spectrum including ground penetrating radar (GPR).

Indeed, following the Falklands conflict of 1983 the British Government funded considerable research and development into smarter ways of locating and neutralising the landmines which scattered the islands, many buried in peat or scree which would prove difficult to detect and clear using conventional metal mine detectors and prodders. This work was halted in 1986 when it was clear that the systems being proposed could not achieve the substantial improvements in clearance rates being demanded by the British Government.

Notwithstanding this absence of a 'silver bullet' there is still a need to find and apply better technologies to demining [9].

3 The Operational Needs

In 2000/01, the Geneva International Centre for Humanitarian Demining (GICHD) was invited by the United Nations to establish a priority list of operational needs that could benefit from improved equipment, processes and procedures. The GICHD's Study of Global Operational Needs [14] which was carried out in partnership with Cranfield University identified a number of generic operational needs and equipment requirements. The purpose of the study was to give guidance to research and development, and provide the user and donor communities with the means to assess more effectively the benefits and cost of technology to mine action programmes. The ultimate aim of the study was to encourage the design, development and manufacture of safer, better and more cost-effective equipment.

The findings and recommendations of the study are still relevant today. Of the 12 capability areas identified by the study two were considered as potentially benefiting greatly from better equipment: the close in detection of landmines, and systems which could more accurately determine the outer edge of mined areas. In particular, the study recommended that such equipments should not only have improved detection accuracy but a much lower rate of false alarms – which leads to inefficiency and can result in complacency of the deminers.

One area of technology where there have been demonstrated improvements in mine detection accuracy and false alarms is in the application of Ground-Penetrating Radar (GPR).

4 Fundamentals of Ground-Penetrating Radar

GPR has been developed over the past couple of decades as a means of detecting buried targets such as landmines. Other applications include the detection of buried utilities such as pipes and cables, as well as archaeological and forensic applications. The technologies also have some similarities to those used for through-wall radar detection and imaging [1, 13], foliage penetration (FOPEN) radar, and for glaciological sounding [12].

Fundamental to all of these applications are the propagation characteristics of electromagnetic radiation through materials such as soil and concrete and at the boundary between air and such materials, and how these characteristics depend on frequency and on material properties. In general it can be appreciated that a lower frequency may give lower propagation loss than a higher frequency, but will in general give poorer resolution, both in range and in azimuth.

Daniels [8] has provided a comprehensive account of the design factors in Ground Penetrating Radar and examples of systems and results. He states that 'GPR relies for its operational effectiveness on successfully meeting the following requirements:

- Efficient coupling of electromagnetic radiation into the ground
- Adequate penetration of the radiation through the ground having regard to target depth
- Obtaining from buried objects or other dielectric discontinuities a sufficiently large scattered signal for detection at or above the ground surface and
- An adequate bandwidth in the detected signal having regard to the desired resolution and noise levels

Table 2 shows the losses for different types of material at 100 MHz and 1 GHz. This shows that the loss is relatively low for dry materials, but that the loss increases substantially with moisture content. It also shows how the losses increase with frequency. However, it should also be understood that the attenuation of an acoustic signal decreases with moisture content, so acoustic (sonar) sensors may in a sense be considered complementary to radar sensors. Fusion techniques to optimally exploit the strengths of both types of sensor may therefore be of interest [11].

Daniels also presents a taxonomy of system design options. The majority of systems use an impulse-type waveform and a sampling receiver, processing the received signal in the time domain. More recently, however, Frequency-Modulated Continuous Wave (FMCW) and stepped frequency modulation schemes have been developed, which allow lower peak transmit powers. Both types of system, though, require components (particularly antennas) with high fractional bandwidths, which are not necessarily straightforward to realise.

Table 2 Material loss at 100 MHz and 1 GHz [8]. (IET, 2004.)

Material	Loss at 100 MHz	Loss at 1 GHz
Clay (moist)	5–300 dB m^{-1}	50–3000 dB m^{-1}
Loamy soil (moist)	1–60 dB m^{-1}	10–600 dB m^{-1}
Sand (dry)	0.01–2 dB m^{-1}	0.1–20 dB m^{-1}
Ice	0.1–5 dB m^{-1}	1–50 dB m^{-1}
Fresh water	0.1 dB m^{-1}	1 dB m^{-1}
Sea water	100 dB m^{-1}	1000 dB m^{-1}
Concrete (dry)	0.5–2.5 dB m^{-1}	5–25 dB m^{-1}
Brick	0.3–2.0 dB m^{-1}	3–20 dB m^{-1}

5 Imaging and Resolution

We can establish some of the fundamental relations for the resolution of an imaging system. In the down-range dimension resolution Δr is related to the signal bandwidth B, thus

$$\Delta r = c/2B \tag{1}$$

where c is the velocity of propagation. High resolution may be obtained either with a short-duration impulse or by a coded wide-bandwidth signal, such as a linear FM chirp, a step-frequency sequence or a pseudo-random digital code, with the appropriate pulse compression processing. A short-duration impulse requires a high peak transmit power and instantaneously-broadband operation; these requirements can to some extent be relaxed in the case of pulse compression.

The rapid increase of attenuation as a function of frequency through most materials (Table 2) demands a low radar frequency. However, high range resolution demands a high bandwidth (Equation (1)). Thus ground-penetrating radars will in general have a high fractional bandwidth:

$$B_F = \frac{f_h - f_l}{\frac{1}{2}(f_h + f_l)} = \frac{B}{f_C} \tag{2}$$

where f_h and f_l are, respectively, the upper and lower frequencies of the radar signal. By convention, a radar with a fractional bandwidth of greater than 25% is characterised as ultra-wideband (UWB) [19, 20]. In the case of an impulse-type radar f_l will tend to zero, so it can be seen from Equation (2) that such radars are inherently ultra-wideband.

The cross-range resolution is complicated by the fact that in many cases the target (at range r) will lie within the near-field of the antenna, i.e.

$$r < \frac{2d^2}{\lambda} \tag{3}$$

where d is the aperture dimension and λ is the wavelength. In the far-field, though, the cross-range resolution is determined by the product of the range and beamwidth θ_B. The beamwidth is determined by the value of d and thus the cross-range resolution Δx at range r is given by

$$\Delta x = r\,\theta_B \approx \frac{r\lambda}{d}. \tag{4}$$

As most antenna sizes are limited by practical considerations, the cross range resolution is invariably much inferior to that in the down range dimension. However, there are a number of techniques that can improve upon this. All of these are ultimately a function of the change in viewing or aspect angle. Thus in the azimuth (cross-range) dimension the resolution Δx is related to the change in aspect angle $\Delta\theta$ as follows:

$$\Delta x = \frac{\lambda}{4\sin(\Delta\theta/2)}. \tag{5}$$

For a linear, stripmap-mode synthetic aperture, Equation (5) reduces to $\Delta x = d/2$, which is independent of both range and frequency. Even higher resolution can be obtained with a spotlight-mode synthetic aperture, steering the real-aperture beam to keep the target scene in view for a longer period, and hence forming a longer synthetic aperture.

Realistic limits to resolution may be derived by assuming a maximum fractional bandwidth of 100%, and a maximum change in aspect angle of $\Delta\theta = 30^\circ$ (higher values than these are possible, but at the expense of complications in hardware and processing). These lead to $\Delta x = \Delta r = \lambda/2$. In the last year or so results have appeared in the open literature which approach this limit [2, 18].

Figure 2 shows that range resolution may be achieved by different methods. In (i) the transmitted signal is an impulse waveform in the time-domain. This requires specialised hardware to generate the high-voltage impulse in the transmitter and to sample the echo in the receiver. In (ii) the transmitted signal is a linear FMCW sweep and the received echo is deramped and processed in the frequency domain. The requirements for the peak transmit power and the digital sampling and processing rate in the receiver are considerably relaxed, but the technique does introduce range sidelobes. These can be lowered by the usual weighting techniques, but nevertheless

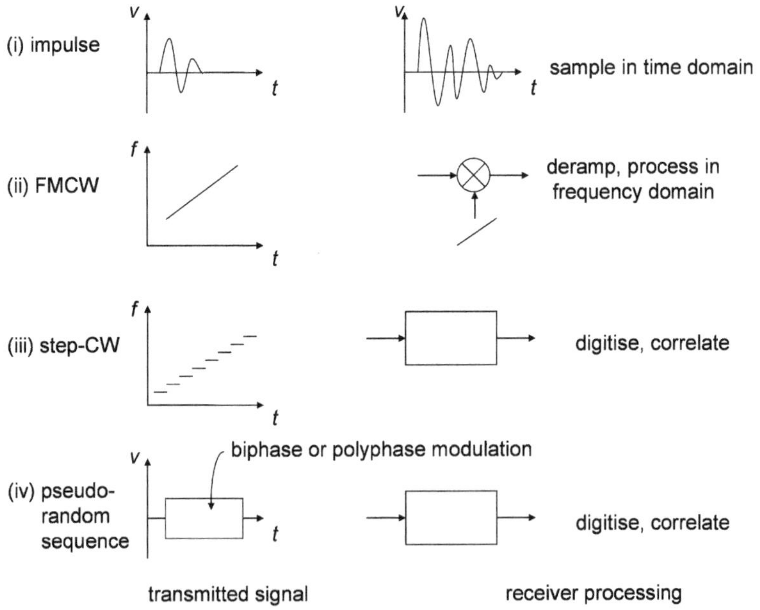

Fig. 2 Form of transmitted signal and receiver processing for different GPR system options. Note that the time axis of (i) is of considerably shorter duration than those of (ii), (iii) and (iv).

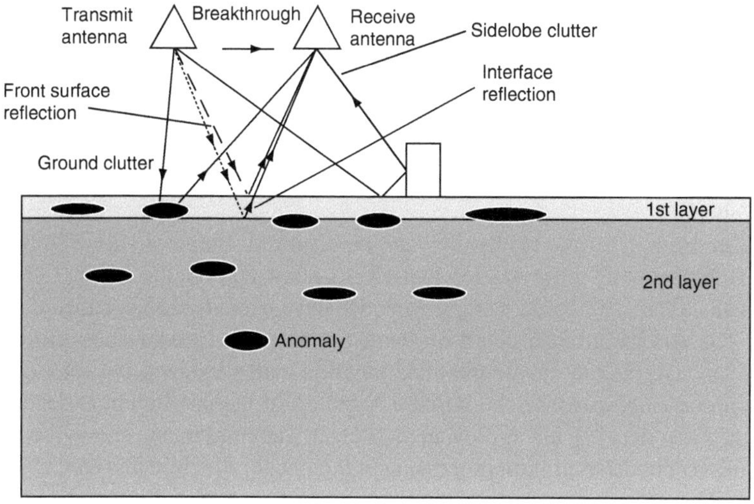

Fig. 3 Physical layout of Ground Penetrating Radar system [8] (IET, 2004.)

the sidelobes from the direct transmit to receive antenna coupling or the strong ground echo may mask target echo features (Figure 3). Similar comments apply to (iii), in which the transmitted signal is a stepped-CW waveform, and (iv) in which it is a pseudo-random biphase- or polyphase-modulated carrier. In both cases the echo is digitised and processed with a matched filter (correlator) in the receiver. In practice the vast majority of GPR systems are of the impulse type.

In contrast, holographic imaging techniques may be used with CW or quasi-CW signals, giving high spatial resolution by exploiting spatial bandwidth rather than frequency bandwidth.

In radar tomography [10] the observation of an object from a single radar location can be mapped into Fourier space. Coherently integrating the mappings from multiple viewing angles enables a three dimensional projection in Fourier space. This allows a three dimensional image of an object to be constructed using conventional tomography techniques such as wavefront reconstruction theory and backprojection where the imaging parameters are determined by the occupancy in Fourier space. Complications can arise when target surfaces are hidden or masked at any stage in the detection process. This shows that intervisibility characteristics of the target scattering function are partly responsible for determining the imaging properties of moving target tomography. In other words, if a scatterer on an object is masked it cannot contribute to the imaging process and thus no resolution improvement is gained. However, if a higher number of viewing angles are employed then this can be minimised. Further complications may arise if (a) the point scatterer assumption used is unrealistic (as in the case of large scatterers introducing translational motion effects), (b) the small angle imaging assumption does not hold and

(c) targets with unknown motions (such as non-uniform rotational motions) create cross-product terms that cannot be resolved.

Finally, image processing techniques (including singularity expansion methods, wavelet transforms, pattern recognition techniques and neural networks) may be used to reduce the effect of clutter and enhance targets. In general these attempt to exploit prior knowledge of the nature of the targets and of the background noise and clutter.

As an example of the results that can be achieved, Figure 4 shows images of a buried antipersonnel mine at a depth of 15 cm, showing both the original image and the results after image processing techniques have been used to enhance the target. The mine was buried at a depth of about 5 cm at an angle of about 30 degrees, in dry sand. In the raw image the mine target is barely evident, but after deconvolution processing, in which the impulse response of the instrument is deconvolved from the radar data [8], the improvement is clear. The third image shows the result of applying Kirchhoff migration processing to the image, which in this case is less successful. These show that, under the right conditions and with the use of appropriate algorithms, significant enhancement is possible.

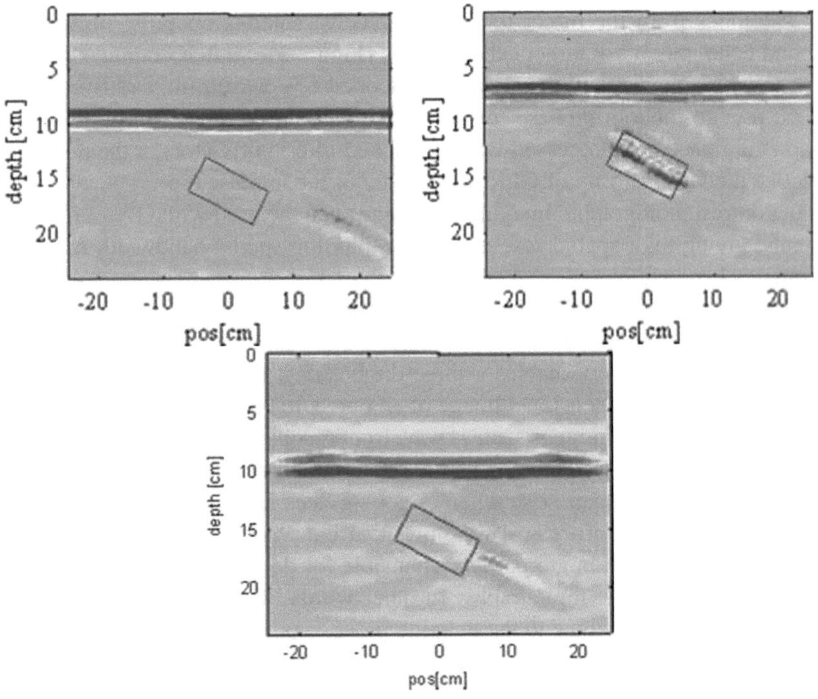

Fig. 4 Oblique antipersonnel mine at an angle of 30 degrees: **(a)** B-scan of raw data; **(b)** after migration by deconvolution; **(c)** after Kirchhoff migration [8]. (IET, 2004.)

Fig. 5 The MINEHOUND instrument (*left*), under test in Sarajevo (*right*) [8]. (IET, 2004.)

6 MINEHOUND

MINEHOUND (Figure 5) is a prototype low-cost, man-portable detector developed for humanitarian demining purposes by ERA Technology, for the UK Department for International Development (DfID). It consists of an ultra-wideband GPR and a metal detector, with the output presented to the operator in audible form, and the signature varies in a characteristic way as the detector is moved over a buried object. Trial results are reported in [8].

7 The *Mineseeker* project

Another example of an advanced radar system for detection of abandoned UXO is the Mineseeker project Figure 6 [6, 21]. The Mineseeker Foundation has the support of some high-profile patrons, and represents a not-for-profit joint venture between the Lightship Group and QinetiQ. The concept uses an ultra-wideband synthetic aperture radar (UWB SAR) developed originally by engineers from DERA Malvern (now QinetiQ), and gimbal-stabilised electro-optic sensors operating in the visible and 3–5 μm IR bands, mounted on an airship platform. The airship has the particular merits of being mobile, stable, low-cost and with long endurance, as well as the ability to carry a substantial payload.

The pulse generator and high-speed digitiser subsystems used in the UWB radar were developed by Kentech, the UWB antennas by researchers at Dundee University, and the synthetic aperture processing and target signature analysis algorithms by Applied Electromagnetics Inc.

Basic parameters of the radar sensor are listed in Table 3 [21].

It has been demonstrated in trials that different mine targets have characteristic signatures, so different mine types may be distinguished from each other and from

Fig. 6 The Mineseeker airship. (Mineseeker Foundation, 2001.)

Table 3 Basic parameters of Mineseeker UWB SAR. (Mineseeker Foundation, 2001.)

Range resolution	5 cm
Azimuthal resolution	0.5 m
Instantaneous bandwidth	>3 GHz
Frequency range	200 MHz to over 3 GHz
Pulsewidth of impulse waveforms	>100 ps
Peak power	1 MW

other false alarm debris (Figure 7). These examples also show the information that may be obtained from the polarimetric signatures of mines and other UXO, though in practice the additional hardware complication of a polarimetric radar makes these techniques very difficult.

MINESEEKER's coverage rate (in terms of location and delineation) of more than 100 m^2/s is claimed, in contrast to 20–50 m^2/day by manual demining.

The preceding are just two examples of practical GPR systems; many more are described in [8].

8 Management of Humanitarian Demining Programmes

Whilst the emphasis here has been on the technology used to detect and neutralise landmines and other UXO, equal prominence should be given to the management of demining programmes, since even the most sophisticated technology is of little use unless deployed in a systematic and properly managed way.

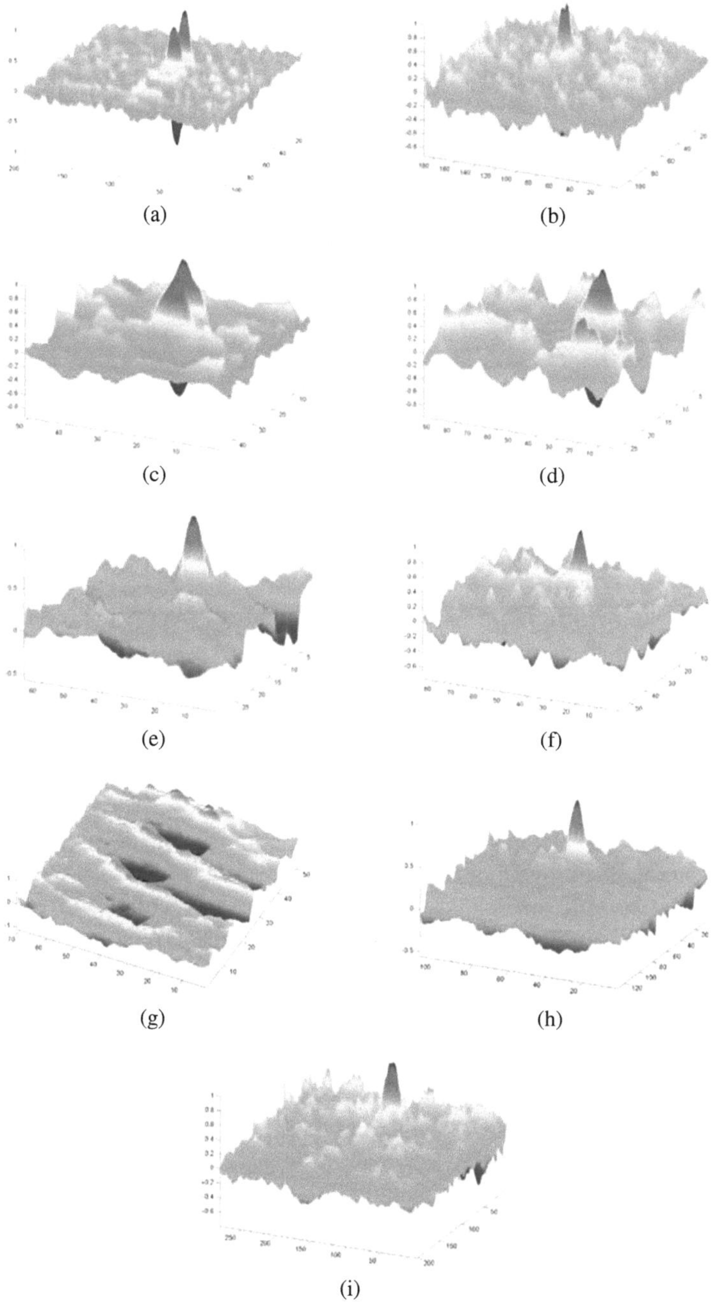

Fig. 7 Signatures of different targets obtained in trials with the Mineseeker UWB SAR: **(a)** surface-laid calibration sphere, HH polarisation; **(b)** surface-laid mortar round (inert), VV polarisation; **(c)** surface-laid RBL755 cluster bomb sub-munition (inert), VV polarisation; **(d)** above-ground PMR2a stake mine (inert), VV polarisation; **(e)** buried TMM1 metal anti-tank mine (inert), VV polarisation; **(f)** buried RBL755 cluster bomb sub-munition (inert), HH polarisation; **(g)** buried mortar round (inert), VV polaristion; **(h)** buried handgrenade (live), VV polarisation; **(i)** buried PMR2a (live), VV polarisation. (Mineseeker Foundation, 2001.)

Fig. 8 Mine action.

Work over more than two decades at the Defence College of Management and Technology, Shrivenham (part of Cranfield University and of the Defence Academy of the United Kingdom, and led by Alastair McAslan) has developed programmes in the management of humanitarian demining, and earlier this year received the Queen's Anniversary Prize for Higher and Further Education, from Her Majesty the Queen, for this work.

'The management of mine action (Figure 8) at the national level is, essentially, about ensuring that programmes, projects and day-to-day mine action activities are carried out effectively, efficiently and safely. This involves defining the requirements through assessment missions and site surveys, prioritising requirements, developing plans, securing funding, implementing projects and confirming that the requirements have been met' [16, 17].

'Resilience' may be defined as understanding the risks to nations and organisations from factors as diverse as terrorism, natural disasters, health pandemics and IT fraud, and hence firstly being able to minimise the risks and effects, and secondly ensuring that the organisation is able to recover as quickly as possible. Demining therefore represents one specific aspect of Resilience.

In March 2008 Cranfield University launched an MSc course in Resilience [22], aimed at professional managers who wish to apply rigorous academic thought to practical problems in their sector, and to acquire the necessary knowledge and skills to analyse threats and build resilient organisations and systems. The course includes an elective module of Managing Post Conflict Challenges, which has been designed for national and international managers operating in mine-affected countries (Figure 9). Students on the course include graduates of the University's national mine action management training programmes.

Fig. 9 Developing national management capabilities consists of training national managers in mine-affected countries to run mine clearance programmes for themselves.

9 Conclusions

All of the foregoing has attempted to show first of all the extreme nature of the UXO detection and disposal problem. Many millions of landmines and other types of ordnance have been deployed in conflicts, with few records of what has been laid and where. Not only do such weapons cause injury and death to innocent civilians, but also they deny the use of substantial areas of land for agricultural and other economic purposes, which may be critical in countries where the threshold of poverty is already low.

Low-frequency ground-penetrating radar represents one of a number of sensors that may be deployed to detect such targets. It is important to understand the strengths and weaknesses of radar techniques for these purposes, and the synergy with other types of sensor. Under favourable (i.e. dry) ground conditions and at relatively low radar frequencies penetration to significant depths can be obtained. However, low frequencies are unable to support wide radar bandwidths, so it is difficult to obtain high resolution at the same time as significant penetration.

Whilst such sensors must always respect the laws of physics, improvements in RF hardware, in digital processing hardware and in processing algorithms mean that steady advances will continue to be made. One promising area is in the complementarity of other types of sensor and hence of data and image fusion techniques to better exploit the strengths of each.

Acknowledgments We express our thanks to Taz Greyling and all other members of the Humanitarian Resilience Group at DCMT, Cranfield University, Shrivenham, and to David Daniels of ERA Technology and Chris Baker of UCL for reading and commenting on early drafts of this manuscript.

References

1. Aryanfar, F. and Sarabandi, K.: Through wall imaging at microwave frequencies using space-time focusing. IEEE Intl. Antennas and Propagation Symposium **3**, 3063–3066 (20–25 June 2004)
2. Benjamin, R., Hilton, G., Litobarski, S., McCutcheon, E. and Nilavalan, R.: 'Post-detection synthetic near field focusing in radar or sonar'. Electronics Letters **35**(8), 664–666 (15 April 1999)
3. Bottigliero, Ilaria: 120 Million Landmines Deployed Worldwide: Fact or Fiction. Fondation Pro Victimis, Geneve (2000)
4. Bridgeland, T. and Morgan, A.: Tunnel-master and Arsonist of the Great War: The Norton-Griffiths Story. Pen and Sword Books, Barnsley (2003)
5. Julius Caesar. De Bello Gallico, p191
6. Crisp, G.N. and Bishop, P.K.:'The Mineseeker airship project'. In: Daniels, D.J. (ed.) Ground Penetrating Radar, second edition, IET Radar, Sonar and Navigation Series, ISBN 0 86341 360 9, pp534–539. Springer, Heidelberg (2004)
7. Croll, M.: The History of Landmines. Pen and Sword Books, Barnsley (1998)
8. Daniels, D.J. (ed.): Ground Penetrating Radar, second edition, IET Radar, Sonar and Navigation Series, ISBN 0 86341 360 9 (2004)
9. EUREL International Conference on Detection of Abandoned Landmines (Edinburgh, October 1996); Second EUREL International Conference on Detection of Abandoned Landmines (Edinburgh, October 1998)
10. Griffiths, H.D. and Baker, C.J.: 'Fundamentals of tomography and radar'. In: NATO Advanced Study Institute Advances in Sensing with Security Applications, Il Ciocco, Italy, 17–30 July 2005, ISBN 1-4020-4286-8 Springer (2005)
11. Heald, G.J. and Griffiths, H.D.: 'A review of underwater detection techniques and their applicability to the land mine problem'. In: Proc. Second EUREL International Conference on The Detection of Abandoned Landmines, Edinburgh, IEE Conf. Publ. No. 458, pp173–176, 12–14 (October 1998)
12. Heliere, F., Lin, C-C., Corr, H. and Vaughan, D.: 'Radio Echo Sounding of Pine Island Glacier, West Antarctica and analysis of feasibility from space'. In: IEEE Trans. Geoscience and Remote Sensing, pp2573–2582 (Aug. 2007)
13. Lin-Ping Song, Chun Yu and Qing Huo Liu: 'Through-wall imaging (TWI) by radar: 2-D tomographic results and analyses'. IEEE Trans. Geoscience and Remote Sensing Pura. Appl. **43**(12), pp2793–2798 (Dec. 2005)
14. McAslan, A. and Bryden, A.: 'Mine Action Equipment: Study of Global Operational Needs'. Geneva International Centre for Humanitarian Demining. Geneva (June 2002)
15. McAslan, A. and Bryden, A.: 'Study of Capability Shortfalls and Equipment Needs in South-east Europe'. European Commission (October 2000)
16. McAslan, A. and Goslin, B.: 'The Mine Action Environment'. Pearson Custom Publishing (August 2004)
17. McAslan, A. and Greyling, T.: 'Humanitarian Resilience: The need to develop sustainable national capacities through education and training'. Principal's Lecture, DCMT Shrivenham (5 Dec. 2007)
18. Morrow, I.L. and Van Genderen, P.: 'Effective imaging of buried dielectric objects'. IEEE Trans. Geoscience and Remote Sensing **40**(4), pp943–949 (2003)
19. Sabath, F., Mokole, E. and Sammadar, S.N.: 'Definition and classification of ultra-wideband signals and devices'. Radio Science Bulletin **313**, pp10–26 (June 2005)
20. http://www.darpa.mil/baa/baa06-04.html
21. http://www.mineseeker.com/
22. http://www.cranfield.ac.uk/students/courses/page1809.jsp

UXO Signal Multi Sensor Detection and Estimation*

Chr. Kabakchiev[1], V. Behar[2], B. Vassileva[2], D. Angelova[2], K. Aleksiev[2], V. Kyovtorov[3], I. Garvanov[3], L. Doukovska[3] and P. Daskalov[4]

Abstract In this chapter, the original advanced algorithms for stepped-frequency GPR imaging are considered. In stepped-frequency GPR, the range profile formation is carried out by reconstruction of a wideband chirp by combining a set of stepped-frequency chirp signals in the time domain. Using the Modelsim simulator, it is shown that the processor VIRTEX II Pro is suitable for implementation of this algorithm. A simple convolution algorithm for simulation of stepped-frequency GPR images from multi-layered subsurface media is described. Different approaches and algorithms for the basic GPR signal and image processing are also considered in this chapter. These algorithms are used for improving the image quality of underground objects, e.g. pipes. It is shown that applying different filters (CFAR, Hough, Kalman, Particle) to GPR image processing is a good decision in the sense of estimation accuracy, probability of target detection and false alarm.

Keywords: UXO signal detection, modeling and estimation, stepped frequency GPR imaging, CFAR for GPR, Hough transform, Canny edge detection, Bayesian, Extended Kalman filtering, Interacting Multiple Model filtering, FPGA implementation

---

[1] Faculty of Mathematics and Informatics, Sofia University "James Bourchier" Str., 5, 1164 Sofia, Bulgaria, e-mail: ckabakchiev@fmi.uni-sofia.bg, ckabakchiev@yahoo.com
[2] Institute for Parallel Processing, Bulgarian Academy of Sciences "Acad. G. Bonchev" Str., bl. 25-A, 1113 Sofia, Bulgaria, e-mail: behar@bas.bg
[3] Institute of Information Technologies, Bulgarian Academy of Sciences "Acad G. Bonchev" Str., Bl. 2, 1113 Sofia, Bulgaria, e-mail: vladimir.kyovtorov@gmail.com, igarvanov@yahoo.com, l.doukovska@mail.bg
[4] Multiprocessor Systems Ltd. Shipchensky prohod Blvd., 63, 1574 Sofia, Bulgaria, e-mail: Daskalov@mps.bg

* This work was financially supported by the Bulgarian National Science Fund (grants MI-1506/2005 and BG051PO001/07/3.3-02-7/08), the National Innovation Fund (grant IF-02-85/2005) with MPS Ltd. Project "Digital Ground Penetrating Radar" and IIT 010089/2007.

J. Byrnes (ed.), *Unexploded Ordnance Detection and Mitigation*,

1 Introduction (UXO Signals and a Multi Sensor Approach)

Ground penetrating radar (GPR) is a well-known method of subsurface exploration, which becomes extremely important for many environmental applications such as unexploded ordnance (UXO) detection and geophysical implementation [8]. It is well-known that most commercial GPR systems are ultra-wideband pulse radars, in which range resolution is determined by the bandwidth of the transmitted pulse. In these GPR, high range resolution is achieved by transmitting very short pulses (or frequency-modulated pulses) to obtain the required bandwidth. The frequency-stepped processing method is a technique developed to overcome the power bandwidth limitations of pulse radars.

In this chapter, GPR range profile formation is carried out by reconstruction of a wideband chirp by combining a set of stepped-frequency chirp signals in the time domain. In order to optimize the parameters of the stepped-frequency algorithm, a simple convolution-based algorithm for simulation of echoes from multi-layered subsurface media has been developed. As a result, a simple algorithm for simulation of frequency-stepped GPR images of multi-layered media has been developed for parameter optimization of the basic GPR signal and image processing [3, 5].

Different approaches and algorithms for the basic GPR signal and image processing are considered and studied in order to improve the image quality of underground objects and enable a recognition of objects and estimation of their parameters. The results of the study described in [1–6, 9, 10] show that different approaches and algorithms for signal and image processing generally lead to similar results. However, in different situations, there can be alternatives. The results described here are obtained in cooperation with MPS Ltd., the Institute of Information Technologies (IIT-BAS) and the Institute for Parallel Processing (IPP-BAS), within the project "Digital Ground Penetrating Radar" financially supported by the National Innovation Fund (IF-02-85/2005-2007).

According to [8], underground objects of interest (e.g. pipes) are very similar to unexploded ordnance (UXO). Therefore, our first conclusion is that algorithms developed for GPR imaging and also for simulation of GPR images can be successfully used in a system for detection of unexploded ordnance (UXO).

Our second conclusion is that the multi-sensor unexploded ordnance detection system (MUDS) approach, usually leading to image improvement, and the parameter estimation of unexploded ordnance (UXO) by using different types of sensors can also be successfully applied to the same GPR sensors with different algorithms for signal and image processing.

2 Stepped-Frequency GPR Imaging

The novelty of the results obtained in [5] is that two stepped-frequency methods intended earlier for SAR applications are used for GPR imaging and implemented on the base of RSPs of Analogy Devices. These methods construct a synthetic high-resolution range profile by transmission of a burst of narrowband LFM pulses with frequency bands separated by a fixed step. The first of them constructs the wideband signal in the time domain as a combination of stepped-frequency narrowband chirps. The other method constructs the wide frequency band of a wideband signal as a combination of the frequency bands of stepped-frequency narrowband chirps. In that case, the range resolution of a synthetic range profile produced by GPR depends on the whole frequency range of the transmitted pulses. Starting from the requirements to implement the stepped frequency processing on the base of RSP AD6624 and AD6624A, four optimal parameter sets of the stepped-frequency processing are proposed for its implementation in GPR. The criterion of optimization was the minimal main lobe width and the minimal sidelobe peaks of the output signal in a synthetic range profile of a homogeneous subsurface medium containing a point target. The first variant of optimal parameter sets corresponds to stepped-frequency GPR operating at 4.6–38.2 MHz, which generates synthetic range profiles with the range resolution of 1–2 m by transmitting 14 narrowband chirps at each GPR position. The simulation results show that the method of constructing a range profile in the time domain is a more appropriate one because it produces the synthetic range profile with lower noise.

2.1 Time-domain processing

The time-domain technique uses a sequence of stepped-frequency narrowband waveforms to produce a high-resolution synthetic range profile. In the time domain, a long wideband chirp is constructed from M narrowband chirps, each of duration T_p, separated in time by a repetition interval T. The central frequencies of narrowband chirps are spaced by step Δf. Since the spectrum of each narrowband chirp is a fraction of a constructed wideband chirp, all transmitted chirps should have the same frequency rate:

$$b_1 = b_2 = ... = b_M = b = \Delta f / T_p. \quad (1)$$

The total bandwidth of a reconstructed wideband pulse is expressed as:

$$\Delta F = f_{max} - f_{min} = \Delta f M. \quad (2)$$

The central frequency of a transmitted narrowband chirp changes as:

$$f_{0,m} = f_c + [m - (1-M)/2]/\Delta f, \; where \; f_c = (f_{min} + f_{max})/2, \; m = 1, ..., M. \quad (3)$$

The transmitted pulse belonging to the same burst can be described by:

$$v_{tx}(t,m) = p(t) exp(j2\pi f_{0,m} t), \; where \; p(t) = A rect(t/T_p) exp(j\pi b t^2). \quad (4)$$

The pulse reflected from a point scatterer located at distance d is a time-delayed version of the transmitted pulse, i.e.:

$$v_{rx}(t,m) = v_{tx}(t-\tau,m),\ m = 1,...,M. \tag{5}$$

The time delay in (5) is $\tau = 2d/V$, and V is the velocity of electromagnetic wave propagation. After quadrature demodulation, the received signal at baseband is given by:

$$v_{bb}(t,m) = v_{rx}(t,m)exp(-j2\pi f_{0,m}t) = p(t-\tau)exp(-j2\pi f_{0,m}\tau). \tag{6}$$

The construction of a synthetic range profile is performed by the following processing steps:

• *Upsampling.* In order to avoid overlaps in the constructed spectrum, the baseband signals have to be upsampled by a factor of M, where M is the number of transmitted pulses.

• *Frequency shift.* The frequency shift of $v_{bb}(t,m)$ is performed in the time domain as:

$$v'_{bb}(t,m) = v_{bb}(t,m)exp(j2\pi\delta f_m t),\ where\ \delta f_m = [m+(1-M)/2]\Delta f. \tag{7}$$

• *Phase correction.* In order to avoid phase discontinuities in the wideband signal, the phase of each narrowband pulse must be corrected by a phase-correcting term, given by:

$$\Phi_m = exp(j\pi bT_p^2[m+(1-M)/2]^2). \tag{8}$$

• *Time shift.* Before coherent summing each narrowband pulse is shifted in the time domain by:

$$\delta t_m = [m+(1-M)/2]/T_p. \tag{9}$$

• *Coherent summing.* In the time domain, the wideband pulse $v'(t)$ is formed by coherently summing all the narrowband signals $v'_{bb}(t,m)$:

$$\begin{aligned} v'(t-\tau) &= \sum_{m=0}^{M-1} v'_{bb}(t-\delta t_m,m)\Phi_m = \\ &= Aexp[j\pi b(t-\tau)^2]\sum_{m=0}^{M-1} rect(\frac{t-\tau-\delta t_m}{T_p}) = Aexp[j\pi b(t-\tau)^2]rect(\frac{t-\tau}{MT_p}). \end{aligned} \tag{10}$$

The bandwidth and duration of the wideband chirp $v'(t)$ are equal to $M\Delta f$ and MT_p, respectively.

• *Pulse compression.* The final operation of constructing a synthetic range profile is performed by filtering the constructed wideband pulse (10). The filter impulse response is formed as the time-reversed conjugate of the wideband pulse, which is constructed from the transmitted narrowband pulses. The signal at the compression filter output is given by:

$$r(t) = |\ FFT^{-1}S(f)V(f)\ |, S(f) = FFT[s(t)], V(f) = FFT[v'(t-\tau)] \tag{11}$$

where $s(t) = conj[v'(t-\tau)]W(t)$ under $\tau = 0$, and $W(t)$ is the weighting function.

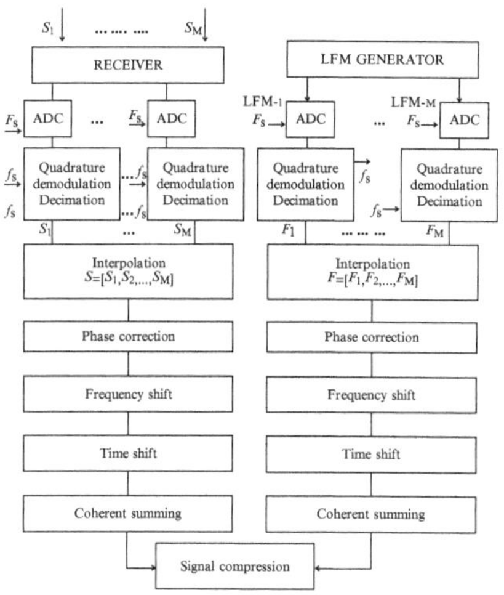

Fig. 1 Time-domain method.

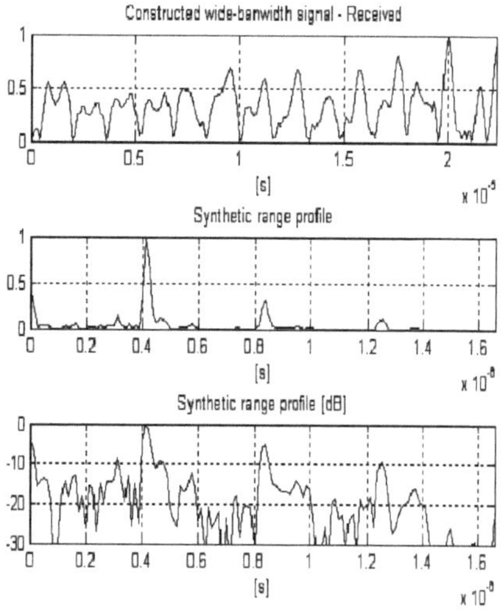

Fig. 2 The synthetic range profile.

2.2 Simulation results

The block scheme for formation of a synthetic range profile using the stepped frequency algorithm in the time domain is shown in Figure 1. According to the block scheme, the left channel for signal processing performs echo-signals while the second channel forms the impulse response of a compression filter. The examples of both, the wideband signal and the synthetic range profile, constructed in the time domain by combining 14 narrowband chirps are shown in Figure 2. Comparison analysis of synthetic range profiles given in [5] shows that the two stepped-frequency processing methods are of equivalent quality. However, it can be seen that the first method is a more appropriate one because it produces the synthetic range profile with lower noise.

3 Simulation of Stepped-Frequency GPR Images

At each transmission of a narrowband pulse, the EM wave radiated from a transmitter antenna travels through the multi-layered media with a velocity that depends on the electrical properties of layers. If the EM wave encounters a boundary between two layers with different electrical properties, a part of the EM energy is reflected or scattered back to the surface, while the rest of the energy continues to travel downward. The radar receiver collects the return signal that contains several returns from various layers of different dielectric properties. There are a variety of methods for simulation of GPR return signals. For a basic first-order simulation, a simple convolution-based modeling technique can be used [6]. More accurate results, taking into account the effects of scattering due to random surfaces and the three dimensional antenna beam pattern can be obtained using advanced methods such as the Finite Difference Time Domain (FDTD) method, at the cost of complexity and computational time.

The novelty of the results obtained in [6] is that a sophisticated convolution-based signal model is proposed for simulation of stepped-frequency GPR images. This model takes into account the basic radar parameters (energy potential, frequency, antenna beamwidth, number of transmitted chirps, wideband of transmitted chirps and so on) and the basic parameters of a multi-layered medium (number of layers, dielectric properties of layers, depth of layers, attenuation) and, therefore, it results in more accurate simulation of stepped-frequency GPR images. The simulation results show that this algorithm can be successfully used for analysis and parameter optimization of the signal processing algorithms in stepped-frequency GPR.

3.1 Echo signal simulation

The synthetic high-resolution range profile is constructed by transmission of narrowband LFM pulses with frequency bands separated by a fixed step. At the m-th transmission of a narrowband LFM pulse, the signal reflected from a medium with L layers can be described as:

$$r(m,t) = r_0(m,t) + \sum_{k=1}^{L} \mu_{m,k}\sqrt{SNR_k}s(m,t) * \delta_k(m,t-\tau_k) + N_0(m,t) \tag{12}$$

where $r_0(m,t)$ is the direct normalized pulse from transmitting to receiving antennas; $s(m,t)$ – the transmitted LFM pulse whose envelope is unity; SNR_k is the signal-to-noise ratio from the interface between layers k and $(k+1)$; $\mu_{m,k}$ – multiplicative noise; $\delta_k(m,t)$ – the impulse response of the interface between layers k and $(k+1)$; τ_k – the two-way time delay of a signal reflected from the interface between layers k and $(k+1)$; L – the number of layers; $N_0(m,t)$ is normalized Gaussian noise with zero mean and unity variation; and "$*$" denotes convolution. Using the basic radar range equation and also taking into account the signal losses in the propagation path from the transmitter to the receiver, the SNR_k (in dB) can be evaluated as:

$$SNR_{k,dB} = \prod\nolimits_{GPR,dB} + \sigma_{k,dB} - 40lg(\sum_{j=1}^{k} z_k) - L_{k,dB}^{REF} - L_{k,dB}^{AT} - L_{R1,dB} - L_{R2,dB} \tag{13}$$

where $\prod_{GPR,dB}$ is the radar energy potential (in dB); $\sigma_{k,dB}$ is the radar cross section of the interface between layers k and $(k+1)$, z_k is the thickness of layer k; $L_{k,dB}^{REF}$ is the signal loss due to signal reflections from the interface between two layers; $L_{k,dB}^{AT}$ is the attenuation loss; $L_{R1,dB}$ is the transmission loss from the antenna to the material; $L_{R2,dB}$ is the retransmission loss from the material to the air. Typically, for many earth materials, both $L_{R1,dB}$ and $L_{R2,dB}$ are about 2.5 dB. The signal losses due to reflection of a signal from the interface between layers k and $(k+1)$ are calculated as

$$L_{k,dB}^{REF} = -20lg(\Gamma_k \prod_{j=1}^{k-1} (T_{j,j+1}T_{j+1,j})) = -20lg(|\Gamma_k| \prod_{j=1}^{k-1} (1-\Gamma_{j,j+1}^2)). \qquad (14)$$

The reflection coefficient Γ_k and the transmission coefficient T_k in (14) are defined as:

$$\Gamma_k = (\sqrt{\varepsilon_{k+1}} - \sqrt{\varepsilon_k})/(\sqrt{\varepsilon_{k+1}} + \sqrt{\varepsilon_k}); T_k = \sqrt{4\sqrt{\varepsilon_k \varepsilon_{k+1}}/[\sqrt{\varepsilon_{k+1}} + \sqrt{\varepsilon_k}]^2} \qquad (15)$$

where ε_k is the permittivity of layer k. The attenuation loss of the material and the radar cross section are calculated as:

$$L_{k,dB}^{AT} = 4\sum_{j=1}^{k} \alpha_j z_j \quad and \quad \sigma_k = \pi(\tan(\Theta/2)\sum_{j=1}^{k} z_j)^2 \qquad (16)$$

where α_j is the attenuation constant in the j-th layer, and Θ is the beamwidth of the transmitter/receiver antenna.

3.2 GPR image simulation

The simulation algorithm for GPR images is shown in Figures 3–4. As can be seen, it includes two main stages: (1) – synthetic range profile formation. This procedure is repeated for each of N positions of a transmitter/receiver system. (2) – B-mode image formation. This procedure uses N synthetic range profiles obtained at a previous stage. According to Figure 3, simulation of M transmissions of LFM pulses results in the signal matrix of M columns, where each column contains the echo

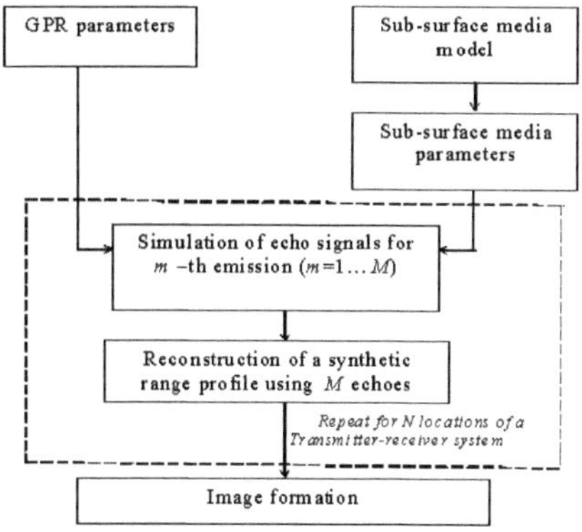

Fig. 3 Simulation of a range profile.

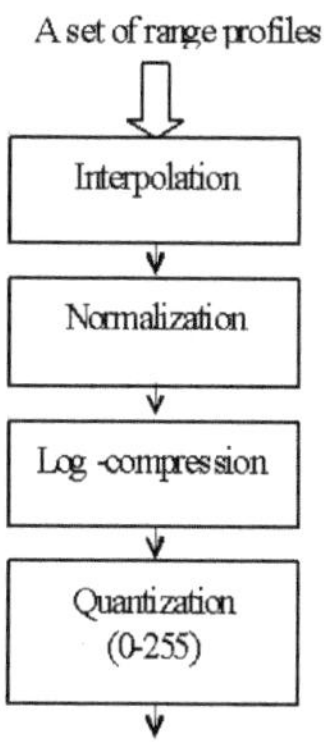

Fig. 4 GPR image formation.

signal received after transmission of a LFM pulse. The simulated signal matrix is used for further construction of a synthetic range profile by one of the two methods, the time-domain method or the frequency-domain method [5]. According to Figure 4, the signal matrix with N columns, containing N synthetic range profiles, is used for GPR image formation. The stage of image formation includes such operations as interpolation, logarithmic-compression, quantization, and color visualization.

3.3 Simulation results

The simulation of B-mode images of a four-layered medium is done by using the convolution-based model described above. The simulated medium includes successive layers of dry sand, green sand, saturated sand and granite with depths of 31, 21, 16 and 24 m, respectively. The electro-magnetic parameters of layers (relative permittivity and attenuation) are $\varepsilon = (4;9;15;9)$ and $\alpha = (0.03;0.1;0.3;0.2)$, respectively.

The following radar parameters are used for calculation of the SNR for each layer: radar energy potential -20 dB; antenna beamwidth -20^0; pulse repetition frequency -300 kHz; number of LFM pulses needed for construction of each synthetic range profile -14, frequency bandwidth of a single LFM pulse -2.4 MHz; total frequency bandwidth – $[4.6 \div 38.2]$ MHz; sampling frequency at RF -80 MHz; sampling frequency at baseband -2.5 MHz. The SNR calculated for a signal reflected from layer 2 is 57 dB, from layer 3–40 dB and from layer 4–21 dB. The corresponding two-way time delay of a signal reflected from layer 2 is 0.4 μs, from layer 3–0.8 μs, and from layer 4–1.25 μs.

The time-domain stepped-frequency method is used to produce synthetic range profiles. The synthetic range profile, constructed in the time domain, and the simulated B-mode GPR image are shown in Figures 5 and 6 respectively.

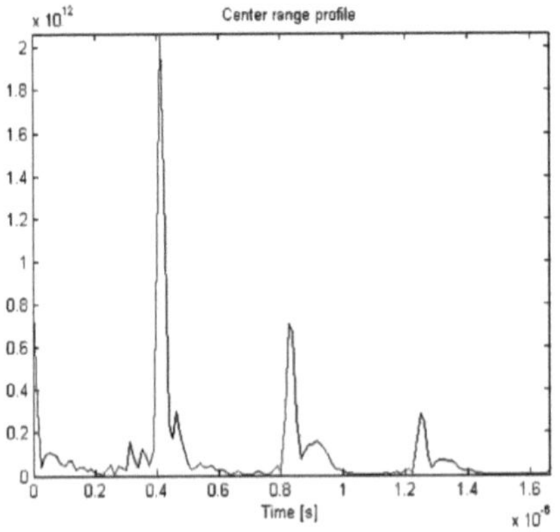

Fig. 5 The synthetic range profile.

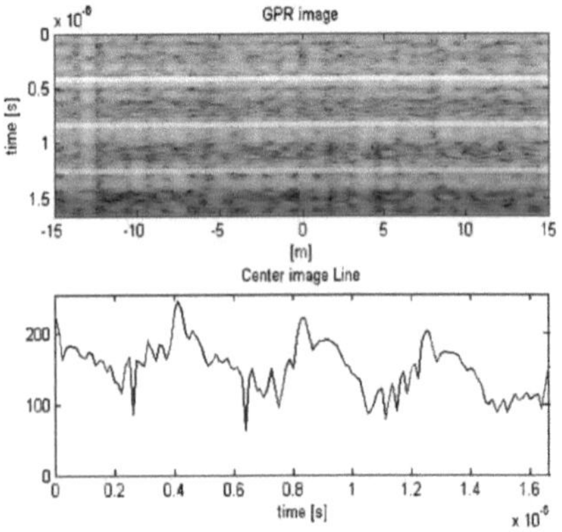

Fig. 6 Simulated B-mode image.

4 GPR Data Basic Processing

The most important basic processing algorithms which are used in [10] have been developed earlier for GPR signal processing. The analysis of GPR data is carried out by processing the data using different filtering techniques and gains.

The most important basic processing algorithms in our case are:

• *Mean filter* (vertical working low-pass filter). This filter acts on each trace independently. The filter performs a mean over a selectable number of time samples for each time step.

• *Running average* (horizontal working low-pass filter). This filter acts on the chosen number of traces. The filter performs a running average over a selectable number of traces for each time step.

• *Stack traces* (compression in horizontal direction). This filter performs a temporal simultaneous stacking of a selectable number of traces.

• *Median filter* (pulse jamming and speckle noise reduction). This filter calculates the median over a selectable time/range area for each time step.

• *Background removal* (spatial high-pass filter which makes visible the shallow objects). This filter performs a subtracting of an averaged trace which is built up from the chosen time/distance range of the actual section.

• *Gain adjustment* (corrects the attenuation losses and makes visible the deep objects). The gain acts on each trace independently. The algorithm parameter (window length) forms a jumping window. The time window samples are normalized in range [0–1]. The experimental results obtained enable one to conclude that the algorithms for the basic signal processing presented in [10] can be successfully used for analysis of GPR images.

4.1 GPR data basic processing – simulation results

In this section some results obtained by the above-mentioned algorithms are shown. The simulated image of a subsurface medium with five layers masked by pulse jamming is shown in Figure 7. It can be seen that after range profile formation, the pulse jamming looks like speckle noise. In order to remove this noise, a median filter can be applied over the selectable time/range area for each time step. In Figure 8, the real radargram acquired by the radar GSSI SIR is contaminated with pulse jamming (Figure 7). The same image "cleaned" by median filtering is shown in Figure 8. The image presents five underground fuel storage tanks.

Benefits of the gain adjustment algorithm are illustrated in Figure 9. The simulated radargram of a subsurface medium with five layers that is reconstructed in frequency is shown in Figure 9 (on the left). The gain adjustment applied to this image is also shown in Figure 9 (on the right). The gain acts on each trace independently. As a result, this algorithm makes the deep objects visible. The time window

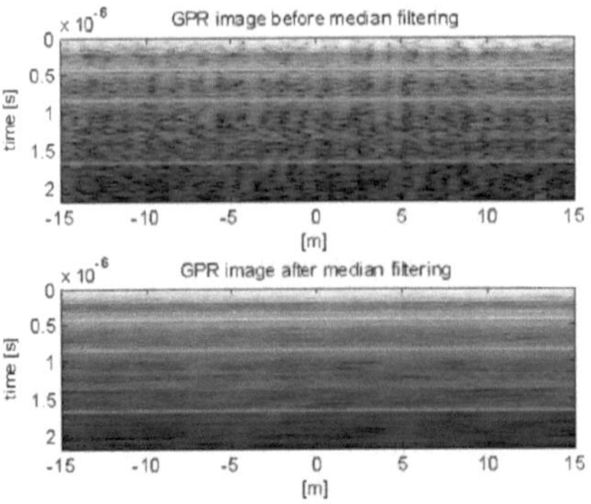

Fig. 7 Median filtering: nr. of time samples = 3; nr. of traces = 21.

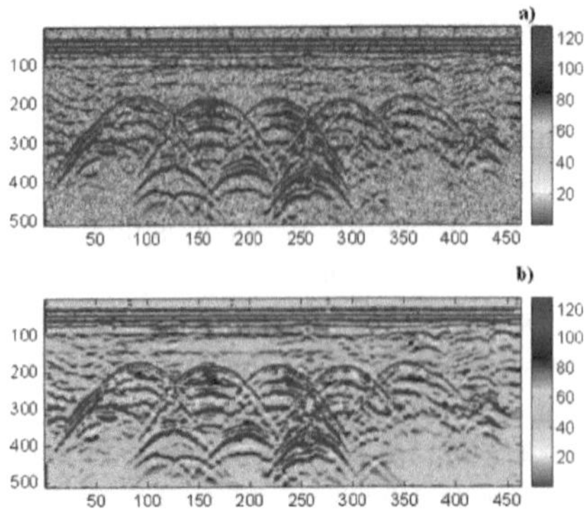

Fig. 8 Median filtering: nr. of time samples = 5; nr. of traces = 5.

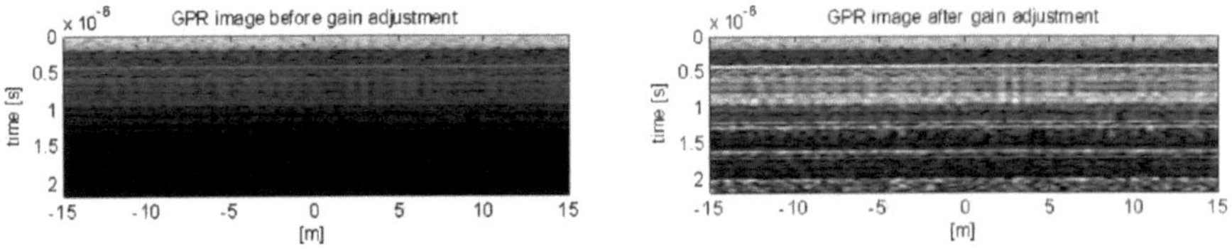

Fig. 9 Gain adjustment corrects the attenuation losses.

samples are normalized in range [0–1]. However this process destroys the original information of the signal. Therefore it is recommended to be applied only for displaying the GPR radargram.

5 CFAR Filter Approach for GPR Processing

A conventional Constant False Alarm Rate (CFAR) detector is often used in primary radar signal processing and is very effective in case of stationary and homogeneous interference. Different approaches proposed in [7] are realized in different structures of CFAR detectors for operating in non-stationary non-homogeneous background and random impulse noise. One of them proposed by Rohling for a multi-target situation is to use the ordered statistics for estimation of the noise level in the reference window. Another approach is to excise high-power samples from the reference window before processing by the conventional cell averaging CFAR detector.

This approach is used by Goldman for design of an excision CFAR detector (EXC CFAR) in order to improve the performance of CFAR detectors in the presence of impulse interference.

It is obvious that two CFAR processors can be used as 2D filters of GPR images. The first of them visualizes images after adaptive thresholding (1 or 0), while the second filter visualizes only amplitudes above the adaptive threshold.

5.1 CFAR filters analysis

In modern radar, signal detection is declared if the signal value exceeds a preliminary determined adaptive threshold. The threshold is formed by current estimation of the noise level in the reference window. In this processor, the target is detected according to the following algorithm:

$$\begin{cases} H_1 : \Phi(q_0) = 1,\ q_0 \geq T_\alpha V \\ H_0 : \Phi(q_0) = 0,\ q_0 < T_\alpha V \end{cases} \tag{17}$$

where H_1 is the hypothesis that the test resolution cell contains echoes from the target and H_0 is the hypothesis that the test resolution cell contains randomly arriving impulse interference only. $V = \sum_{i=1}^{N} x_i$ is the noise level estimate. The constant T_α is a scale coefficient, which is determined in order to maintain a given constant false alarm rate (CFAR).

The presence of randomly arriving impulse interference in both the test resolution cells and the reference cells can cause drastic degradation in the performance of such a CA CFAR processor.

To overcome the heavy noise environment where the detection is performed, a CFAR processor with Binary Integration (CFAR BI) is proposed. This signal processor can be considered as N single dimensional CA CFAR processors working in parallel. The binary integration processor employs a two-step thresholding technique for target detection. Firstly, a preliminary decision is made about each pulse of the pulse train reflected from a target. Pulse detection is declared if the first adaptive threshold is exceeded in the test cell. For this aim, the conventional CFAR detector can be used. Secondly, the number of samples, where the first threshold is exceeded, are counted and the obtained number of detections is compared with the second threshold. Target detection is declared if the second threshold is exceeded. The results in [7] show that the CFAR BI detector is more effective in conditions of intensive randomly arriving impulse interference.

CFAR processors with post detection integrators are proposed for the case of a homogeneous environment and chi-squared family of target models (CFAR PI). The possibility for parallel processing of samples in the reference window can be realized by a parallel computing architecture of the target detection algorithm. This post detection integration (PI) CFAR processor consists of a single pulse matched filter, square-law envelope detector, linear post detection integrator, noise level estimator and comparator.

5.2 CFAR filters – simulation results

One real GPR image containing a waste water pipe and a land mine under a thin layer of wet sand is presented in Figure 10. Three images after the CA CFAR

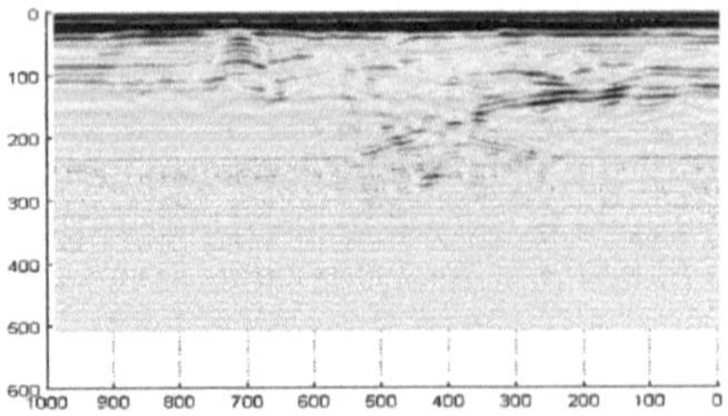

Fig. 10 Geo-radar profile (waste water pipe and a land mine).

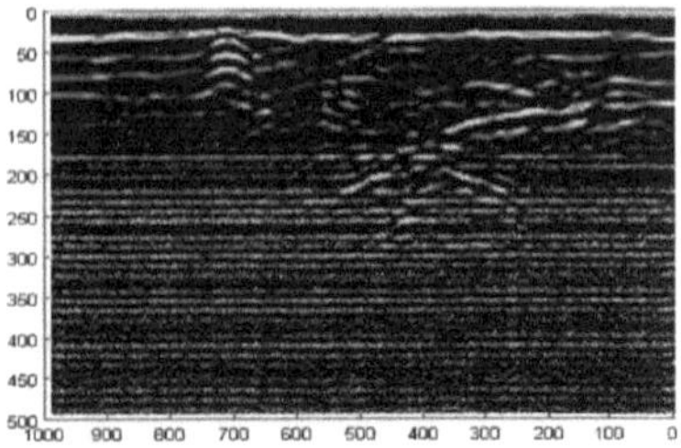

Fig. 11 Geo-radar profile after CA CFAR filtering ($T = 1$).

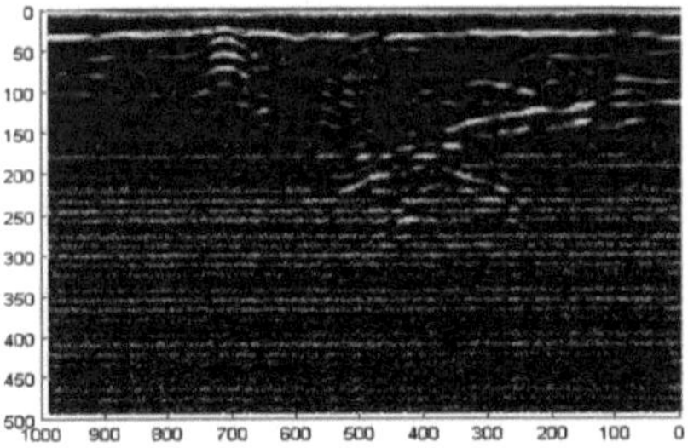

Fig. 12 Geo-radar profile after CA CFAR filtering ($T = 1.1$).

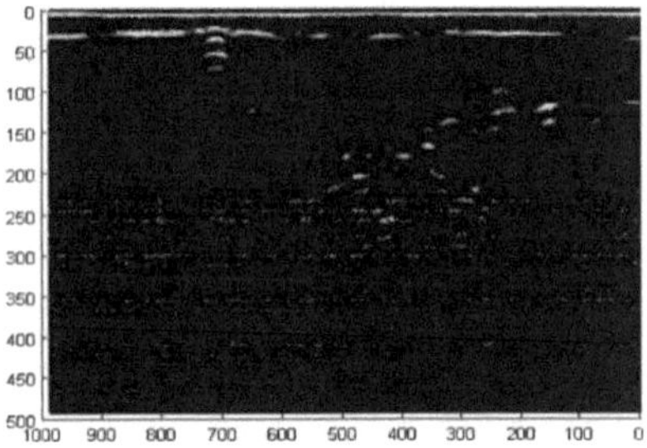

Fig. 13 Geo-radar profile after CA CFAR filtering ($T = 1.3$).

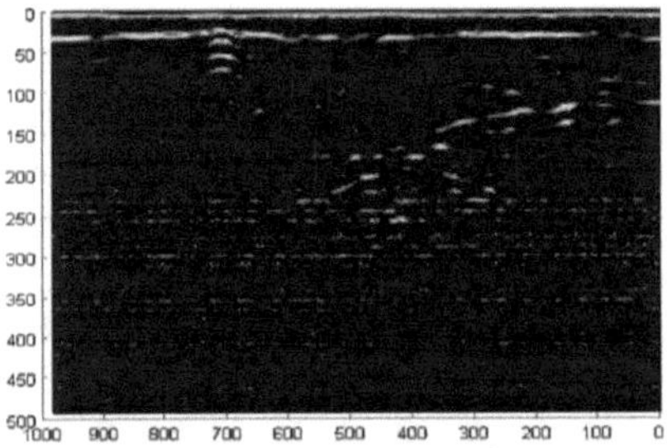

Fig. 14 Geo-radar profile after CFAR PI filtering ($N = 16, M = 16, T = 18$).

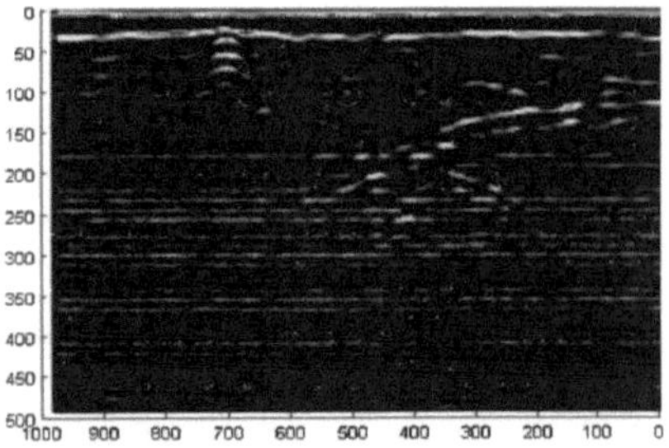

Fig. 15 Geo-radar profile after CFAR BI processing with binary rule 10/16).

filtering are shown on Figures 11–13. They are performed by a 16-element moving window in depth and scale constants $T = 1, 1.1, 1.3$.

When the scale factor increases the borders between layers become less visible. For higher values of T the presence of foreign substances (land mines, pipes) is more perceptible. When CFAR PI filtration is applied to the image from Figure 10, the result in Figure 14 is obtained. The performance is done with a rectangular window of size (16 X 16) and $T = 18$.

After CA CFAR BI processing of the image from Figure 10, the result is shown on Figure 15. The binary integration with rule 10/16 leads to the results depicted on Figure 15. In this case the reference window is of size (16 X 16) and $T = 6$.

6 Hough Approach in GPR Processing

The Hough Transform (HT) is regarded as a template matching method for feature detection. The conventional HT approach is usually used for straight line detection and linear objects localization. However, the HT can be successfully used for ellipse or circle detection and even for arbitrary form detection. As a consequence, the HT algorithm can be applied to buried mines detection transforming all image pixels by automatic detection of circular shapes [1].

6.1 Hough transform based hyperbola detection

The standard hyperbola equation doesn't meet all GPR constraints. To deal with the antenna speed fluctuation and the anisotropic wave propagation an additional parameter is included in this equation. The slightly modified equation takes the form:

$$y(t_i) = \sqrt{k^2(x_r(t_i) - x_0)^2 + d_{min}^2}. \tag{18}$$

A parameter k may express the difference between the antenna speed and the basic speed v_{0x}:

$$y(t_i) = \sqrt{[\frac{k(x_r(t_i) - x_0)}{t_i - t_k}(t_i - t_k)]^2 + d_{min}^2} = \sqrt{[kv_{0x}(t_i - t_k)]^2 + d_{min}^2}. \tag{19}$$

k may also express the variations of the velocity of wave propagation when spreading through the subsurface medium:

$$\begin{aligned} y(t_i) = \sqrt{k^2(x_r(t_i) - x_0)^2 + d_{min}^2} = k\sqrt{(x_r(t_i) - x_0)^2 + (\tfrac{d_{min}}{k})^2} = \\ = k\sqrt{(x_r(t_i) - x_0)^2 + D_{min}^2} = kv_{0x}(t_i - t_k) \end{aligned}$$

where v_{0x} is the accepted wave propagation velocity through the earth.

There are two approaches for hyperbola maximum localization by HT. The *first* considers two consecutive standard (for straight line detection) HTs, followed by a logical analysis of detection of the corresponding lines. The idea is to approximate a hyperbola with two straight lines (Figure 16) and find them with the standard HT. These straight lines have the restricted space position (depending on the parameter variation), and the algorithm requires limited computer resources. The main drawback is that the accumulator doesn't contain information about coordinates of points. As a result many hyperbolas will be detected, but most of them will be false. A complex combinatorial algorithm has to be applied to reject ghost hyperbolas. The second drawback concerns the horizontal hyperbola's part. This part is usually formed by the most powerful echoes from the target of interest with the highest

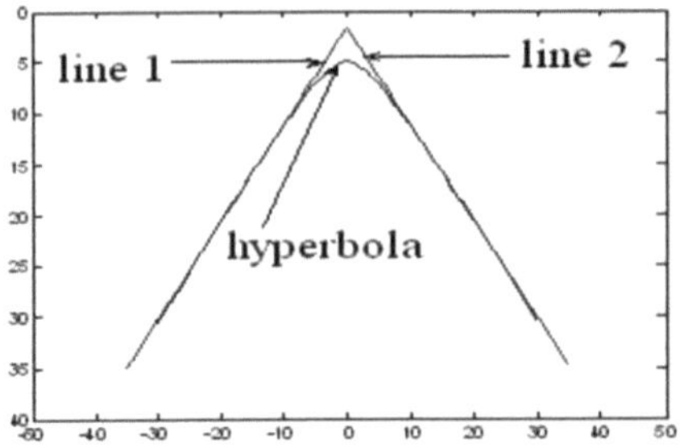

Fig. 16 Hyperbola approximation with two straight lines on hyperbola.

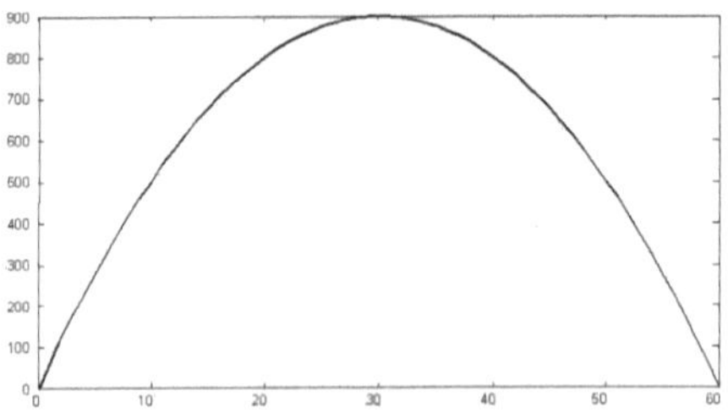

Fig. 17 Nonlinear asymmetrical weighting (for the case of 60 received votes).

signal-to-noise ratio. That is the main reason to regard this part as the most reliable source of information about the target position. But in the first approach it isn't considered at all.

The *second approach* applies the HT directly to Equation (19). In this case the HT uses 3D parameter space. The three parameters are x_0, y_0, k. The larger parameter space presumes more time for searching the peak and requires more memory. Still the second approach is accepted as a more perspective one. The hyperbola strip width is matched to errors, generated by the antenna speed fluctuation and by the variable velocity of propagation. The hyperbola strip is convolved with a Gaussian filter to weight the votes, falling from points, lying on the central (for this strip) hyperbola, and the votes from points lying aside it (Figure 17).

6.2 Algorithm realization and simulation results

The generalized HT algorithm requires transformation of each image point from the image (feature) space to the parameter space and accumulates their votes. Usually the GPR images include not less than 0.5 M pixels, every pixel with 216 or 28 intensity levels. The problem is to find such pixels of similar intensity lying on a hyperbola (or near to it) that differ from the neighboring pixels. It is obvious that the computer processing of a whole set of image points is a tedious task, requiring serious computer resources. This problem is solvable, but the algorithm will be intensity dependent, which is an undesired characteristic of every image processing algorithm. To reduce the initial set of potential points belonging to hyperbolas, filtering algorithms are applied. Real-time solution of the task includes several steps: the GPR input; 2D FFT filtering; Canny edge detection; HT hyperbola detection; visualization.

The strongest and almost constant echo-signals are received from borders between different subsurface layers (Figure 18a). These echo-signals play a role as powerful low frequency noise and should be removed from the image. The high frequency noise is also present in the image and looks like one or a few grouped

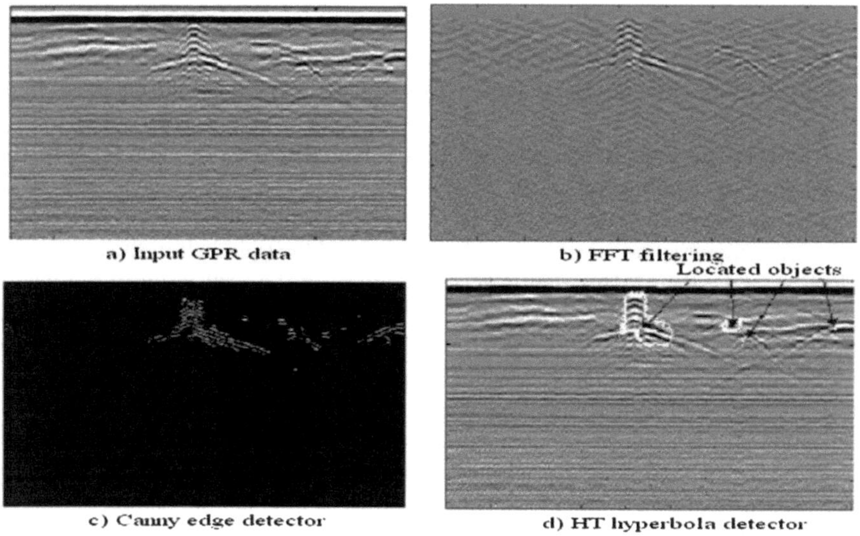

Fig. 18 HT algorithm applications.

pixels in the image, strongly differentiating from the surrounding pixels. To reject them a 2-D band-pass frequency filtering is applied over a raw GPR image. The lowest and the highest several frequencies are rejected, including the constant or DC Fourier component. The filter band-pass frequencies are matched to both the antenna speed and the echo-signal attenuation. The band-pass frequency filtering is realized in three steps: (1) Fast Fourier Transform (FFT); (2) Weighing the Fourier components; (3) Inverse FFT. The image may be preprocessed in order to limit the bandwidth. For example, Gaussian smoothing can be applied in advance.

For GPR data, the most suitable Fourier components of a 2-D frequency filter are chosen as follows: For the highest frequencies, the last 5 Fourier components in both directions, horizontal and vertical, are nulled. For the lowest frequencies, the first 30 Fourier components in the horizontal direction are nulled and the first 10 components in the vertical direction are nulled (rectangular window). Using frequency domain filtering, excellent robustness against correlated and frequency dependent noise is achieved (Figure 18b). Careful analysis of the output image shows an appearance of weak Gipps effects near the image borders, but without the influence on the edge detection algorithm.

The Canny edge detector is used as an image contour detection algorithm. The Canny method finds edges by looking for local maxima of the gradient of the pixel intensity. The gradient is calculated using the derivative of a Gaussian filter. The method uses two thresholds for detecting the strong and the weak edges, and includes the weak edges in the output only if they are connected with strong edges. This method is therefore less sensitive to noise than the others, and it can localize true weak edges. The Canny edge detector is used mainly to reduce the number of pixels of interest by two orders of magnitude. The final result is very promising – the number of non-zero pixels in GPR images is reduced from 0.5 M to several thousands. The values of chosen parameters are: 0.3 – for lower threshold, 0.33 – for higher threshold and 3 – for standard deviation of the Gaussian filter (Figure 18c).

The HT algorithm is realized as follows. Two windows, one for the left half of a hyperbola and the second for the right half of a hyperbola are generated. Both windows are applied to the image at the output of a Canny edge detector. As a result, two accumulator spaces are obtained for the left and the right half of the hyperbola. How to merge them? This step is very important for the final result of the whole filtering. Practice proves that robust results are obtained only if there is symmetry of votes for both parts.

The proposed algorithm realizes the common accumulator space by multiplication of contents of both accumulator spaces, element by element. It is clear that this operation will amplificate accumulators in both parameter spaces with the equivalent number of votes and will weaken accumulators with asymmetric distribution but with the equivalent sum of votes to the previous case in both parameter spaces (Figure 18b). Peak detection is performed after thresholding. The located objects are displayed on Figure 18d.

7 A Bayesian Algorithm for Object Detection in GPR Data

A number of sophisticated techniques for background signal reduction and object detection have been proposed, accounting for the nonstationary and correlated nature of GPR signals. They usually incorporate complex models and time-consuming learning stages for model parameter adjustment. The aim of this investigation is to use simple models with robust processing algorithms.

A GPR data processing algorithm relying on simple background and target models is suggested by Dr. Carevic, cited in [3]. It is based on the "variable dimension filtering approach" to target tracking. Background estimation, target detection and target-background separation are performed within a common Kalman filter-based computational procedure. This algorithm is successfully applied to reduce the background interference signals and to detect shallow buried targets. However, in the case of large state dimensions the target signal estimate can be unsatisfactory. Also, additional information is needed for identifying target extent.

The novelty of the results obtained in [2, 3] is that the constructive elements of a Kalman filtering approach are extended with the advantages of hybrid Bayesian estimation. The set of GPR data is processed in two consecutive steps. At the first step, a part of the algorithm of Dr. Carevic is realized: a Kalman filter (KF) estimates the background signal, time-varying noise characteristics and detects possible targets. The estimated noise parameters are utilized at the second step, where an Interacting Multiple Model (IMM) algorithm is applied. Using multiple models and efficient Bayesian mechanism for information fusion, the IMM algorithm assesses more precisely the target signal and target extent. The IMM posterior model probabilities assist in the decisions of the Kalman filtering procedure, increasing the probability of target detection.

Change detection methodology provides efficient tools for automatic on-line (or off-line) signal segmentation. The cumulative sum (CUSUM) tests are computationally simple and robust procedures, giving relevant results in the cases of slowly time-varying signals before and after the abrupt change. A sequential CUSUM test is developed and investigated using the innovation properties of the Kalman filter as the next stage of the target recognition system. The experiments show promising results in terms of estimation accuracy, probability of target detection and false alarm probability [3].

7.1 Processing algorithms

B-scan (or radargram) data contain the received GPR signals $u(n,k)$, where $n = 0,...N-1$ denotes the signal time samples and $k = 0,1,...,$ corresponds to the spatial position of the receive antenna. In the framework of state space representation, the radargram data are divided into P non-overlapping horizontal strips with depth m. Based on appropriate models, a set of P KFs is run in parallel on each data strip. Next, a set of P IMM filters is implemented, using the KFs' output parameters. The estimators work independently of each other, but exchange information for more reliable decision making. The goal of this combined KF-IMM algorithm is to detect and estimate target signals by fragmenting the data into target and background regions. The algorithm can be summarized by the following two steps.

Step I: A KF for background estimation and target detection. Using a "quiescent state model" [3] and the GPR measurements, the KF recursively produces a background state estimate with its associated state covariance. The properties of measurement residual (innovation) are employed to detect the targets and to adapt the filter to time-varying background parameters. The detection algorithm uses a χ^2 test and innovation-based statistic (normalized innovation squared (NIS) [3]) to detect the presence of possible targets. Under the hypothesis that the target is not present ("target-free" hypothesis), the NIS has a χ^2 distribution with m degrees of freedom. If NIS exceeds a threshold, determined by some level of significance, a procedure for target detection is initiated. If the "target-free" hypothesis is rejected for at least K_1 consecutive spatial positions (traces) and for at least K_2 of the total of P strips, the target is considered to be detected and its size is determined proportionally to the values of K_1 and K_2.

Noise identification. The correct knowledge of process and measurement noise statistics is a prerequisite for consistent KF operation. In general, the noise statistics are not known or partially known. The measurement error covariance can be estimated on line or selected a priori according to some rules or practical considerations. In the present realization it is determined through the variances of the radargram data, calculated along the traces for each strip. Due to soil inhomogeneities, the background signal slowly varies with trace numbers k and soil layers $p = 0, 1, ..., P$. The filters are adapted to this feature by using time-varying process noise characteristics. Two background adaptation procedures are implemented and experimented here. According to the first one, the noise covariance is updated recursively by a scaling factor. At each trace number k, the scaling factor is modified according to logic, managed by the NIS values and a set of thresholds, selected according to χ^2 distribution. A hybrid estimation technique for noise identification is also realized and experimented.

Step II: An IMM filter for target signal estimation and target-background separation. Using multiple models, accounting for different data regions (background or background plus target), the IMM algorithm has a potential for robust data segmentation. The IMM design configuration incorporates three models. The first one corresponds to the hypothesis that only a background is present. The process noise covariance is obtained by the background adaptation procedure, implemented at Step I. The target is modeled approximately as a stochastic bias with different magnitudes. The next two models match to the hypotheses that a bias is available in addition to the background. The input noise covariances are selected in such a way that the signal jumps caused by the targets can be detected and estimated. The IMM filter produces a combined state estimate as a weighted sum of model-matched estimates. The posterior model probabilities segment the radargram data into target and no-target areas. Comparison with the KF detection output can help for more reliable target identification.

A great variety of change detection tests are proposed and investigated in the statistical literature. Here, the CUSUM test is appended to the KF algorithm (discussed in Step I) in order to increase the probability of target detection.

A Kalman filter with CUSUM test for B-scan data segmentation. It is found in the investigations devoted to GPR data processing, that the residual energy obtained by removing background components from the GPR signal is more reliable for change detection. Based on this inference, we apply the CUSUM algorithm to the difference between the radargram data and background state estimate. If the evolution of this substacted signal is described by a simplified linear model, a KF can be implemented to yield the subtracted signal state estimate. If the target is not present, the measurement residual is zero mean, Gaussian and white. The signal anomalies, caused by the objects, alter the parameters of the Gaussian distribution. Thus, the task of target onset detection is transformed to the problem of change detection in the Gaussian distribution. Practically, the change point detection is approximately considered as the detection of changes in the mean value of the Gaussian distribution. A simple recursive two-sided CUSUM test is realized and experimented. It confirms the decisions of the KF-based detector and provides additional information for target onset. The CUSUM test is also applied to the transposed GPR image. Thus, the test approximately outlines the borders between the layers and can be used for the purposes of ground layers segmentation.

7.2 Experimental results

Algorithm performance is studied over a series of real radargrams, acquired by GSSI SIR Systems. The design parameters are chosen as follows: the depth of the horizontal strips is $m = 50$ and the number of strips is $P = 6$. The parameters of the KF detection algorithm are selected as follows: $K_1 = 15$ and $K_2 = 2$. The combined KF-IMM algorithm detects two objects in the image, presented in Figure 19a. Outputs of the KF and IMM detectors are presented in Figures 20a, b, respectively. Based on this information, the objects' positions are approximately determined, as can be seen in Figure 19b. The first object (a sewerage pipe under consideration) is positioned over three consecutive layers ($p = 2,3,4$) and its presence is confirmed by both detectors. Since the second object (positioned on strips $p = 0,1$) is a clutter object, additional information about the pipe size is needed to discard it. The KF-IMM algorithm detects two real objects in the image presented in Figure 21a. The CUSUM test, implemented after the KF procedure, validates the presence of the objects (Figure 21b).

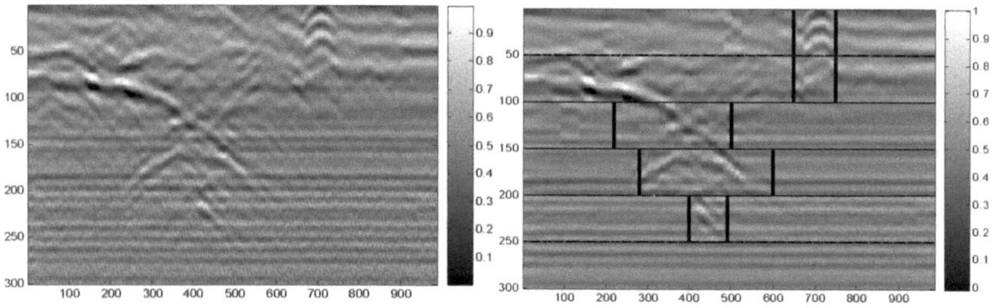

Fig. 19 **(a)** Original GPR image and **(b)** estimated objects' position by KF-IMM.

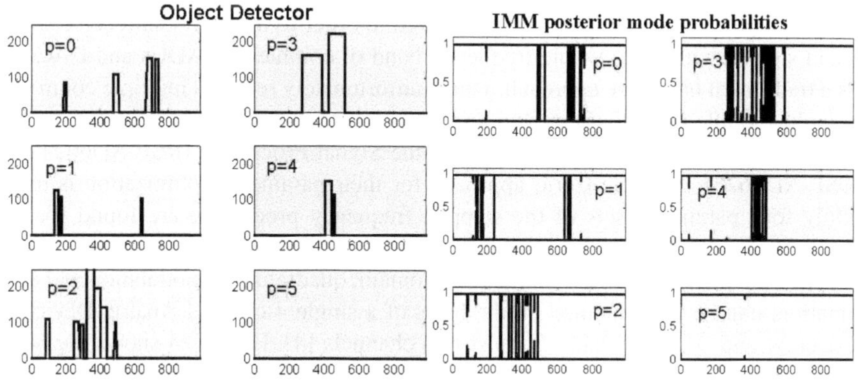

Fig. 20 **(a)** KF object detector and **(b)** IMM detector by KF-IMM.

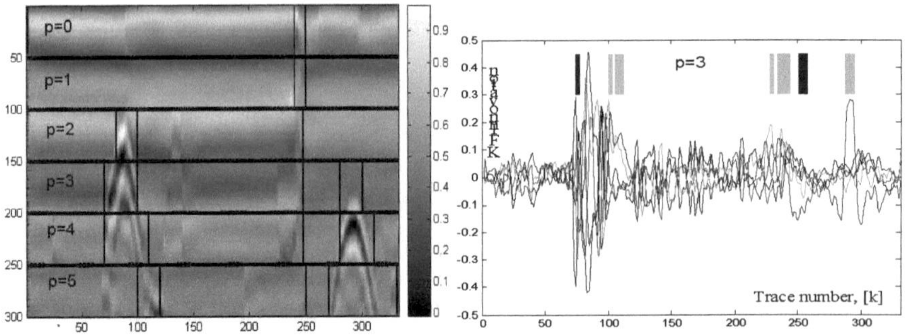

Fig. 21 (**a**) KF-IMM algorithm detects two pipes (**b**) CUSUM detector confirms the KF results at $p = 3$.

7.3 Implementation of low-frequency GPR signal algorithms using a conventional narrowband digital transmit-receiver systems

The earlier described stepped-frequency approach with time domain reconstruction reveals a possibility to obtain high-range resolution images with a conventional narrow-band transmitter/receiver digital system for GPR implementation. A survey was performed to test the maximal frequency bandwidth by using a traditional narrow-band transmit/receiver system composed of commercial signal processing devices: ADC(AD6644), DAC(AD9772), receiver, synthesizer, on-line signal processor running on a PC. In that way the whole ADSs and DACs bandwidth (60–100 MHz) can be filled up with a set of narrow-band Receiver/Tranceiver Signal Processors (RSP AD6624 and TSP AD6623 – 2.2 MHz) [4]. A multi-module and a multi-channel digital system composed of narrowband receivers (RSPs) and transmitters (TSPs) (Figures 22, 23) is developed in order to transmit and receive wide-band signals within the whole frequency band of commercial ADCs and DACs. It is a traditional hardware approach, which unfortunately requires multiple control of a multi-module digital system and, evidently, involves high financial expenses [4]. Considering the limitation parameters of the Signal Processors (RSP AD6624 and TSP AD6623), a Monte-Carlo approach for their parameter optimization is used. Only four parameter sets of the stepped frequency processing are found for the implementation in GPR. The theoretical calculations show that the role of a GPR stepped-frequency algorithm in the time domain, quadrature demodulation and decimation, can be implemented on the basis of a single 4-channel Analog Devices's AD6624 and AD6623 using at most two channels [4]. Figure 24 shows the MPS signal processor based on two DSP signal processors and conventional industrial box, which encapsulates the signal processing hardware and is used in MPS Ltd. for a GPR system.

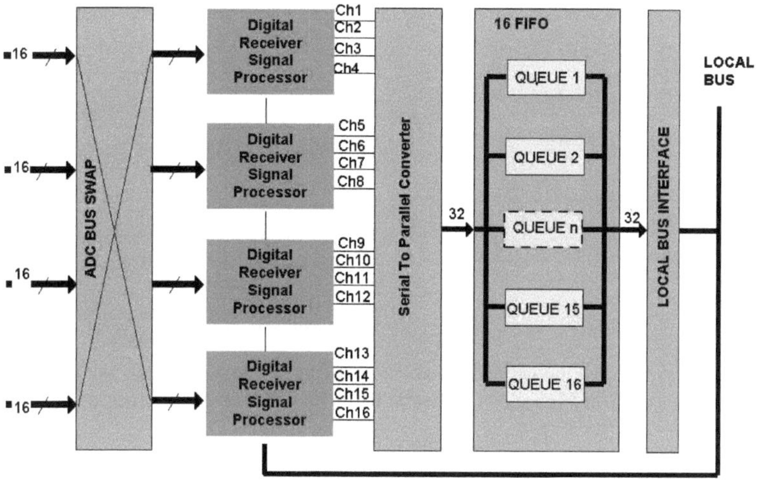

Fig. 22 Block diagram of the digital receiving system based on four-channel receive signal processors (RSP) AD6624.

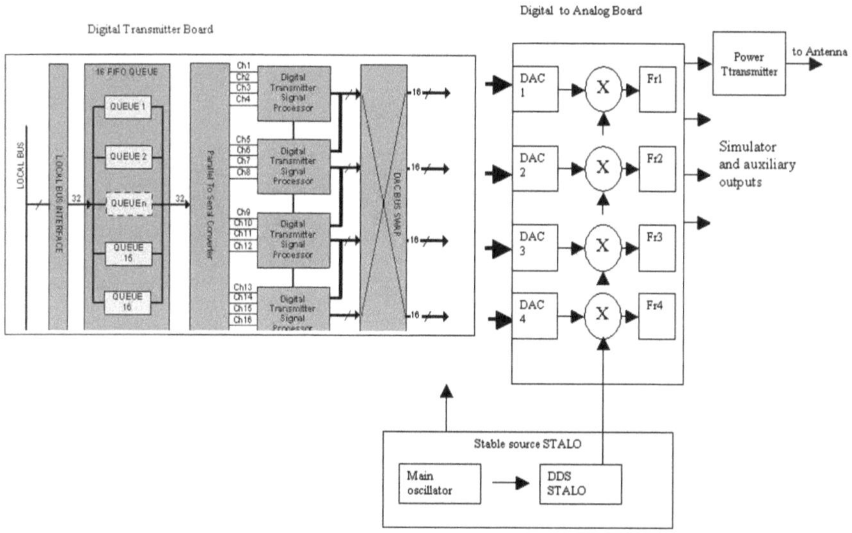

Fig. 23 Block diagram of the digital transmitting system based four-channel transmit signal processors (RSP) AD662.

7.4 FPGA implementation of a low-frequency GPR signal algorithm

This paragraph reveals a possibility for implementation of the stepped-frequency algorithm with time-domain reconstruction on a hardware platform in real time i – a step closer to real implementation. The hardware reconfigurable platform XUPV2P,

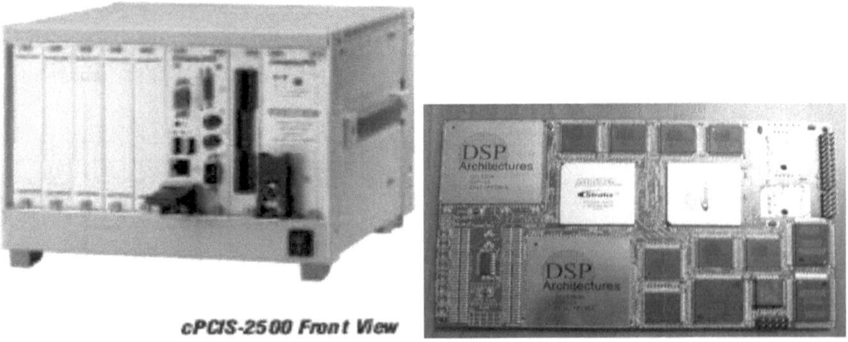

Fig. 24 The conventional industrial box, which encapsulates the signal processing hardware in a MPS Ltd. GPR system and the signal processor based on two DSP signal processors.

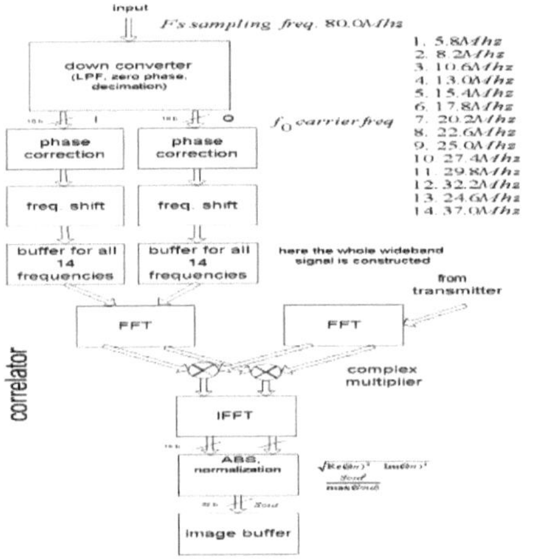

Fig. 25 Block diagram of the receiving structure).

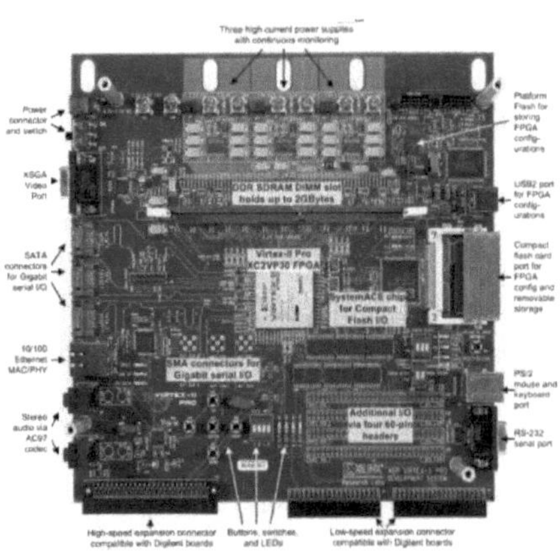

Fig. 26 XUP^{TM} Virtex-II Pro Development System.

based on VIRTEX II Pro technology is used (Figure 26). A block diagram of the algorithm suitable for a reconfigurable hardware implementation is presented in Figure 25. All computational kernels from the algorithm are designed as separate hardware blocks, and verified individually and stacked together. Considering the previously described stepped frequency algorithm [5], a block diagram of the receiver was made. It consists of: down conversion; interpolation; phase correction; frequency shifting; buffering the whole constructed signal; correlation; envelope detection, normalization and image storing. The block diagram of the receiver is shown in Figure 25. The number of transmitted pulses is $M = 14$. The sampling frequency of the signal is 80 MHz, the sampling frequency of the video signal – 2,25 MHz, minimal frequency carrier $f_{min} = 4.6$ MHz, maximal frequency carrier $f_{max} = 38.2$ MHz, the step in frequency is $d_f = 2.4$ MHz. The frequency sweep rate is $b = \Delta f / Tp$. Tp is the time duration of a narrow-band chirp, in our case –1.6 ms.

The down converter is implemented according to the specifications of the Digital Down Converter (DDC) V1.0 (Xilinx IP^{TM}) [9]. It encompasses the following processing: Quadrature Amplitude Demodulation, Low Pass Filter and decimation by 32. The input signal consists of 5376 samples. The tests were performed considering following parameters: System frequency rate: 100 Hz; Input signal frequency: 80 MHz; Input data width: 16 bits; Output data width: B8=18 bits; Spurious dynamic range of the digital synthesizer: 40 dB; Frequency resolution: 0.5 MHz; Phase angle: fixed; Output mixer width: 20 bits. The finite impulse response (FIR) filter is included in the synthesis of the digital down converter, the decimation rate is 16; the FIR filter length is 16 and the result precision is 12. The time domain reconstruction follows. It consists of phase correction and frequency shifting. Next a buffer for signal reconstruction (coherent summing) follows. It consists of a standard storage buffer based on memory block core [2]. Considering the signal processing principles the correlator consists of multiplication between received and transmitted signals in the frequency domain. Therefore we put two 64-point FFT transforms, one each for the received signal and for the transmitted signal. Next an IFFT is needed to come back to the time domain (Figure 25) [8]. An envelope detector and a signal normalization follow (Figure 25). The envelope detector consists of two multipliers and a sqrt block, which is based on the CORDIC v.3.0 architecture [9]. The transceiver consists of a look-up table, which contains the signals for transmitting. The number of signals is 14 and each of them consists of 128 samples. The transmitted signal is formed by the Tukey window before sending it to the Digital to Analogue Converter.

Simulation results. The simulation results are obtained via the *Modelsim*^{TM} simulator [9]. A VHDL code was written, and studied through the *Modelsim*^{TM} simulator. After the performed simulation, the constraints for real time imaging were defined. The correlation is performed for 108 μs. The total synthesis estimation parameters are: number of slices = 8937; BRAM = 30; Mult18 x 18 = 62. After the simulation performing the real time constraints took approximately 400 μs. According to the synthesis report, the usage of the processor was almost 75%.

8 Conclusions

The simulation results based on the Monte-Carlo approach enable us to conclude:

• The stepped-frequency GPR processing method for range profile formation in the time domain, operating at 4.6–38.2 MHz, generates synthetic range profiles with the resolution of 1–2 m by transmitting 14 narrowband chirps at each GPR position.

• A new convolution-based algorithm for simulation of stepped-frequency GPR images from multi-layered media can be successfully used for analysis and parameter optimization in stepped-frequency GPR.

• The basic algorithms for GPR signal and image processing such as mean filters (vertical working low-pass filter), running averages (horizontal working low-pass filter), stack traces (compression in horizontal direction); median filters (pulse jamming and speckle noise reduction), background removals (spatial high-pass filter) that make visible the shallow objects and gain adjustment algorithms (correct the attenuation losses and make visible the deep objects) are effective algorithms for GPR image processing.

• Applying CFAR filters and Hough filters to GPR image processing is a good decision.

• The designed multiple models Particle Filter (PF) for contour determination and segmentation in GPR images has shown encouraging results in terms of convergence and accuracy, at the cost of acceptable computational complexity.

• Applying Kalman filters to GPR data processing gives promising results in terms of estimation accuracy, probability of target detection and false alarm.

• The processor VIRTEX II Pro is suitable for implementation of the stepped-frequency processing algorithm for synthetic range profiling in the time domain.

Generally speaking, the approaches and algorithms considered in this chapter can be successfully used for UXO signal processing and multi-sensor (channel) UXO signal processing.

References

1. Alexiev K., "Object Localization in Ground Penetrating Radar Images", Comptes rendus de l'Academie Bulgare des Sciences, Tome 60, No 11, 2007, pp. 1237–1244.
2. Angelova D., P. Konstantinova, L. Mihaylova, "Contour Tracking in 2D Images Using Particle Filtering", Proceedings of the IRS'2007, 5–7 September, Cologne, Germany, pp. 515–519.
3. Angelova D., "A Bayesian Algorithm for Object Detection in GPR Data", Proceedings of the IRS'2008, 19-23 May, Wroclaw, Poland, pp. 301–304.
4. Babalov S., V. Vassilev, P. Daskalov, Chr. Kabakchiev, "DPCR-Development Platform for Digital Communications and Radars", Proceedings of the IRS'2004, Warsaw, Poland, pp. 131–136.
5. Behar V, Chr. Kabakchiev, "Stepped-Frequency Processing in Low Frequency GPR", Proceedings of the IRS'2007, 5–7 September, Cologne, Germany, pp. 635–639.
6. Behar V, Chr. Kabakchiev, "A Simple Algorithm for Simulation of Stepped-Frequency GPR Images of Multi-Layered Media", Proceedings of the IRS'2008, 19–23 May, Wroclaw, Poland, pp. 289–292.
7. Garvanov I., V. Behar, Chr. Kabakchiev, "CFAR Processors in Pulse Jamming", Proceedings of Conference "Numerical Methods and Applications" – NMA02, Borovets, Bulgaria, LNCS 2542, 2003, pp. 291–299.
8. Daniels D., Ground Penetrating Radar, 2nd edition, The Institution of Electrical Engineers, London, 2004.
9. Kyovtorov V., Chr. Kabakchiev, V. Behar, G. Kuzmanov, I. Garvanov, L. Doukovska, "FPGA Implementation of Low-Frequency GPR signal algorithm using Frequency Stepped Chirp Signals in the time domain" Proceedings of the IRS'2008, 19–23 May, Wroclaw, Poland, pp. 297–300.
10. Vassileva B., Chr. Kabakchiev, "Singular Spectrum Analysis (SSA) for GPR Data Processing", Proceedings of the IRS'2008, 19–23 May, Wroclaw, Poland, pp. 293–296.

Advanced Multifunctional Sensor Systems

Lena Klasén*

Abstract This work addresses the role of multifunctional sensor systems in defence and security applications. The challenging topic of imaging sensors and their use in object detection is explored. We give a brief introduction to selected sensors operating at various wavelength bands in the electromagnetic spectra. Focus here is on sensors generating time or range resolved data and spectral information. The sensors presented here are imaging laser radar, multi- and hyper-spectral sensors and radar systems. For each of these imaging systems, we present and discuss analysis and processing of the multidimensional (n-dimensional) data obtained from these sensors. Moreover, we will discuss the benefits of using collaborative sensors, based on results from several ongoing Swedish research projects aiming to provide end-users of such advanced sensor systems with new and enhanced capabilities. Major applications of this kind of systems are found in the areas of surveillance and situation awareness, where the complementary information provided by the imaging systems proves useful for enhanced systems capacity. Typical capabilities that we are striving for are, e.g., robust identification of objects being possible threats on a sub-pixel basis from spectral data, or penetrating obscurant such as vegetation or certain building construction materials. Hereby we provide building blocks for solutions to, e.g., detecting unexploded ammunition or mines and identification of suspicious behavior of persons. Furthermore, examples of detection, recognition, identification or understanding of small, extended and complex objects, such as humans, will be included throughout the chapter. We conclude with some remarks on the use of imaging sensors and collaborative sensor systems in security and surveillance.

Keywords: Full 3-D imaging, gated viewing, image analysis, image processing, imaging sensors, laser radar, multi- and hyper-spectral sensors, multi-sensor systems, radar systems, multidimensional data, synthetic aperture radar

*Information Coding, Department of Electrical Engineering,
Linköping University, SE-581 83 Linköping, Sweden,
e-mail: lena@orlunda.e.se

J. Byrnes (ed.), *Unexploded Ordnance Detection and Mitigation*,

1 Background Motivation

Safety and security applications bring several challenging problems at hand. This especially becomes apparent when facing the complex task of surveillance in order to detect and identify any possible threat. Such tasks can, for example, be to detect a person applying an improvised explosive device (IED) on a bus whilst he is being recorded by a surveillance camera, or to identify a person placing out surface laid mines in a remote and desert area without any surveillance capabilities. Thus, both suspicious objects and abnormal behavior of humans are of interest. To accomplish capabilities to handle such tasks, we truly need a variety of tools, e.g. spanning from surveying large areas to providing evidence to be used in the criminal justice system.

The importance of images in security and safety applications needs not to be questioned. Video cameras producing streams of image sequences usually builds up the surveillance systems of today. But many additional problems arise from the surveillance system technologies in use. The most commonly used short-range, passive surveillance systems are not optimal to capture the events and objects in a scene for further analysis and processing, but these systems will still be in use for many more years. Reviewing recordings from these systems e.g. surveillance video, is a time demanding task. It is also very difficult to detect all objects by the human visual system. Another major problem that the existing surveillance techniques provide, and that seriously limits the possibilities of identification in the criminal justice system, is the lack of images rather than lack of analysis methods [10, 11, 24]. The task in a forensic situation, for example, is often to handle situations where the image sequence comes from a single camera, or multiplexed cameras where the image streams are recorded on the same media. Furthermore, camera parameters and the characteristics of the imaging devices and recording conditions are usually unknown or limited, as the circumstances seldom provide calibration procedures to be performed. Moreover, there are many examples of applications where human assisted analysis is no alternative and there is a need for automatic or semi-automatic processing. Hence, we foresee a lot of challenging issues if we want to be able to detect and identify all kinds of events and objects that could cause a threatening situation.

The scientific areas of sensor technique and sensor data processing have evolved significantly. By using sophisticated and existing sensor systems and algorithms, several problems of conventional surveillance systems can be solved. Nowadays there are a large number of new sensors and image processing techniques for tracking and analyzing moving persons or detecting small objects, see e.g. [25]. We introduce somewhat more unconventional sensors for means to present complicated information in a way that can be easily, correctly and quickly understood. Complementary sensors addressed here are gated viewing, full 3-D imaging laser radar, multi- and hyperspectral sensor and radar systems. These imaging systems brings new capabilities such as to penetrate vegetation, clothing, building materials, and can be used despite of poor weather conditions or at long ranges.

But, the nature of the threats against us in our society constantly increases in complexity. Consequently, there are several situations to be handled and that need

Fig. 1 Example of multisensors for urban monitoring, [4,35].

even more complicated sensor systems. A possibility to provide better capability is to make use of the additional data provided by complementary imaging sensors. So, in addition to the individual sensors and algorithms, the combination of passive and active sensors are used. This brings flexibility and the capability to enhance our possibility to "see" the threats that we usually are unaware of or believe are unlikely to occur. Not only do we need the capability to "see" the threats, we can also do so without being "seen" ourselves, illustrated in Figure 1.

The work addressed in this chapter emanates from several ongoing activities at the Defence Research Agency FOI on the subject of automatic target identification for command and control in a netcentric defence. The driving force for the research activities at FOI is strongly motivated by requirements that emanates from defence applications and law enforcement. Although, the main applications areas of interest in our research are found in security and safety, there are many other possibilities. Hence, we give some examples of successful imaging systems that in combination with image processing and analysis techniques provide means to e.g. to improve surveillance capacity.

Finally, some concluding remarks on the use of imaging sensors for applications in security and surveillance rounds off the chapter.

2 Imaging Sensors

We have got the sensors – but what can they accomplish? What we usually strive for is recommendations and specifications for future sensors systems, and we want the computer do the "dirty work" for us in the process of identifying objects, events and phenomena in image sequences by the use of image analysis and image processing techniques. These methods provides a complement to the human visual system so that we can use the visual information in a better way.

A key issue is provide good quality data – rather than trying to enhance and analyze poor data. This does not necessarily mean that the image quality needs to be of good visual quality. On the contrary, data collected might not make sense when presented to an operator but are very useful in an automatic image analysis process. The importance, though, is that the data quality is of high quality. This, in turn requires knowledge about the sensor in use, regardless being conventional or being newly developed. Furthermore, we need knowledge about the problem at hand, the depicted scene and the objects of interest. Thus, a useful rule of thumb is to get it right from the start.

The focus here is on laser radar systems (in Section 3), multi- and hyper spectral systems (in Section 4) and radar systems (in Section 5) that are sensors generating complementary time resolved or range data and spectral information, in contrast to CCD- and IR-cameras that passively images a broad spectra of the visual or infrared range. After a brief introduction on each of those imaging sensors we present methodologies and applications by the use of image processing and analysis techniques. One important computer vision task is the understanding of complicated structures representing threats, crimes or other events. Here a major part of the problem originates from the difficulty of understanding and estimating data describing the events taking place in the imagery.

The main objective for using advanced sensor systems is to provide descriptors related to the problem of understanding complex objects from images, such as mines and vehicles (in Section 6) or humans (in Section 7). These descriptors can, for example, be used in a recognition or an identification process. Detection, recognition, identification or understanding of small (covered by a few pixels or sub-pixel sized), extended (covered by many pixels) and complex objects from images provides us with a variety of difficult but challenging problems. Here we use the term *complex* to denote an object that can simultaneously move, articulate and deform, while *detection* is referred to as the level at which object are distinguished from background and other objects of no interest, i.e., clutter objects such as trees, rocks, or image processing artifacts. *Recognition* is used to distinguishing an object class from all other object classes and *identification* is used to distinguishing an object from all other objects.

For any method, either supporting an operator or a fully autonomous method, the whole chain must be taken into consideration, from the sensor itself to what the sensor can comprehend. This includes sensor technology, modeling and simulation of the sensor, signal- and image processing of the sensor data, evaluation and validation of our models and algorithms e.g. by experiments and field trials with well known ground truth data to finally obtain the desired data. The outcome can e.g. be further used for data- and information fusion at higher system levels, such as alerting an operator of the position of a detected suspicious object that, e.g., could be a surface laid mine.

To investigate the performance bounds to reveal the role of the system parameters and benefits of sensor performance, we model and simulate each of the individual sensors. Modeling requires knowledge of the atmosphere, object and background characteristics and there is a need for characterization at the proper wavelengths. But, if we get it right, we can use our models to simulate larger scale sensor system, including different types events, scenarios, object types, sensor types and data processing algorithms. Hereby we have a good platform to analyze the performance of systems of higher levels, as exemplified in Section 6.

3 Laser Imaging

Laser imaging range from laser illumination systems enabling active spectral imaging to range gated and full 3-D imaging systems. Coherent laser radars will also provide Doppler and vibration information. We will concentrate on 3-D imaging systems. Real time 3-D sensing is a reality and can be achieved by stereo vision, structured light, by the various techniques for estimating depth information or by range imaging. Laser radar, in contrast to passive imaging systems, provides both intensity and range information, see e.g. [26, 27, 34, 41, 47]. The 3-D image can be derived from a few range gated images or from each pixel directly coded in range and intensity using a focal plane array or a scanning system with one or few detector elements. Each pixel can generate multiple range values. The range information provides several advantages and has impact on many military and also civilian applications. For example, 3-D imaging laser radars have the ability to penetrate scene elements such as vegetation, windows or camouflage nets. The latter is illustrated in Figure 2.

3-D imaging systems are predicted to provide the capability of high resolution 3-D imaging at long ranges at full video rate, supporting a broad range of possible applications.

3.1 Laser radar systems

The majority of the early laser radar systems are based on mechanically scanning the laser beam to cover a volume. The 3-D "image" (or point cloud) is then built up by successive scans where each laser pulse (or laser shot) will return intensity and multiple range values corresponding to the different scene elements within the laser beam footprint. In many systems, the full return waveform is captured for each laser shot and stored for further processing. Other systems capture parts of the returning waveform (e.g. first or last echo). The range information provides several advantages when compared to conventional passive imaging systems such as CCD and infrared (IR) cameras. The current development of laser radars, from scanning systems to fully 3-D imaging systems, provide the capability of high resolution 3-D imaging at

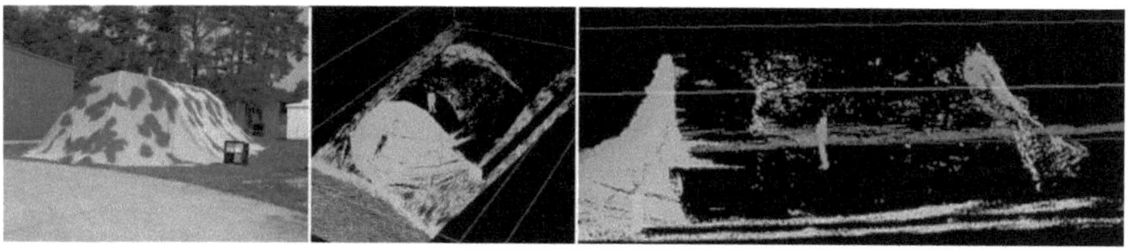

Fig. 2 A camouflage net scanned by a laser radar system (*rightmost pictures*), revealing a person inside.

long ranges with cm resolution at high video rate. For example, 3-D imaging laser radars have the ability to penetrate scene elements such as vegetation and windows.

The range resolution and the spatial resolution (cross range) depends on the properties of the receiver and are important in system performance measurements. The received laser effect can be described by the laser radar equation as

$$P_m = P_s \eta_s \frac{A_\Delta}{\pi \left(\Phi R/2\right)^2} \frac{A_m}{R^2} \eta_m t_A^2 , \tag{1}$$

where P_m is the received laser power $[W]$, P_s laser power $[W]$, η_s transmission of transmitter optics, η_m transmission of receiver optics, Φ laser beam divergence $[rad]$, R distance transmitter-target $[m]$, A_Δ object effective area $\left[\frac{m^2}{sr}\right]$, A_m area of receiver $\left[\frac{m^2}{sr}\right]$ and t_A represents the atmosphere transmission. The range resolution varies with different types od laser radar sensors. The spatial resolution depends on the spatial resolution of the imaging sensor, but also on the atmospherical conditions and the distance to the target.

There are several concepts for scanner-less 3-D laser radar systems. The technology which seems to draw the largest attention in 3-D imaging for military applications is 3-D sensing flash imaging FPAs, which here is in focus. The remaining techniques are detailed in [34, 41]. A laser flood illuminates a target area with a relatively short pulse (1–10 ns), [45, 46]. The time of flight of the pulse is measured in a per pixel fashion. The position of the detecting pixel yields the angular position of the object element, and the time of flight yields the range. Hence, with a single laser shot, the complete 3-D image of an object is captured.

3.2 Modeling and simulation

To model a scene we need to know the characteristics of the system itself and also gain knowledge about the various scene elements. This especially holds for any object we want to detect. For a long time, theories for laser beam propagation and reflection have been developed and adjusted. Many of these theories have been useful to simulate and evaluate parts of a complex laser radar system, but modeling of a complete system was not possible in the early stage. The laser radar technology has become more expensive and a system model was desired to reduce the cost of laser system development and to expand the amount of training data for signal processing algorithms.

The simulation of the reflected waveform from a laser radar system is based on the ray-tracing principle and, inspired by [15], divided into four sub problems. Each sub system contains several parameters controlling the simulation. The abstraction level of the simulation is often a trade-off between complexity and efficiency. Too complicated models would require parameters not understandable by the average user and too simple models would not simulate enough conditions to produce accurate results. The *laser source* is specified by the wavelength and the temporal and spatial distribution of the light intensity. The *atmosphere model* is simplified and controlled only by the aerial attenuation and the turbulence constant, C_n^2, as a function of the altitude. The *target* is a scenario of polygon models and their corresponding reflection properties at the current laser wavelength. Finally, the *receiver* is modeled electronically as a standard receiver from [15]. Since many of these subproblems contain complex analytic mathematical expressions, especially when combined, we choose to make the calculations discrete, both in the temporal and spatial dimension. Another problem is the trade-off between the computational speed and accuracy. Based on our experience, a reasonable resolution in the spatial domain lies about 0.1 mrad, and in the temporal domain 0.1–1 ns.

The laser radar system model by FOI combines the theories for laser propagation and reflection with the geometrical properties of an object and the receiver characteristics such as noise and bandwidth. Our simulation model has been further developed over the years, through gated viewing (GV) systems and aerial scanning laser radar, up to the forthcoming 3-D focal plane arrays (3-D FPAs). There are several publications by FOI on this subject, see for example [9, 13, 19, 37–40, 43, 44, 48]. Another example is [42] also described in [25], that includes atmosphere modeling in terms of e.g. aerosols and turbulence, image processing, object recognition and estimating performance of different gated viewing (range imaging) system concepts. Moreover, we addressed the object/background contrasts of the reflectance value at eye safe wavelengths to investigate the recognition probabilities in cluttered backgrounds. An advantage with laser systems is the ability to penetrate vegetation. A tool is also developed at FOI for the purpose of estimating the laser returns as a function of distance to the sender/receiver, e.g. useful for detection of hidden vehicles as shown in Figure 3.

3.3 *Object recognition*

The development of algorithms at FOI for object recognition includes methods that aim to support an operator in the target identification task and also more autonomous algorithms. This work is described in [7,8,20,26,27,41–43]. To obtain point clouds at long ranges, data achieved by an experimental GV system [25, 42] out to 14 km was used, in combination with a method for reconstruction of the surface structure [7]. This system, however, initially operated at 532 nm that is not eye safe. Thus, the simulation model was essential to estimate the performance of a system operating at an eye safe wavelength, which now is built. Examples of range gated imaging at 1.5 m is found in [47].

A major advantage is that a 3-D cloud often can be directly viewed without any processing. Furthermore, by visually searching a point cloud by varying the viewing distance and angle, objects that are not immediately obvious to the human eye can become easy to detect and recognize, see Figure 4. Fusing data from multiple viewing angles enhances this possibility which becomes an effective method to reveal hidden targets.

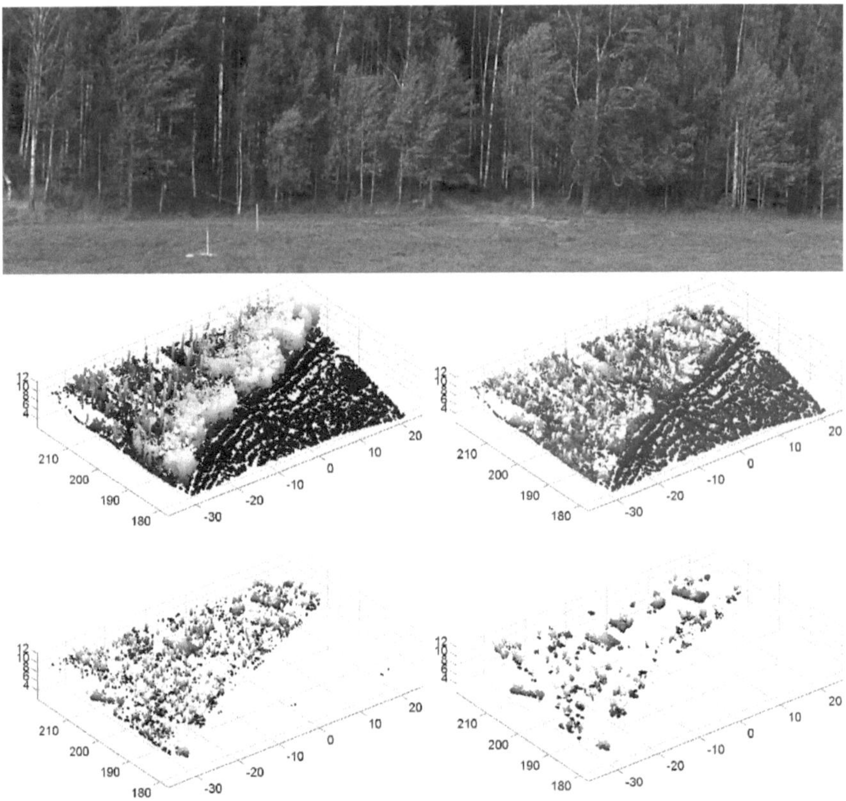

Fig. 3 The scene for the laser measurement (*upper row*). The raw data from the laser radar system (*middle row to the left*) and the processed bare earth data (*middle row to the right*). All data less than 0.3 m above estimated ground (*bottom row to the left*) and finally the tree streams and noise clutter has been removed, reveling the vehicles (*bottom row to the right*).

Fig. 4 To the *left* is a laser scanned terrain area viewed from a frontal view. In the *middle* is a close up of the point cloud viewed at a different aspect angle to better reveal the target. To the *right* is a 3-D model of the vehicle, also created from scanned laser radar data of high resolution.

Laser radars also have the ability to penetrate Venetian blinds provided there are tiny openings, and thus have the ability to see into buildings. A method for matching 3-D sensor data with object models of similar resolution is detailed in [6]. For GV data, a combination of a method for 3-D reconstruction and a 3-D range template matching algorithm is developed.

The current problem tackled is methods on extracting object points based on detection from hyperspectral data. In parallel, there are ongoing works addressing methods based on multi sensor approaches for detection of hidden objects, surface laid mines [49] where the objects can be in vegetation [1, 3, 14] and urban environments [4, 5], further described in Section 6. The exchange of information between different sensors, such as CCD, IR, SAR, spectral and laser radar, can provide solutions to problems that are very difficult to solve by using raw data from one single sensor only. Consequently, our work on 3-D imaging sensors for object recognition is incorporated in several multi-sensor approaches.

4 Multi- and Hyperspectral Imaging

Multi- and hyperspectral electro-optical sensors sample the incoming light at several (multispectral sensors) or many (hyperspectral sensors) different wavelength bands, see e.g. [2, 12]. Compared to a consumer camera that, typically, uses three wavelength bands corresponding to the red, green and blue colors, hyperspectral sensors sample the scene in a large number of wavelength (or spectral) bands, often several hundred. Images providing spectral information give the possibility to detect and recognize objects from the spectral signatures of the object and the background, without regarding spatial patterns. The methods used for object detection differ strongly depending on the characteristics of the used sensor and of the expected object and its surrounding background. For example, pattern recognition techniques are used for detection, classification, and recognition of extended objects (covering many pixels). Multi- or hyperspectral images sequences provides means to detect objects of sub-pixel size as well. Although, it is important so specify the system performance from the situation at hand e.g. from matching the object- and background signatures to the spectral bands of the camera (bandwidth, number of bands etc.). Moreover, the spectral bands can be beyond the visible range, i.e. in the infrared domain, which opens up a variety of new applications [12].

Here we briefly describes methods for detecting extended or small targets in multispectral images. In this context we limit the discussion to treat spectral information only, i.e., spatial correlations are not considered. There are two main types of object detection methods. Object detection is, in the first context, is about finding pixels whose spectral signature do not correspond to some model of the background spectral signature but do correspond to a object model, if available. The spectral signature of the target is not assumed to be known, instead spectral anomalies with respect to the background are searched for. The process of detecting unknown targets is called anomaly detection. The second case is when a target model is available, which we call signature-based object detection.

4.1 Anomaly detection

Anomaly detection, detailed in [2] provides new capabilities in object detection where the aim is to detect previously unknown objects as shown in Figure 5.

Anomaly detection is the case when we do not know the spectral signature of the target, and we want to find pixels that significantly differs from the background. We use a background model B, a distance measure $d(\cdot)$, and a threshold t. We regard a pixel x as an anomaly if $d(B,\mathbf{x}) > t$. Thus, a model for the background signature is needed, as well as an update scheme, i.e., a degree of locality of the model. For example, we could use a local model (estimating the background signature from a local neighborhood only), a global model (using the entire image), or a combination. Then, to measure the distance from each pixel signature to the background model, we need a distance measure. The choice of distance measure is restricted, or even determined, by the model used for the background and thus the assumptions about background spectral distribution. Finally, we need to set the threshold t.

A signature-based algorithm for target detection searches for pixels that are similar to a target probe. The target probe is a model of a certain target signature T, i.e., the spectral signature of the target or target class is known. Basically, we measure the distance from a pixel signature to the target model and to the background model, and choose the smaller. That is, we can classify pixel x as a target pixel if $d(T,\mathbf{x}) < d(B,\mathbf{x})$.

The detection methods require spatial and spectral models for targets and background. The spatial model is used to define background areas to classify any object areas. The spectral modeling is to represent the properties of the object and background classes in use, There are several possible methods, with the common goal to measure a distance from an object class to the modeled background class in order to classify in what category the pixels belongs to.

Combining anomaly detection with signature based detection can improve detection performance. Moreover is the detection useful as input e.g. to a 3-D laser radar for identification.

Fig. 5 Detection of military vehicles by a hyperspectral camera. The targets are in the open and hidden in the terrain and the targets are detected by the signal processing algorithm applied to the data. One of the vehicles, which is under camouflage, is enlarged.

5 Imaging Radar Systems

Among the many possible radar systems available and found in the literature see e.g. [50], we will address only a few; SAR and imaging radar systems for penetration of certain materials.

5.1 Resolution in a radar system

The concept of resolution of a radar system is usually defined as the width of the impulse response when the signal energy has decreased to half. The impulse response can be divided into two dimensions, range and azimuth. The range resolution is determined by the transmitted bandwidth (B) as $X_r = \frac{c}{2B}$ where c is the speed of light and $B = \frac{1}{T}$ where T is the length of the transmitted radar pulse; i.e. a short pulse has a large bandwidth equalling a small resolution cell in range. In reality, the bandwidth is often created by some kind of frequency modulation of the transmitted pulse in order to increase the mean power in the system. The return signal is then compressed in an inverse filter in the system receiver. In azimuth, the resolution is determined by the attributes of the antenna. A radiation beam is created with an opening angle depending on the antenna size vs. the wavelength. The opening angle of the beam will be $\varphi = \frac{0.88\lambda}{d}$ where λ is the wave-length and d is the aperture of the antenna. This implies that the azimuth resolution (measured as the distance in azimuth between two point targets, which can be resolved by traditional radar) will depend on the range between the radar and the target area, i.e. the azimuth resolution performance will decrease with range. For most imaging applications the antenna will soon become impractically large when trying to keep a good image resolution at great distances.

5.2 Synthetic Aperture Radar (SAR)

The SAR-technology, [50] is a signal processing method for increasing the azimuth resolution of a radar system. The first patent was issued already 1951 for Carl A. Wiley at the Good-year Corporation in the USA but was not widely used until the modern digital technology became available. SAR has the fantastic characteristics of being like a camera featuring all-weather capabilities and range independent image resolution.

With SAR-technology the azimuth resolution is generated in the signal processing and will be independent of the range from the sensor to the target. The trick is to use a small antenna placed on a moving platform, e.g. an aircraft. The small antenna will generate a wide beam of radar illumination. The beam must cover the complete area of interest, and the signal is received in amplitude and phase during the fly-by of the platform. By using different mathematical methods, e.g. Fourier methods, the phase history (Doppler shift) of the signal can be analyzed and a synthetic antenna aperture equal to half the length of the flight track, the synthetic aperture L, can be generated.

FOI has, since many years, a diverse research program for low frequency radar development for ground and airspace supervision. We have developed the foliage penetration CARABAS system operating in the VHF band (20–90 MHz).

The system is a unique tool for providing information on targets concealed under foliage. It combines unprecedented wide area stationary target detection capacity with the capability of penetration of vegetation and camouflage. The VHF band used, allows target detection at a low surface resolution enabling the large surveillance capacity. The new LORA system, operating in the UHF band (200–800 MHz), is also capable of moving target detection and will be used as a generic research tool.

The research at FOI on SAR provides methods for generation of high resolution radar images. In fact the resolution on ground is independent of the distance from the radar to the target area. In urban environment there is the problem of detecting small objects due to the very strong backscattered signal from buildings and other large structures. The target signal will be obscured by the background clutter in the image. By separating the transmitter and receiver in the radar system and hence creating a bi-static situation this problem can be reduced. Furthermore, by placing receivers on the ground, receiving opportunities are opened for "tomographic" 3-D imaging of the internal structures of the buildings. This is a relatively new field of research that in all probability will enhance the situation awareness in future urban surveillance.

Among the many publications available, we also recommmend [16, 17, 22, 30, 33, 51].

5.3 Radar for penetration of materials

Another very promising upcoming technology [23] is the ability to penetrate certain materials, such as clothes and construction materials, with radar. This capability lets us penetrate materials that we cannot visually see through with the human eye. This opens up possibilities in military situations but also in law enforcement and rescue situations.

Researchers at FOI have developed imaging radar systems, capable of delivering through-the wall measurements of a person. Figures 6 and 7 shows the radar images when measuring a person through three different inner wall types at 94 GHz.

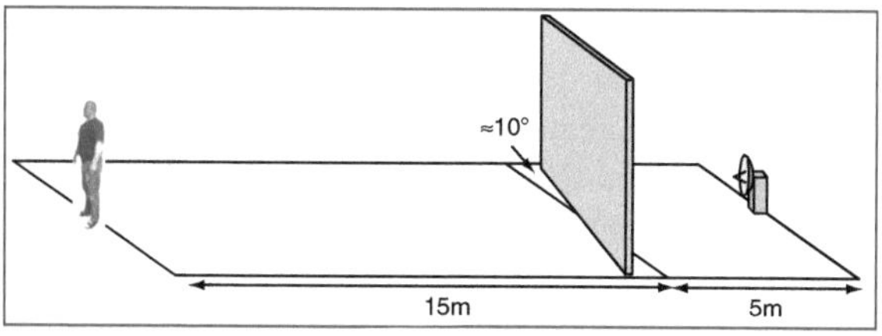

Fig. 6 Localization of a person behind a wall by measurements carried out at FOI with an in-house developed imaging radar system.

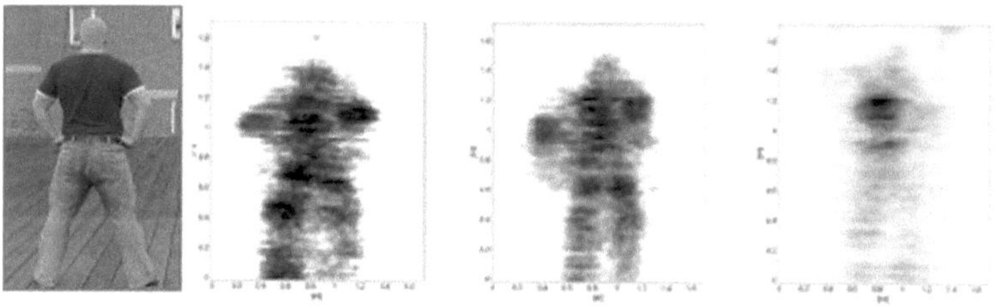

Fig. 7 Radar images when measuring a person through three different inner wall types at 94 GHz are shown. A 12.5 mm thick plasterboard (*left*). Two 12.5 mm thick plasterboards separated by a 45 mm air slit (*middle*). A 12.5 mm thick chipboard (*right*).

6 Multisensor Approaches

As mentioned, the complex task of surveillance to detect and identify any possible threats brings the need for multifunction and multisensor systems to have the flexibility to meet the environmental subsystems at hand, see e.g. [1, 3, 28, 29].

6.1 Detection of surface laid mines

Methods for detecting surface laid mines on gravel roads are being investigated in a national research program at FOI. Among other basic issues, is the idea in [8] that human-made objects are expected to appear more structured than surrounding background clutter. Another key issue is to base any detection method on the phenomenology of the surface laid mines, striving for to select the right combination of sensors to provide optimal data as input to the detection algorithms.

Using data from laser radar has shown some promising results [21]. This method basically relies on a fusion of intensity and hight features obtained from laser radar data. Although intensity usually is useful as a feature for separating mines from background data, is will not be enough for desired system performance. A gravel road is a relatively flat surface and hence the height above the ground plane is a feature that improves the separation of mines from the road. However, for more complex environments, such as forest, the height feature worsens the separation of the mine from the background, which motivates a search for other features. In [49, 53], 3-D data received from the laser radar is used to extract features relevant for mine detection in vegetation. These features varies with the nature of the vegetation. By involving data from an infrared (IR) sensor, synchronized with the 3-D laser radar data, additional features can be extracted. These features are evaluated to determine what combination that gives robust anomaly detection. A method based on Gaussian mixtures is proposed. The method tackles some of the difficulties with Gaussian mixtures, e.g., the selection of number of initial components, the selection of a good description of the data set, and the selection of which features that are relevant for a good description of the current data set. The method was evaluated with laser radar data and IR data from real scenes.

6.2 Urban monitoring

In recent years, significant research related to tasks in an urban environment has started, see e.g. [35]. Many sensor systems are, for instance, able to handle detection, but for classification and especially for identification, there are still many unanswered questions. Additional research is needed e.g. in sensor technology, data processing and information fusion. Consequently, there is a broad spectrum of challenging research topics. Here we present some resent examples from the ongoing research activities at the Swedish Defence Research Agency FOI that can contribute to the Swedish Armed Force's ability to operate in the urban terrain.

It is of importance to handle monitoring of the urban environment in a broad perspective, spanning from the everyday civilian surveillance situation to a full-scale war, bearing in mind that the border between law enforcement and military operations is somewhat fuzzy especially when considering terrorist activities. During military operations, surveillance systems are useful for detection of trespassing, tactical decision support, training and documentation to mention a few. The demand for fast and reliable information sets high requirements for data processing, spanning from fully automatic processes to visualization of data to support an operator. In the end, decision-makers from low rank soldiers to high commanders must be given the support required for different situations. Visual surveillance systems already exist and are increasingly common in our society today. We can hardly take a walk in the center of a modern city without being recorded by several surveillance cameras, even less so inside shops. The rising numbers of surveillance sensors, although being very useful, also introduces problems. Problems arise on how to get an overview of the surveillance data, and how to preserve the personal integrity of the people being watched by the sensors.

Overview is one of the greatest obstacles in a surveillance system with a large number of sensors. The most common type of surveillance sensor is video camera networks or other types of cameras. Images and video give rich information about the world, but are difficult to interpret automatically. Therefore, it is most common that the images are interpreted by a human operator of the surveillance system. The human operator of a surveillance system is not seldomly showered with a large number of images of micro events that are difficult to position in space and in time. However, there are upcoming technologies to handle this. In 2004 FOI defined a number of urban surveillance situations. The purpose was to exploit an approach to create a framework for surveillance of urban areas. From these scenarios, we built up a concept for future large area monitoring where situation awareness is critical. Subsequently, on May 13 2004, we launched a field campaign in an urban environment, "The Norrkping riot". A number of our different sensors, being both off the shelf products and experimental set ups, provided useful data. The sensor data were fused by projecting them onto a 3-D model of the area of interest. By combining technologies, methods for data analysis and visualization we introduced new concepts for surveillance in an urban environment, and suggestions on how to realize these concepts using technology developed at FOI.

This concept is built around a 3-D model of the urban area to be surveyed. In this virtual environment, the cameras from the real environment are represented by projectors that project the camera views onto the 3-D model. This approach has several advantages. The context in which each camera is placed is visualized and becomes obvious. The spatial relation between different cameras hereby becomes obvious. Imagery from several cameras can be studied simultaneously, and an overview of the entire area is easily acquired. Even if the idea is not completely new, it is not widely used, and it improves the general situation awareness tremendously. In the 3-D model, all available sensor data can be visualized in such a way that their context and mutual relations are immediately visible. We have developed a research platform for visualization of the surveyed area. The platform is a visualization tool built at FOI on open source software that visualizes 3-D models and projects textures from input video, and is controlled using either a user interface or by commands over a network.

The actual key to making this into an operational system is that the 3-D model can be automatically generated, [5].

The key issue with the multiple heterogeneous sensors concept is to make use of the benefit brought by new capabilities by new and cooperating sensor systems. Besides conventional acoustic, seismic, electro-optical and infrared sensors, this can e.g. include range gated imaging, full 3-D imaging laser radar sensors, multispectral imaging, mm-wave imaging or the use of low frequency radars in urban environment. Assume, for example, that we have a sensor that can localize gunfire. The position of the sniper can then immediately be marked in the 3-D model, which gives several interesting possibilities. If the shooter is within the field of view of a camera, he is pointed out by marking the location of the shot in the 3-D model. The shooter can then be tracked forwards and backwards in time, searching for pictures suitable for identification and also warn others in the area. Regardless if the shooter is within the field of view of a camera or not, the shooter's field of view can be marked in the 3-D model. The marked area is a risk area that should be avoided and warned for. The same functionality can be used in a deployment scenario, aiding the placement of sensors, snipers and people. Other examples are passage detection sensors, sensors that track or classify vehicles, sensors that detect suspicious events or behavior.

6.3 Sensor networks for detecting humans

A network of acoustic sensor nodes can also be used to locate gunshots, and also track sound sources. For example, technology used in military applications for tracking ground vehicles in terrain can be modified to fit in with an urban scenario. The output of the sensor network is synchronized with all other information in the system and user specified or general areas can be displayed in the 3-D model with a classification tag to indicate the type of event, see [4].

Passage detection sensors can be used for determining when people and/or vehicles enter a surveyed area and the other sensors should be activated. Several types of passage detectors are commercially available. Ground alarms for example, that react on pressure, i.e., when someone walks on the sensor (that consequently should be placed slightly below the ground's surface). Further examples are fibre-optic pressure-sensitive cables, laser detectors that react when someone breaks an (invisible) laser beam and seismic sensors, e.g., geophones that register vibrations in the ground. All of these were used in the "Norrköping Riot" supporting the imaging sensors in situations where these suffer from drawbacks, further described in [4].

6.4 Multisensor simulation

A multisensor simulation (MSS) tool is developed at FOI, systematically incorporating and synchronizing results from a very large number of sensor research projects. Detailed terrain-models, e.g. from laser radar data [5], is an important building block. As is our results from estimating and simulating the signatures of objects and scene elements in the operating wavelengths of the sensors in use. Hereby we achieve high realism and quality in signals and signatures. Included is object models for estimation of realistic target signatures. The MSS lab also integrates a variety of sensor simulators and signal- and image processing via HLA interface. Finally, we have developed a tool for verification and validation of the simulated sensor system, mainly based on the sensor platform, weather condition, sensors, environment, and the function needed to accomplish a certain task.

Providing high accurate signatures to physically based simulation of the scene elements in a realistic, high resolution 3-D environment model had resulted in a very promising resource for various applications. An example of using the MSS lab is to predict and analyze the performance of a mission by an unmanned airborne vehicle that performs automatic target recognition, as seen in Figure 8.

Fig. 8 Simulation of a mission by an unmanned airborne vehicle that performs automatic target recognition. A high resolution 3-D model from laser data is used, modeled as seen by sensors operating in the visual range (*upper left*), IR range (*lower left*), respectively, and by a SAR (*upper and lower right*).

7 Detecting Humans and Analyzing Human Behavior

An important issue, especially in security applications, is to address humans, which are complex to detect, identify or to analyze behavior and intention of either a particular individual or a group, [4]. Another strong motivation to our research at FOI is the need for methods to separate our troops from combatants, non-combatants and even temporal combatants. The latter can for example be a civilian picking up a an IED from his back-pack in a mole, throwing it and injuring people. Likewise, integrity preserving surveillance is a new and important area, stressing the importance of providing technologies that serve the community, not act against it. This will be discussed below.

7.1 Preserving integrity

We have introduced the term integrity preserving surveillance to denote various technologies enabling surveillance that does not reveal people's identities. The implication for integrity preserving surveillance is that people generally do not like to be watched and/or identified, and, furthermore, the use of surveillance cameras is often restricted by law. Integrity preserving surveillance systems put high demands on functionalities like robust classification and tracking of people and vehicles. The scenario below explains some of its potentials. We want to deploy a surveillance system in certain areas in a city. The problem is that we know that this is unpopular among the city's inhabitants, and the solution can be an integrity preserving system. The system maps, as described above, the videos on a 3-D-model of the areas, but replaces people and vehicles with blobs or symbols. The original and authentic videos are encrypted and stored at an institution that the local population have trust in. The processed videos can even be publicly displayed, for example on a web server. The semantic data used for image processing is also used for behavior analysis and warning, e.g. in case of suspicious activities.

7.2 Automatic analysis of humans

Most environments that are interesting to survey contain humans. Currently, automatic analysis of humans in sensor data is limited to passage detectors and simple infrared motion detectors. More complex analysis, like interpretation of human behavior from video, is likely to be performed by human operators. With the recent rapid development in computing power, image processing and computer vision algorithms are now applicable in an entirely different way than a few years ago, especially those for looking at humans in images and video. The benefits of automating analysis of human behavior are mainly robustness. If the video surveillance data is analyzed by a human, a certain error ratio is to be expected due to the human factor, i.e., fatigue and information overload. By automating parts of the process, the human operator can concentrate on interpretation based on the refined information from the human-aware system.

A basic capability of a human-aware system is the ability to detect and locate humans and other moving objects in the video images. This could either be used in a stand-alone manner in the same way a trespassing sensor is used, or for initializing tracking or recognition systems. A method for detection of human motion in video, based on the optical flow pattern, has been developed at FOI. For the purpose of masking out individuals or groups of people from a surveillance video sequence, in order to reveal their activities to a human observer but not their identity, we present each individual in the image masked out with a separate color. An advantage with this technique is that it greatly enhances the human understanding of the activity in the scene.

Our work now is focused on analyzing human motion, see Figure 9. This is to train a system to recognize what can be considered as normal, e.g. that a waist paper basket is emptied every day about 10 o'clock. Hereby, we can detect any deviation from what we have classified as normal, e.g. that a person puts a suspicious object in the same waist paper basket, at ten in the evening, that later on explodes.

Fig. 9 An illustration of the process of localization and classification of humans and vehicles to recognize human motion. Foreground and background separation (*upper row*), separating the foreground into distinct objects (*middle row*) and activity recognition from shape (*bottom row*).

Hence, the goal is to understand human motion and human interaction from images, to be able to detect anomalies. We also want to be able to understand an classify actions, which has to be considered in the current role and environment. In the area of analysis of humans in video, the focus has moved from tracking of humans in video [18], via articulated tracking and tracking in 3-D [25, 31], towards analysis of human motion on a higher level [52]. Due to the increased computational power, focus has also shifted from logic-based methods to probabilistic methods that learn from training data. Tools from probability theory and machine learning has enabled the development of efficient and robust methods for, e.g. 3-D articulated tracking [31], sign language recognition [36], face expression recognition [32] and methods for biometric analysis of humans.

8 Concluding Remarks

Here we have given some insight in FOI's research on sensor technologies and methods for advanced multifunctional sensor systems. The driving force is brought by the defence capability needs for operations in the urban environments. Urban environment is difficult to monitor, being built up by complex structures and situations to monitor. Small object like mines and IEDs are difficult to find and identify. Moreover, humans are perhaps even more complex to detect, identify or to analyze behavior and intention of either a particular individual or a group. However, we foresee that the ongoing research and technical development of new imaging technologies are important contributions to the Swedish Armed Force's ability to perform several tasks in various terrain and conditions. By developing techniques and methods for object identification and situation analysis, we can provide tools and specifications for future systems.

Examples of new imaging technologies are 3-D imaging laser radars, multi- and hyperspectral imaging and new trends in the radar region of the electromagnetic spectra, such as bi-static SAR. These systems have the ability to penetrate e.g. vegetation, clothing material and certain building structures. It also provides detection and recognition of small or extended target. With the recent rapid development in computing power, image processing and computer vision algorithms are also being developed for applications such as looking at humans in images and video. Moreover, we have emphasized the importance of having proper knowledge and information on the close environment (weather, turbulence etc.), that brings factors that can seriously degrades the performance unless handled correctly. Thus, we need to look at the whole problem at hand in close connection to the sensor/sensors in use. We have also given some application examples on new and approved capabilities from using combined sensor and methods.

Conclusively, using advanced multifunctional sensors and considering the whole chain, from the sensor itself to what the sensor can comprehend, we can provide means to a variety of new and complementary capabilities of importance, such as detecting object and abnormal behavior of humans.

References

1. J. Ahlberg, L. Klasén, C. Grönwall, M. Ulvklo and E. Jungert (2003) Automatic target recognition on a multi-sensor platform, *Proceeding of Swedish Society of Automated Image Analysis Symposium on Image Analysis*, pp. 93–96
2. J. Ahlberg, T. Chevalier, P. Andersson and L. Klasén (2004) Target detection in multispectral imagery, *Proceeding of Swedish Society of Automated Image Analysis Symposium on Image Analysis*, pp. 50–53
3. J. Ahlberg, T. Horney, E. Jungert, M. Folkesson, C. Grönwall, L. Klasén, K. Silvervarg and M. Ulvklo (2004) An information system for automatic target recognition, *Multisensor, Multisource Information Fusion: Architectures, Algorithms, and Applications VIII, Proceedings of the International Society for Optical Engineering*, SPIE Press, vol. 5434

4. J. Ahlberg and L. Klasén (2005) Surveillance systems for urban crisis management, *Proceeding of Swedish Society of Automated Image Analysis Symposium on Image Analysis*
5. S. Ahlberg, M. Elmqvist, Å. Persson and U. Söderman (2004) Three-dimensional environment models from airborne laser radar data, *Laser Radar Technology and Applications VII, Proceedings of the International Society for Optical Engineering*, SPIE Press, vol. 5412
6. P. Andersson (2003) Automatic target recognition from laser radar data. Applications to gated viewing and airborne 3D laser radar *FOI-R–0829-SE*, Defence Research Agency, Sweden
7. P. Andersson, L. Klasén, M. Elmqvist, M. Henriksson, T. Carlsson and O. Steinvall (2003) Long range gated viewing and applications to target recognition, *Proceeding of Swedish Society of Automated Image Analysis Symposium on Image Analysis*, pp. 89–92
8. P. Andersson and G. Tolt (2007) Detection of vehicles in a forest environment using local surface flatness estimation in 3-D laser radar data, *Proceeding of Swedish Society of Automated Image Analysis Symposium on Image Analysis*, pp. 6572
9. L. Andrews and R. Phillips (1998) Laser Beam Propagation Through Random Media, SPIE Press
10. J. Bijhold, Z. Geradts and L. Klasén (2004) Forensic imaging 2001–2004: A review, *Proceedings of the 14th Interpol Forensic Science Symposium*
11. S. Bramble, D. Compton and L. Klasén (2001) Forensic image analysis, *Proceedings of the 13th Interpol Forensic Science Symposium*
12. X. Briottet, Y. Boucher, A. Dimmler, A. Malaplate, A. Cini, M. Diani, H. Beckman, P. Schwering, T. Skauli, I. Kasen, I. Renhorn, L. Klasén, M. Gilmore and D. Oxford (2006) Military applications of hyperspectral imagery, *Targets and Backgrounds XII: Characterization and Representation*, SPIE Press, vol. 6239
13. T. Carlsson, O. Steinvall and D. Letalick (2001) Signature simulation and signal analysis of 3-D laser radar *FOI-R–0163-SE*, Defence Research Agency, Sweden
14. T. Chevalier, O. Steinvall and H. Larsson (2007) Performance of laser penetration through forest vegetation, *Laser Radar Technology and Applications XII, Proceedings of the International Society for Optical Engineering*, SPIE Press, vol. 6550
15. S. Der, B. Redman and R. Chellappa (1997) Simulation of error in optical radar range measurements, *Appl. Optics* 36(27):6869–6874
16. X. Dupuis, P. Dreuillet, L. M. H. Ulander and A. Gustavsson (2006) Multi-pass and multi-date at P and L bands: Ground penetration and change detection, *Proceeding of EUSAR 2006*
17. (2006) P.-O. Frölind, L. M. H. Ulander and D. Murdin fast factorised backprojection algorithm for processing of SAR data, *Proceeding of IoA International Conference on Synthetic Aperture Sonar and Synthetic Aperture Radar*, pp. 168–175
18. D. M. Gavrila (1999) Visual analysis of human movement: A survey. CVIU 73(1):82–98
19. C. Grönwall, T. Carlsson and F. Gustavsson (2003) Performance analysis of measurement error regression in direct-detection laser radar imaging, *Proceedings of International Conference on Acoustics, Speech and Signal Processing*, April 6–10, 2003, Nice, France
20. C. Grönwall, T. Chevalier, Å. Persson, M. Elmqvist, S. Ahlberg, L. Klasén and P. Andersson (2004) Methods for recognition of natural and man-made objects using laser radar data, *Laser Radar Technology and Applications VII, Proceedings of the International Society for Optical Engineering*, SPIE Press, vol. 5412
21. C. Grönwall and A. Linderhed (2007) Statistical approaches to mine detection using optical sensors, *Proceeding of Swedish Society of Automated Image Analysis Symposium on Image Analysis*, pp. 7376
22. A. Gustavsson, L. M. H. Ulander, M. Karlsson and M. Lundberg (2005) Employing UAVs to augment ISR capabilities, *Proceeding of RVK 05, Radiovetenskap och kommunikation, Sweden*, pp. 221–226
23. A. Jänis, S. Nilsson, L.-G. Huss, M. Gustafsson and A. Sume (2004) Through-the-wall imaging measurements and experimental charachterization of wall materials, *Military remote sensing*, London, 27–28 October 2004
24. L. Klasén (1999) Forensic Image Analysis, (ed.) R. S. Frank and H. W. Peel, *Proceedings of the 12th Interpol Forensic Science Symposium*, pp. 261–302

25. L. Klasén (2002) Image sequence analysis of complex objects – Law enforcement and defence applications, *Linköping Studies in Science and Technology, Thesis No 762, 2002, University of Linköping, Sweden*
26. L. Klasén, T. Chevalier, H. Larsson, P. Andersson and O. Steinvall (2004) 3D imaging by laser radar and applications in object recognition, *Proceeding of Swedish Society of Automated Image Analysis Symposium on Image Analysis*, pp. 1–4
27. L. Klasén, P. Andersson, H. Larsson, T. Chevalier and O. Steinvall (2004) Aided target recognition from 3-D laser radar data *Laser Radar Technology and Applications VII, Proceedings of the International Society for Optical Engineering*, SPIE Press, vol. 5412
28. L. Klasén (2005) Urban Warfare to see without being seen from a sensor field of view. *Swedish J. Military Technol.*, vol. 3–4
29. L. Klasén (2007) Analysis of multidimensional data from advanced sensor systems, *Proceeding of Swedish Society of Automated Image Analysis Symposium on Image Analysis*
30. M. Lundberg, L. M. H. Ulander, W. E. Pierson and A. Gustavsson (2006) A challenge problem for detection of targets in foliage, *Algorithms for Synthetic Aperture Radar Imagery XIII, Proceedings of the International Society for Optical Engineering*, SPIE Press, vol. 6237
31. T. B. Moeslund and E. Granum (2001) A survey of computer vision-based human motion capture. CVIU 81(3):231–268
32. N. Sebe, M. Lew, I. Cohen, Y. Sun, T. Gevers and T. S. Huang (2004) Authentic facial expression analysis, *Proceeding of International Conference on Automatic Face and Gesture Recognition*
33. J. R. Rasmusson, M. Blom, B. Flood, P.-O. Frölind, A. Gustavsson, T. Jonsson, B. Larsson, G. Stenstrm and L. M. H. Ulander (2007) Bistatic VHF and UHF SAR for urban environments. *Radar Sensor Technology XI, Proceedings of the International Society for Optical Engineering*, SPIE Press, vol. 6547
34. R. D. Richmond (2004) Eye safe laser radar focal plane array for three-dimensional imaging *Laser Radar Technology and Applications VII, Proceedings of the International Society for Optical Engineering*, SPIE Press, vol. 5412
35. H. Sidenbladh, J. Ahlberg and L. Klasén (2005) New systems for urban suirveillance, *FOI-R–1668-SE*, Defence Research Agency, Sweden
36. (1998) T. Starner, J. Weaver and A. Pentland real-time American sign language recognition using desk and wearable computer based video. PAMI 20(12):1371–1375
37. O. Steinvall (1997) Theory for laser systems performance modelling, *FOA-R–97-00599-612–SE-SE*, Defence Research Establishment, Sweden
38. O. Steinvall (2000) Waveform simulation for 3-D sensing laser radar, *FOA R 00 01530 612, 408 SE*, Defence Research Agency, Sweden
39. O. Steinvall and T. Carlsson (2001) Three-dimensional laser radar modelling, *Laser Radar Technology and Applications IV, Proceedings of the International Society for Optical Engineering*, SPIE Press, vol. 4377, pp. 23–34.
40. O. Steinvall (2002) Effects of target shape and reflection on laser radar cross sections, *Appl. Optics* 39(24):4381–4391
41. O. Steinvall, T. Carlsson, C. Grönwall, H. Larsson, P. Andersson and L. Klasén (2004) Laser based 3-D imaging new capabilities for optical sensing, *FOI-R–0856-SE*, Defence Research Agency, Sweden
42. O. Steinvall, L. Klasén, T. Carlsson, P. Andersson, H. Larsson, M. Elmquist and M. Henriksson (2004) Grindad avbildning – fördjupad studie *In Swedish FOI-R–0991–SE*, Defence Research Agency, Sweden
43. O. Steinvall, L. Klasén, C. Grönwall, U. Söderman, S. Ahlberg, å. Persson, M. Elmqvist, H. Larsson, D. Letalick, P. Andersson, T. Chevalier and M. Henriksson (2004) 3 D laser sensing at FOI – Overview and a system perspective, *Laser Radar Technology and Applications VII, Proceedings of the International Society for Optical Engineering*, SPIE Press, vol. 5412
44. O. Steinvall, H. Larsson, F. Gustafsson, P. Andersson, T. Chevalier, Å. Persson, U. Söderman, S. Ahlberg, D. Letalick and L. Klasén (2004) Characterizing targets and backgrounds for 3 D laser radars *Military remote sensing*, London, 27–28, October 2004, vol. 5613, pp. 51–66

45. O. Steinvall (2005) Review of laser sensing devices and systems *Technologies for Optical Countermeasures II; Femtosecond Phenomena II; and Passive Millimetre-Wave and Terahertz Imaging II, Proceedings of the International Society for Optical Engineering*, SPIE Press, vol. 5989
46. O. Steinvall (2007) Laser systems and technology for surface mine detection and classification – A literature updateand performance discussion, *FOI-R–2269–SE*, Defence Research Agency, Sweden
47. O. K. Steinvall, P. Andersson, M. Elmquist and M. Tulldahl (2007) Overview of range gated imaging at FOI, *Infrared Technology and Applications XXXIII, Proceedings of the International Society for Optical Engineering*, SPIE Press, vol. 6542
48. O. Steinvall, T. Chevalier, P. Andersson and M. Elmqvist (2007) Performance modeling and simulation of range-gated imaging systems, *Infrared Technology and Applications XXXIII, Proceedings of the International Society for Optical Engineering*, SPIE Press, vol. 6542
49. G. Tolt, D. Westberg and C. Grónwall (2008) A sensor fusion model for detection of surface laid mines, *Proceeding of Swedish Society of Automated Image Analysis*
50. L. M. H. Ulander, H. Hellsten and G. Stenström (2003) Synthetic-aperture radar processing using fast factorized back-projection. *IEEE Transactions on Aerospace and Electronic Systems* 3(39):760–776
51. L. M. H. Ulander and T. Martin (2006) Bistatic clutter suppression in low-frequency SAR, *Proceeding of EUSAR 2006*
52. L. Wang, W. Hu and T. Tan (2003) Recent developments in human motion analysis. Pattern Recognition 36(3):585–601
53. D. Westberg (2007) A sensor fusion method for detection of surface laid land mines, Master Thesis, LITH-ISY-EX07/4021SE, Linköpings University, Sweden

Applications of Luminescence to Fingerprints and Trace Explosives Detection

Ismail Mekkaoui Alaoui*

Abstract Fingerprints and trace explosives detection requires great sensitivity, which is provided by luminescence and appropriate physical and chemical treatments. Ninhydrin, 1,2-indanedione and other chemicals react with the amino acids present in the fingerprint residue. The chemically treated samples, on which the prints are to be detected, are excited with the blue lines 476.5 and 488 nm of an Argon laser, and the sample's fluorescence is observed under orange filters. The detection of common explosives including trinitrotoluene (TNT) may also be carried out using luminescence techniques. Trace explosive and fingerprint detection require sensitivity due to the minute amount of matter left and available on the samples to be detected. Detection sensitivity can be gained by taking advantage of luminescence techniques. To increase the sensitivity of such detection luminescent chemicals are used, and to distinguish among compounds in a mixture of explosives, time-resolved imaging techniques may suppress any unwanted and background luminescence. Explosives are tagged with europium complexes showing long lived luminescence (0.4 ms) and appropriate for time-resolved imaging. The europium luminescence excitation utilizes a laser operating at 355 nm. Comparison between photoluminescence fingerprints and trace explosives detection will be presented and discussed: common difficulties will be exposed.

Keywords: Laser, luminescence, time-resolved, explosives, detection, fingerprints, 1,2-indanedione, europium

*Department of Physics, Faculty of Sciences Semlalia
Cadi Ayyad University, BP. 2390 Marrakech 40000, Morocco,
e-mail: mekkaoui@ucam.ac.ma

J. Byrnes (ed.), *Unexploded Ordnance Detection and Mitigation*,

1 Introduction

Luminescence is a process in which energy is emitted from a material at a different wavelength from that at which it is absorbed. Luminescence covers photoluminescence, electro luminescence, chemiluminescence, etc. We are interested here in photoluminescence (fluorescence and phosphorescence). The principle of photoluminescence transitions is sketched in Figure 1. Fluorescence is a phenomenon in which electron de-excitation occurs almost spontaneously (emission lifetime of a microsecond or less), and in which emission of a photon from a luminescent substance ceases when the exciting source is removed. In fluorescent materials, the excited state has the same spin as the ground state. In phosphorescence, light emitted by an atom or molecule persists some time after the exciting source is removed (emission lifetime of a microsecond or more). It is a quasi stable electron excitation state involving a change of spin state (intersystem crossing) which decays only slowly.

The emission decay curve is usually expressed by a single exponential when only one species is emitting.

$$N(t) = N(0)\exp-\left(\frac{t}{\tau}\right), \tag{1}$$

where τ is the lifetime of the emitting species at a given wavelength λ.

Photoluminescence, which includes fluorescence and time-resolved luminescence, has been applied to the detection of explosives [12], and has been used for fingerprint detection since 1980 [10]. This technique helps in identifying the explosive element from the molecules released by exposing a high power laser beam on it or in its vicinity. The excited molecules give off photons (light) of the characteristic wavelength of the material when the light source is removed.

The photoluminescence detection of trace explosives and fingerprints on objects share the same main problems: often there are only minute quantities available or left at the crime scene, and the surface to be screened may fluoresce under laser excitation overwhelming the sample's signal. For explosive detection, there is a broad range of effective explosives that need to be screened for, and most current detection technologies require close proximity to the person or object being screened. In fingerprint and trace explosive detection we have the problem of the surface on which the print/explosive is left: some surfaces are difficult to handle and/or luminesce

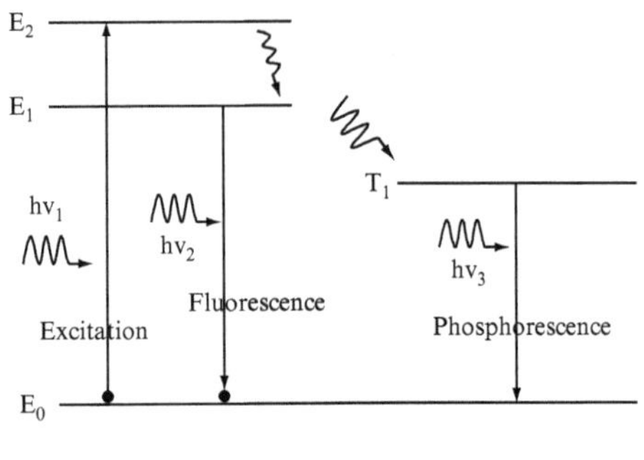

Fig. 1 Principle of photoluminescence transitions.

intensely under laser excitation. Trace explosive and fingerprint detection require sensitivity due to the minute amount of matter left and available on the samples to be detected. Time resolved luminescence is used to suppress the unwanted luminescence background, and may be used to distinguish between explosives that need to be screened.

CW argon-ion lasers are used as excitation of the samples. They deliver continuous wave signals. For time-resolved luminescence purposes, we need to modulate the excitation of the samples as desired depending on the luminescence decay time and also on the background decay lifetime. Two ways of laser modulation were used depending on the decay time range of the compounds: mechanical light chopper for relatively long lifetimes (on the order of milliseconds), and electro-optic modulation for relatively short lifetime decays (on the order of nanoseconds or microseconds). For Eu-RP compounds (0.4 ms), a mechanical chopper is sufficient when operating at 169 Hz (6 ms).

The rare earth (Eu^{3+}) salts are known for their narrow and weak absorption bands in the UV region coupled with emission bands which have narrow half-widths and long luminescence lifetimes (on the order of ms) in the visible region. The radiative transitions of these elements can be enhanced, when Eu^{3+} is bonded to appropriate organic ligands, via intramolecular energy transfer from the organic ligands to the rare earth ions when excited with the right excitation [19]. Rare earth-RP complexes show emission enhancement of the rare earth ions [1]. RP is the reaction product of glycine with ninhydrin. So far, the only excitations that lead to energy transfer from the organic ligands, RP, to the Eu^{3+} are in the near UV range (200–400 nm) [2].

2 Fingerprint Detection by Photoluminescence

2.1 Procedures not requiring time-resolved luminescence

Fingerprint reagents, such as ninhydrin followed by ZincChloride treatment [6], DFO [15], 1,2-indanedione, [9], fluoresce under Ar laser and develop fingerprints without requiring time resolved luminescence imaging. The reaction product of 1,2-indanedione with glycine [2] emits yellow under blue-green laser excitation. The emission spectrum is a broad band having its maximum around yellow (560–575 nm). The samples, on which the prints are to be detected, are excited with the blue lines 476.5 and 488 nm of an Argon laser, and observed under orange filters. 1,2-indanedione is a single-step fluorescence way on porous surfaces; on smooth surfaces, the potential of 1,2-indanedione to detect fingerprints depends on how much luminescence is coming from the sample's surface.

2.2 Cases requiring time-resolved luminescence

When the routine procedures described above fail to give good results, time-resolved luminescence imaging may be the solution. There are a large number of intensely luminescent surfaces for which conventional detection techniques fail. The basic principle of the time-resolved luminescence technique is described in previous papers [14] (Alaoui, 2007). Undeveloped fingerprints are then treated with chemicals having much longer luminescence lifetimes. The imaging device will only detect this long-lived luminescence and eventually suppress the background.

3 Photoluminescence Trace Explosive Detection

Explosives are chemicals (molecules) containing at least two nitro groups (NO_2). When exited with appropriate laser light, the photoluminescence spectrum through a spectrometer will have some peaks characterizing the molecule. Depending on the nature of the transitions the peaks can be more or less sharp. In case of allowed transitions the peaks are usually easy to get while in prohibited transitions the peaks are narrow and hard to find. These electron transitions (peaks) serve as a "fingerprint" for identifying substances. The effectiveness of any trace explosive analysis is dependent on three distinct steps: sample collection, sample analysis, and comparison of results with known standards. All three steps are essential to detect explosives that are present in a crime scene, for example.

It was found recently [7] that many explosives (TNT, nitroglycerin, etc.) share a common photoluminescence peak at 705 nm when excited with a UV laser source emitting at 325 nm. Common non-explosive substances showed no 705 nm peak under the same experimental conditions [8]. Schllhorn's research team [16] developed a portable explosives detector based on photoluminescence. The device shines ultraviolet, infrared and visible light onto two sample areas at the same time and then calculates the difference in reflectance between them for each part of the spectrum. If explosives are present at one sample area but not the other, the characteristic signature of reflected light should show up clearly in this difference measurement. This enables easy identification of the explosive by comparing the signature with a database stored in the detector. Three ways of detecting trace explosives using photoluminescence are presented in the following subsections.

3.1 Photoluminescence versus colorimetric trace explosive detection

Many reactions with trace explosive residue produce a product that is colored and luminescent. In using a luminescent product for detection, one gains at least one to three orders of magnitude in detection sensitivity [13]. To compare colorimetric versus photoluminescence detection of RDX ($C_3H_6N_6O_6$), the following procedure was used by Menzel's team. RDX was solvated in a small amount of acetone, then in a larger amount of methanol and spotted on chromatography paper. The spots were allowed to air dry for about half an hour. Then reagents from an Explosive

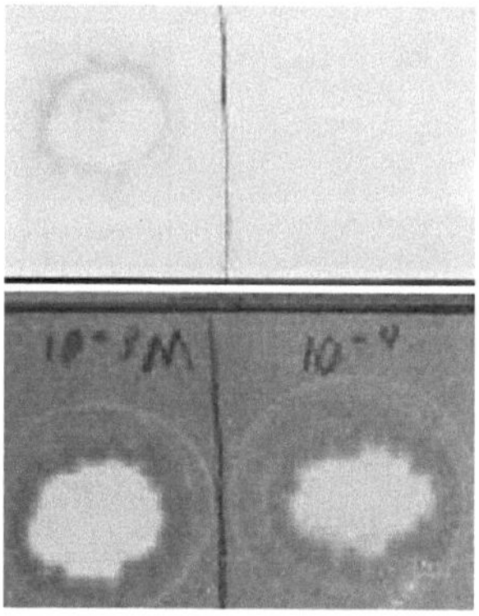

Fig. 2 Colorimetric versus photoluminescence detection of RDX.

Testing Kit were spotted on the RDX. The reaction products were immediately visible. The colorimetric product was light purple and visible in room light. The luminescent product was viewed under blue-green laser light using red and orange filters and appeared as a reddish color (Figure 2). At low concentrations (10^{-4} M and lower), the photoluminescence trace explosive detection is superior to the colorimetric method due to its sensitivity.

3.2 Trace explosive detection by time-resolved luminescence

The luminescence time-resolved approach records the luminescence intensity in a specific time at a given delay after the excitation pulse, where both delay and gate width are carefully chosen based on the characteristic decay of the explosive signal and the background luminescence, on which the explosive is to be detected. The photoluminescence trace explosive detection approach may employ Eu^{3+} lanthanide tagging for time-resolved detection due to the fact that the photoluminescent properties of its compounds depend slightly on the nature of the ligands [3]. The observed decay time of its main peak (616 nm) is 0.4 ms, which is suitable for time resolved imaging (Figure 3). The ground state for europium trivalent ions is an ^7F state. The lowest excited states inside the 4f-shell for Eu^{3+} are $^5D_0$ (about 17,267 cm^{-1}), the main emissive level, and $^5D_1$ (about 19,025 cm^{-1}).

To detect trace explosives on a high background surface, Eu^{3+} compounds (long lifetime) are needed, so that time-resolved imaging could be utilized. In the study done by Menzel's team [13], several $EuCl_3$ complexes were tested on a variety of different explosives (NG, RDX, and two kinds of smokeless powder). The explosive was rubbed on filter paper, then the lanthanide complex was spotted on the explosive. A control, with solvent and no explosive, was also spotted with the lanthanide complex. The luminescence intensities were then compared between the samples and the controls. There was a change in luminescent intensity and a slight color shift for these tests when viewed immediately (under both near and deep UV) after

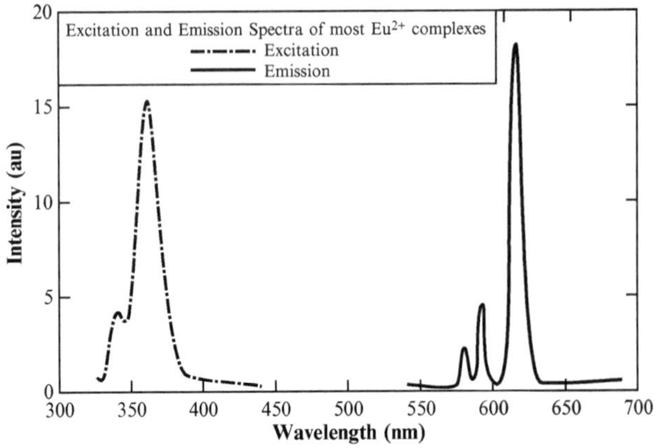

Fig. 3 Excitation and emission spectra of Eu^{3+} compounds.

spotting. There are possible reactions taking place, which is a good sign: Eu^{3+} is probably forming complexes with the explosive molecules. But further study and research would be necessary for Eu^{3+} and time-resolved luminescence to become a trace explosives detection method.

3.3 Trace explosive detection with luminescent polymers

Luminescent polymers were used (Toal and Trogler, 2006) as sensors for TNT. When a molecule of an explosive binds to a polymer it can 'turn off' the luminescence of the polymer. This change can be used to sense very low concentrations of explosives. The Toal team has made a silicon-containing polymer that glows blue or green under illumination with a UV light, and dims in the presence of TNT. This could be used to detect trace explosives left by the fingerprints of a bomb maker. Usually, fingerprint residue does not hold much matter, as stated in the introduction. Other studies using photoluminescent metallole-containing polymers [18] reported detection of trace explosives limits observed to be as low as 5 ng for TNT, and 20 ng for DNT (2,4-dinitrotoluene).

4 Conclusion

Even though the chemistry is different, photoluminescence fingerprints and trace explosive detection share many common difficulties: scarcity of the materials left at the crime scenes, sensitivity, high power excitation, and high background signal. When the luminescence signal from explosive materials is strong, luminescence can be used as a main technique for explosive materials detection and identification.

In the case of a weak signal and/or high background luminescence under a laser excitation source, time resolved luminescence and other techniques (Raman spectroscopy and laser-induced luminescence spectroscopy) may solve the problem. In any case luminescence may be used as a complimentary property; it serves as a "fingerprint" for identifying explosive materials. The distinct photo luminescent peak at 705 nm (if confirmed by further studies), common to all explosive materials, opens new windows: luminescence may become an excellent and extremely sensitive trace explosive identification tool.

Acknowledgments Part of this work has been done at the Center for Forensic Studies (E.R. Menzel, Director) and the Physics Department, Texas Tech University.

References

1. Alaoui, I. M., and Menzel E. R. (1993) Spectroscopy of Rare Earth Ruhemann's Purple Complexes. J. Forensic Sci., 38(3), 506–520.
2. Alaoui, I. M. (1995) Non-Participation of the Ligand First Triplet State in Intra-molecular Energy Transfer In Europium and Terbium Ruhemann's Purple Complexes. J. Phys. Chem., 99, 13280–13282.
3. Alaoui, I. M., and Menzel, E. R. (1996) Constituent Effects on Luminescence Enhancement in Europium and Terbium Ruhemann's Purple Complexes. J. Forensic Sci. Int., 77, 3–11.
4. Alaoui, I. M., Menzel, E. R., Farag, M., Cheng, K. H., and Murdock, R. H. (2005) Mass spectra and time-resolved fluorescence spectroscopy of the reaction product of glycine with 1,2-indanedione in methanol. Forensic Sci. Int., 152, 215–219.
5. Alaoui, I. M. (2007) Time-resolved luminescence Imaging and Applications, ASI-Imaging for Detection and Identification, 243–248.
6. Herold, D. W. and Menzel, E. R. (1982) Laser detection of latent fingerprints: ninhydrin followed by zinc chloride. J. Forensic Sci., 27, 513–518.
7. Hummel, J. (2004) Photoluminescence spectroscopy: New technique for detecting explosives. www.buzzle.com/editorials/10-11-2004-60363.asp.
8. Hummel, R. E., Fuller, A. M., Schllhorn, C., and Holloway, P. H. (2006) Detection of explosive materials by differential reflection spectroscopy. Applied Phys. Lett., 88, 23.
9. Joulli, M. M. and Petrovskaia, O. (1998) A better way to develop fingerprints. ChemTech, 28(8), 41–44.
10. Menzel, E. R. (1980) Fingerprint Detection with Lasers. Marcel Decker, New York.
11. Menzel, E. R., and Menzel, L. W. (2004) Ordinary and time-resolved photoluminescence field detection of traces of explosives and fingerprints. J. Forensic Ident. 54, 560–571.
12. Menzel, E. R., Bouldin, K. K., and Murdock, R. H. (2004) Trace explosives detection by photoluminescence. ScientificWorldJournal 4, 55–66.
13. Menzel, E. R., Menzel, L. W., and Schwierking, J. R. (2004) A photoluminescence-based field method for detection of traces of explosives. ScientificWorldJournal 4, 725–735.
14. Mitchell, K. E., and Menzel, E. R. (1989) Time resolved luminescence imaging: Application to latent fingerprint detection, Fluorescence Detection III, SPIE Proceedings, E. R. Menzel, Ed., 1054, 191–195.
15. Pounds, C. A., Ggrigg, R., and Mongkolaussavaranta, T. (1990) The use of 1,8-diazafluoren-9-one (DFO) for the fluorescent detection of latent fingerprints on paper. J. Forensic Sci., 35(1), 169–175.
16. Schllhorn, C., Fuller, A. M., Gratier, J., and Hummel, R. E. (2007) Developments on standoff detection of explosive materials by differential reflectometry. Appl. Opt., 46, 6232–6236.

17. Toal, S. J., and Trogler, W. C. (2006) Polymer sensors for nitroaromatic explosives detection. J. Mater. Chem., 16, 2871–2883.
18. Toal, S. J., Sanchez, J. C., Dugan, R. E., and Trogler, W. C. (2007) Visual Detection of Trace Nitroaromatic Explosive Residue Using Photoluminescent Metallole-Containing Polymers. J. Forensic Sci., 52(1), 79–83.
19. Weissman S. I. (1942) Intramolecular energy transfer, the fluorescence of complexes of europium. J. Chem. Phys., 10, 214–217.

Electromagnetic Methods for UXO Discrimination

Kevin O'Neill[*,1] and Juan Pablo Fernández[2]

Abstract The subsurface remote-sensing technology currently used in the United States for UXO decontamination is relatively crude, consisting of DC (static) magnetometry. Ultrawideband electromagnetic induction (EMI) is emerging as a technology with reasonable discrimination potential. EMI devices operate in the magneto-quasistatic (MQS) band, usually between tens of Hz and perhaps a couple hundred kHz, and engage a substantially different phenomenology than that of wave electromagnetics. Over the relevant space scales, soil, fresh water, and rock are effectively lossless in the MQS regime, which encourages EMI application.

Here we review the relevant EMI physics and phenomenology and then discuss state-of-the-art EMI discrimination methods like the Standardized Excitation Approach (SEA). This can be used in signal matching to decide if an unseen target belongs to a catalogued set. It can also quickly provide many examples of realistic input to train statistical learning algorithms such as Support Vector Machines (SVM). SVMs can also use SEA parameters themselves as discriminators. Most realistic UXO-sensing scenarios are clutter limited. We examine computational upward continuation as a clutter-mitigation strategy with a rational physical basis.

Keywords: UXO, discrimination, electromagnetic induction, EMI, magneto-quasistatics, MQS, magnetometry, standardized excitations approach, SEA, excitation mode, pattern matching, magnetic dipole, dipole moment, magnetic diffusion, ground response, magnetic charge, statistical learning algorithms, support vector machine, SVM, slack variable, classification

[*,1]Thayer School of Engineering, Dartmouth College and U.S. Army Corps of Engineers, ERDC-CRREL, Hanover, NH 03755, U.S.A., e-mail: Kevin.A.O'Neill@dartmouth.edu
[2]Thayer School of Engineering, Dartmouth College, Hanover, NH 03755, U.S.A., e-mail: jpfb@dartmouth.edu

J. Byrnes (ed.), *Unexploded Ordnance Detection and Mitigation*,

1 Introduction

Surveying and cleanup of sites with potential contamination by unexploded ordnance (UXO) is an extremely high priority environmental objective in the United States, yet one that is very challenging. The problem is complicated by its sheer size (millions of hectares); hundreds, probably many hundreds of sites; diverse and often problematical geological and terrain conditions; and great diversity in the sizes and types of UXO. A comparable or larger scale problem exists at the international level, beyond military training grounds to the sites of past conflicts. The problem comes into focus when one notes that, while some very large UXO may be 10 m deep underground, by far most UXO are within the top meter of soil, mandating very shallow surveying. Further, we cannot yet sense what we are in fact most concerned about, namely the explosive within intact shells. Therefore we have to sense metal and only thereby characterize the object.

The first priority in UXO surveying is detection—making sure that some sufficiently clear signal is obtained from essentially all UXO in the field. Unfortunately, in the service of this objective, our sensors record responses from virtually everything in the environment capable of producing a signal. Site remediators frequently excavate hundreds of objects for each UXO that is found [4]. The resulting costs are frequently prohibitive. Thus the second crucial requirement is that of discrimination. Signal anomalies identified in broader surveying must be subjected to close investigation to distinguish the nature of the responding object and to judge how likely it is to be a UXO. This chapter focuses on discrimination, as opposed to detection.

Electromagnetic sensors of some kind are currently the most logical choice with which to sense buried metallic bodies. Ground penetrating radar (GPR), while used in many applications of geophysical, environmental, and infrastructure surveying, has not generally been successful at distinguishing UXO. The combination of ground surface reflection, signal loss over depth, and signal clutter due to both metallic fragments and dielectric heterogeneities is simply too great. While holding some potential as an adjunct to other sensing modes for close discrimination, even ultrawideband (UWB), fully polarimetric GPR in the 10–810 MHz range coupled with extensive processing is still challenged in the discrimination realm [6, 14]. The most common sensing mode by far in actual practice in the United States is static (DC) magnetometry. Magnetometers detect perturbations of the earth's field caused by ferrous objects. While relatively reliable at least as detectors of steel, magnetometers produce a rather crude picture by virtue of both current practice and inherent information content in the signals. Discrimination capability is quite limited, though progressing [16].

Between DC magnetometry and GPR lies electromagnetic induction (EMI) sensing. One may hope that EMI sensors combine the best of magnetometry and GPR. Like the former, they are immune to dielectric heterogeneities and, on our scale of observation, the ground is essentially transparent to them. State-of-the-art sensors are UWB, with frequencies of operation possibly from tens of Hz up to tens or, rarely, hundreds of kHz, and some devices register vector response fields. Received signals have high information content, being sensitive to object distance, shape, orientation, and composition. Altogether, whether alone or in tandem with other kinds of sensors, EMI appears to offer the greatest immediate promise for discrimination of buried UXO.

2 Relevant Electromagnetic Theory and Phenomenology

To understand both the challenges and the potential of UWB EMI technology, it is vital to gain some grasp of the fundamental physics and phenomenology in that domain, particularly if one is approaching from what may be the more familiar realm of electromagnetic waves.

2.1 Basic relations; waves vs. diffusion and potential fields

All the relevant phenomena are governed by Maxwell's equations [9,20]. Faraday's Law pertains specifically to induction, stating that a time varying magnetic flux is linked to circulation of **E**:

$$\nabla \times \mathbf{E} = \left\{ \begin{array}{cc} -\dfrac{\partial \mathbf{B}}{\partial t} & \text{Time domain (TD)} \\ i\omega\mu\mathbf{H}[e^{-i\omega t}] & \text{Frequency domain (FD)} \end{array} \right\}, \tag{1}$$

where **E** is the electric field (V/m) and **B** (T) is the magnetic flux density, equal to the magnetic field **H** (A/m) times the medium magnetic permeability μ (H/m). Integrating the normal component of this equation over a surface S, e.g. the planar region within the loop in Figure 1, produces an integral version of (1):

$$\oint_\Gamma d\gamma \cdot \mathbf{E} = -\frac{d}{dt}\int_S dS\, B_n = -\frac{d\Phi}{dt}. \tag{2}$$

The line integral around the edge of the surface, Γ, constitutes essentially a voltage, an electromotive force. If there is a conductor surrounding the surface, this implies

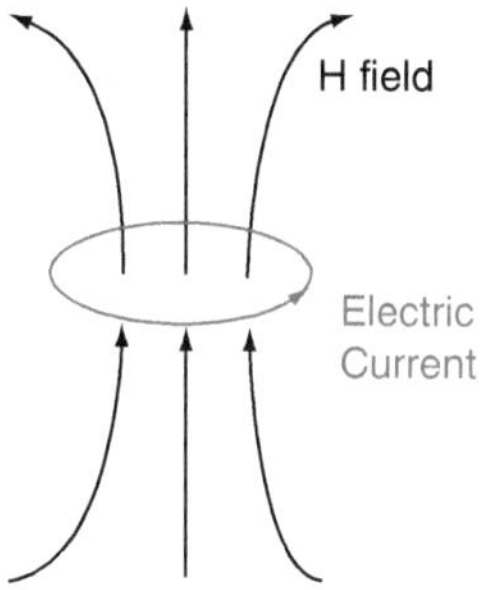

Fig. 1 Schematic of a magnetic dipole formed by an electric current loop, with associated magnetic field lines passing through the loop.

an electric current loop. Depending on what part of the system is forced, such a current loop may induce a changing magnetic flux Φ, as in our EMI transmitters; or, conversely, an imposed change in magnetic flux may induce a current loop, as when our transmitted magnetic fields encounter a metallic object (see Figure 1).

An infinitesimal magnetic dipole of moment **m** produces an **H** field [20]

$$\mathbf{H}(\mathbf{r}) = \frac{3\hat{\mathbf{r}}\hat{\mathbf{r}} - \mathbf{I}}{4\pi r^3} \cdot \mathbf{m}. \tag{3}$$

where **m** is the dipole moment of the current loop. For this ideal dipole approximation to apply, the distance r from the center of the loop need only be greater than a couple times the loop diameter.

As all our EMI transmitters and responding objects of interest form finite dipoles, the relation in (3) is fundamental, at least as an approximation. In particular, the $1/r^3$ spatial dependency of signals produces crucial limits on applicability and resolution of the technology. Except rarely, where noted, we will assume in what follows that the sensor's transmitters and receivers are co-located. In this case, (3) applies over the same distance, r, in both directions between transmitter and responding object, for a total signal falloff proportional to $1/r^6$. As we shall see below, ground lossiness does not afflict EMI with signal loss in the same way that it does GPR, where it is a serious problem. Instead, the $1/r^6$ signal falloff in EMI is due purely to the inherent geometry of the quasistatic fields. There is little that can be applied to counteract it.

Ampère's Law relates the curl of the magnetic field to various currents:

$$\nabla \times \mathbf{H} = \left\{ \begin{array}{c} [\mathbf{J}_{\mathrm{sc}}] + \mathbf{J}_{\mathrm{env}} + \dfrac{\partial \mathbf{D}}{\partial t} \\[2ex] [\mathbf{J}_{\mathrm{sc}}] + \sigma \mathbf{E} - i\omega\varepsilon\mathbf{E} \end{array} \right\}. \tag{4}$$

The source current $\mathbf{J}_{\mathrm{sc}}$ is taken to be non-zero only in isolated, concentrated regions (e.g., a wire loop as in Figure 1), and we do not analyze it further. The conduction current density in the environment, $\mathbf{J}_{\mathrm{env}}$ (A/m^2), is related to the electric field via the electrical conductivity σ (S/m); and $\partial\mathbf{D}/\partial t$ is the "displacement current," where $\mathbf{D} = \varepsilon\mathbf{E}$ and ε is the permittivity of the medium (F/m). The exposition that follows explores the ways in which the magnitudes of the last two terms on the right, relative to one another and to the quantity on the left, fundamentally determine the nature of the electromagnetic phenomena at hand.

The divergence law for magnetic fields states, in effect, that there are no isolated magnetic charges (poles):

$$\nabla \cdot \mathbf{B} = 0 = \nabla \cdot \mu\mathbf{H}. \tag{5}$$

In magnetically homogeneous regions, **H** will be divergence-free as well. Equation (5) also means that the magnetic field lines in Figure 1 actually form closed loops. As a convenient fiction, we may introduce nonzero equivalent magnetic charges q_m on the right in (5), outside of regions of application, in order to generate mathematically the fields within regions of interest (ROI). This is valid as long as the source distributions imply fields that satisfy the governing equations within the ROI as well as the appropriate conditions on its boundary. A q_m-based approach is used in the clutter-suppressing upward continuation system described below.

Using (5) for **H** together with (1) in the curl of (4) produces a Helmholtz-type equation:

$$\nabla^2 \mathbf{H} = \left\{ \begin{array}{ll} \sigma\mu \dfrac{\partial \mathbf{H}}{\partial t} + \mu\varepsilon \dfrac{\partial^2 \mathbf{H}}{\partial t^2} & \text{(TD)} \\ (i\omega\sigma\mu + \omega^2\mu\varepsilon)\mathbf{H} = -k^2\mathbf{H} & \text{(FD)} \end{array} \right\}, \tag{6}$$

where k has different meanings depending on the parameter range that applies. When the second derivative with respect to time dominates then (6) becomes a wave equation, possibly with a significant loss term. In that case, k is a true wave number. When the first derivative with respect to time dominates, then (6) becomes a diffusion equation and k is no longer a true wave number.

The relative magnitude of the two time-derivative terms in (6) is probably best apprehended from the ratio of their corresponding frequency-domain expressions, namely $\sigma/\omega\varepsilon$. In the GPR frequency range (10^7–10^9 Hz) we may assume that the fields form waves in air ($\sigma \sim 0$) and that penetration of metallic reflectors is negligible. They serve as perfect reflectors. The dielectric constant of soil, $\kappa = \varepsilon/\varepsilon_0$, ranges from about 6 for dry soil to a maximum of about 30 for soil completely saturated with water. For ground, σ ranges from a low of about 10^{-3} S/m (particularly for dry and granular soil) up to about 1 S/m for media saturated with seawater. Altogether, for GPR we have

$$\frac{\text{Diffusion}}{\text{Wave}} \sim \frac{\sigma}{\omega\varepsilon} = \left\{ \begin{array}{ll} 10^{-3}, & \text{lowest } \sigma \text{, highest } f \\ 10^{2}, & \text{highest } \sigma \text{, lowest } f \end{array} \right. . \tag{7}$$

In principle, either waves or diffusion may dominate. In the absence of saltwater in the soil, a more typical maximum of the ratio in (7) is on the order of unity. Thus wave phenomena are typically dominant or at least highly significant.

Magneto-quasistatics is defined by the condition that the displacement current $\partial \mathbf{D}/\partial t$ is negligible. This can occur when the ε term on the right hand side of (6) is overshadowed by the σ term. It is the case over essentially the entire EMI band in soil and metal.

$$\frac{\text{Diffusion}}{\text{Wave}} \sim \frac{\sigma}{\omega\varepsilon} = \left\{ \begin{array}{l} \text{Soil} \quad \Rightarrow \left\{ \begin{array}{l} \sim 10^1\text{, lowest } \sigma\text{, highest } f \\ \sim 10^8\text{, highest } \sigma\text{, lowest } f \end{array} \right\} \\ \text{Metal} \quad \sigma \sim 10^7 \text{ S/m}, \quad \Rightarrow \quad \sigma/\omega\varepsilon \gg 1. \end{array} \right. \tag{8}$$

While diffusion can be very important, waves are essentially always negligible. *In the EMI realm there are no true reflections, diffractions, resonances, etc., of the sort expected in the wave regime.*

As explained below, it is possible for both terms on the right hand side of (4) to be negligible relative to the derivatives in the term on the left. In this case, the magnetic field is irrotational and can be represented simply in terms of the gradient of a scalar potential Ψ [9, 20]:

$$\nabla \times \mathbf{H} = 0 \quad \longrightarrow \quad \mathbf{H} = -\nabla\Psi. \tag{9}$$

Combining this with the divergence-free condition on **H** produces a simple Laplace governing equation in terms of Ψ:

$$\nabla \cdot \mathbf{H} = 0 \quad \longrightarrow \quad \nabla^2\Psi = 0. \tag{10}$$

A scalar potential may be generated most simply via scalar sources, *i.e.* equivalent magnetic charges q_m situated outside the ROI within which the divergence free condition applies to **H**:

$$\Psi(\mathbf{r}) = \int_{S_0} dS' \, \frac{q_m(\mathbf{r}')}{4\pi \left|\mathbf{r} - \mathbf{r}'\right|}. \tag{11}$$

By construction, Ψ obtained from (11) from any set of q_m outside the ROI will satisfy the equations in **H** within the ROI, as per (9) and (10). To obtain the particular, realistic field required in any circumstance, one uses the gradient of (11) to enforce the standard boundary conditions in **H** [9, 20] on the boundary of the ROI. This is the strategy we will employ for upward continuation.

2.2 Character of the EMI regime: metal

Metallic targets are significantly penetrable over much of the EMI band (Figure 2). From the point of view of discrimination of unknown targets, this is both fortunate and unfortunate: metal type matters. Figure 3, left, shows the frequency response to a uniform excitation **H** field by a hypothetical 20-cm diameter metal sphere, a case for which there is an analytic solution [23]. The response is construed in terms of a component in phase with the excitation field (real-valued part) and a part in phase quadrature with it (imaginary part). Around the high-frequency limit, penetration of the excitation field is so slight that essentially only surface currents exist and asymptotic behavior is reached, independent of material type. In accordance with Lenz's law, these surface currents circulate in such a way as to oppose the primary field, hence the negative real value of the response there.

At the low-frequency end of the spectrum, approaching magnetostatic conditions, penetration of the object by the excitation field is complete but there are no induced currents because $\partial\mathbf{B}/\partial t$ is negligible. If the metal is permeable ($\mu > \mu_0$), as in the example in the figure, a magnetization (polarization) response appears in the absence of macroscopic induced currents. It is aligned with the excitation field (positive sign) and is due to microscopic magnetic dipole structures within the material. For non-permeable materials ($\mu = \mu_0$), the inphase response will be zero at the low-frequency limit, descending from there toward the high-frequency asymptote. Between the low- and high-frequency regimes, over most of the band, one encounters some mixture of magnetization and induced (macroscopic) current

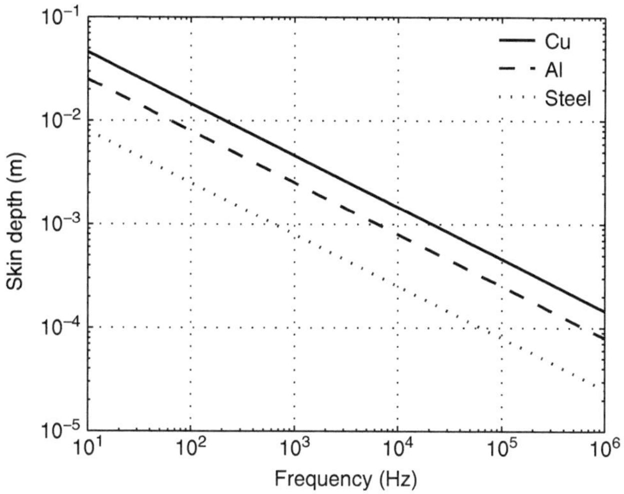

Fig. 2 Skin depth vs. frequency for various common metals over the EMI band.

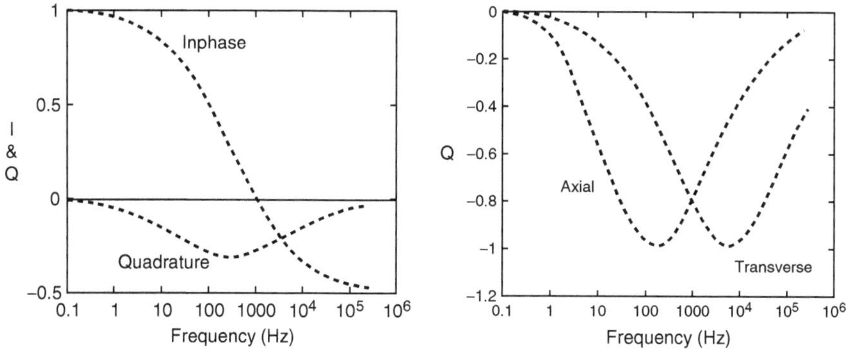

Fig. 3 *Left*: Frequency response of a sphere with properties of steel, $\sigma = 4 \times 10^6$ S/m, $\mu_r = 100$. *Right*: Normalized transverse and axial responses by a prolate spheroid of the same material.

responses within the object. The quadrature component is due to induced volume currents and is delayed relative to the excitation field because of the finite conductivity of the material. This effect peaks somewhere within the band.

Given that **D** and its derivatives are negligible in magneto-quasistatic fields, one can manipulate the equations above to show that induced currents must be divergence-free. There can be no accumulations of free charge, and all induced currents must form continuous, closed loops. The current loops induced by impinging primary (excitation) fields thus effectively form finite-dimensional magnetic dipole structures. For the special case of a homogeneous sphere, the induced currents and polarization throughout its volume produce a secondary field outside the object that is exactly the same as would be produced by an infinitesimal magnetic dipole at its center.

While essentially all metallic objects produce some variant of the relaxation-type curves in Figure 3, left, the particulars are case-dependent. Material type, size, proportions, and orientation all influence the location of the quadrature peak as well as many specific details in the relations between the two components. Figure 3, right, shows normalized quadrature response spectra of a hypothetical steel prolate spheroid, 20 by 5 cm, with the same material parameters as the sphere on the left and with its axis oriented parallel and then perpendicular to the excitation field. A new analytical solution is available for these cases [1]. Note that the peak in this component shifts much higher in frequency for the transverse orientation. Examining spectral features such as this is the basis for frequency-domain (FD) discrimination.

These spectral features also correspond to response patterns in time when the object is subjected to an imposed change in the surrounding magnetic field. Responses to any such excitations can only decay through the time following the change, but with patterns of magnitude, temporal gradient, etc., dependent on the object's particulars. Time-domain (TD) discrimination examines the features of such received temporal decay curves. Note that, while responses proceed through time in TD EMI sensing, signal time does not correspond to distance to a responding entity, as in radar. On our EMI time and space scales there is no true wave-type propagation, and all distances sensed respond effectively at the same time.

2.3 Character of the EMI regime: air

One can show that the magnitude of $\partial \mathbf{D}/\partial t$ in air in the EMI band is negligible compared to the terms in $\nabla \times \mathbf{H}$. Because the electrical conductivity of air is also negligible, neither displacement nor conduction currents are significant, and the right hand side of (4) is zero. With an irrotational **H** field, the simple scalar Laplace equation (10) governs. Viewed another way, as long as there is time variation of the fields, they are waves on some scale, but not on our scale of observation ($\sim$1 m). Wavelengths range from some kilometers at the top of the band up to perhaps 10,000 km near the bottom. Over our scale of observation there is no discernible phase difference between one point and another; no delay. The fields have the structure of static fields with time dependence imparted only by boundary conditions/forcing functions.

2.4 The EMI regime in soil

Here σ is finite and one can show that conduction currents will typically have a larger effect than displacement currents. At the same time, the effect of σ is negligible: Skin depths are much greater than the scale of observation (Figure 4).

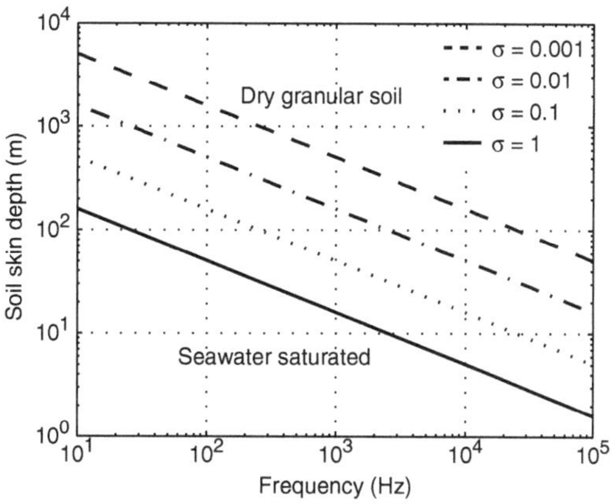

Fig. 4 Skin depths for different soil conductivities, assuming a representative dielectric constant of 16.

In this sense, the soil is transparent to signals in the EMI band. In the absence of seawater, the upper two lines probably furnish reasonable bounds under common circumstances. With induced soil currents negligible and displacement currents even smaller, once again the **H** field is irrotational and a scalar Laplace potential equation governs. Further, note that as tangential **E** fields are continuous between soil and buried metal objects, the ratio of induced current magnitudes in each will be approximately the ratio of their conductivities, which is $\sim 10^9$. Currents induced in the metal dominate the response signal within the soil, given that we only sense a soil volume that is not many orders of magnitude greater than that of a target of interest. Of course, this situation may not be the case in the absence of metal targets and when large volumes or depths of ground are sensed, e.g. ~kilometers in deep geophysical prospecting. Overall, in our case, for all practical purposes the transmitted field reaches a metallic buried object essentially unaffected by the ground, and system responses both within the soil and above ground at the receiver are dominated by the response of the target.

While induced soil currents do not produce significant signal responses, magnetically permeable soil can still, however, produce a significant half-space response, including rough surface effect, which we indeed see in field work. The relative permeability μ_r is unity in free space; for soil it is typically construed in terms of the (volumetric) magnetic susceptibility as $(1+\chi)$. Even though χ magnitudes are generally on the order of 10^{-3} or less [4], this can still be enough to produce a half-space response that is notable relative to that of buried metal targets. Some UXO sites are particularly problematic, such as volcanic terrains (e.g., Maui and Kaho'olawe in Hawai'i). The nature and character of soil susceptibility are currently an area of active research. Our recent experience suggests that, for common soil, the instantaneous response (real-valued χ, constant over the band) is large relative to a trailing relaxation response, with only the latter affecting TD instruments [26].

2.5 The EMI realm, summary

In the MQS EMI band, on our scale of observation:

- There are no waves nor attendant wave phenomena (reflection, diffraction, refraction...). To emphasize the difference, we speak of the transmitted field impinging on a target as the "primary" (as opposed to "incident") field, and the field from the object's response as the "secondary" field.
- One can map the vector, UWB subsurface response over an area of ground surface in terms of inphase and quadrature components in the FD, overall using perhaps five decades of frequency, or using decay time points in the TD.
- Metal targets are penetrable, completely so at the very bottom of the band, with possibly negligible penetration (surface currents only) at the very top of the band. Different metal types respond differently, and the magnetic field within the metal operates by diffusion.
- In both soil and air, *both* conduction and displacement currents, *i.e.*, both diffusion and wave effects, are negligible. Magnetic fields are irrotational and can be expressed in terms of a scalar Laplacian potential. The only significant induced currents are those in metal objects, which dominate the secondary fields in the soil and air in their vicinity.
- In contrast to GPR, there is no significant delay, travel time, or phase difference over space – fields have the structure of magnetostatic fields, with time dependence imposed by boundary conditions/forcing functions. In TD EMI, elapsed signal time corresponds to duration of signal decay in an object at a given depth. It does not correspond to depth of responding entities. All depths respond essentially at once and a single picture emerges over each portion of the ground surface.
- Soil lossiness has a negligible effect. In that sense the ground is transparent to EMI signals, though half-space-type magnetic responses from permeable soils are seen.

3 Standardized Excitation Approach Forward Modeling in EMI

For use in many kinds of discrimination algorithms, we benefit from rigorous but fast simulations of EMI responses by objects of interest, taking into account the particular sensor characteristics. The Standardized Excitation Approach (SEA) formulation based on fundamental spheroidal modes can calculate the sensor responses produced by a geometrically complex, materially heterogeneous object, accounting for near and far field effects and all internal interactions [7, 19, 21, 25]. The total response to any excitation is constructed simply as an appropriate superposition of responses to defined excitation modes. These modes form a sufficient basis to express any excitation. As the simulations are extremely fast relative to detailed numerical solution of the governing equations, they can be run many times in the course of an inversion or classification computation.

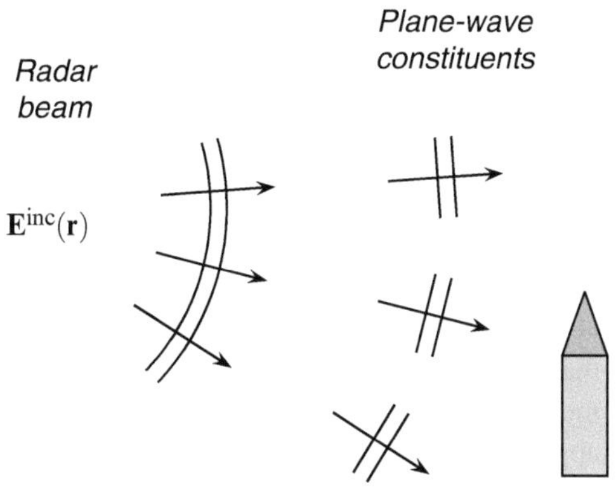

Fig. 5 Schematic of a radar beam incident upon a UXO, indicating the beam's decomposition into constituent plane waves.

Consider an analogy to decomposition of radar beams into constituent plane waves (Figure 5). At each frequency, an incident radar beam can be decomposed into a bundle of plane waves, all of the same frequency but with different vector wave numbers $\mathbf{k}_j$ depending on the direction of propagation of each:

$$\mathbf{E}^{\rm inc}(\mathbf{r}) = \sum_j \beta_j \mathbf{E}_j^{\rm inc}(\mathbf{r}), \qquad \mathbf{E}_j^{\rm inc}(\mathbf{r}) = e^{i\mathbf{k}_j\cdot\mathbf{r}}. \tag{12}$$

If, either by computation or experiment on a particular object, one catalogues its response to a unit magnitude of each of these constituent $\mathbf{E}_j^{\rm inc}$, then one can easily construct the total response to the bundle of them constituting the complete beam. One can do this for any sensor-target arrangement as quickly as one can solve for the β_j in (12).

To parallel this procedure in EMI, for magnetic fields, one decomposes the primary field into fundamental or "standardized" excitations,

$$\mathbf{H}^{\rm PR}(\mathbf{r}) = \sum_j b_j \mathbf{H}_j^{\rm PR}(\mathbf{r}), \tag{13}$$

and constructs the total response $\mathbf{H}^s$ as the corresponding sum of responses to each excitation mode:

$$\mathbf{H}_j^{\rm PR}(\mathbf{r}) \to \mathbf{H}_j^s(\mathbf{r}), \qquad \mathbf{H}^s(\mathbf{r}) = \sum_j b_j \mathbf{H}_j^s(\mathbf{r}). \tag{14}$$

Cataloguing the response to each excitation mode means solving for some set of parameters, S_k^j, for each fundamental (jth) input, *i.e.*, setting $\mathbf{H}_j^s(\mathbf{r}) = \sum_k S_k^j \mathbf{G}_k(\mathbf{r})$. One can obtain the S_k^j from data in controlled measurements on an object of interest by calculating the b_j for various sensor-object configurations via (13), then using these in the combination of (13) and (14):

$$\mathbf{H}^s(\mathbf{r}) = \sum_j b_j \sum_k S_k^j \mathbf{G}_k(\mathbf{r}). \tag{15}$$

With the b_j known and a sufficient number of measurements of the left-hand side of (15), one can solve for the necessary S_k^j. For a given object or object type, this need only be done once. While the data or particular beam composition may be a function of the sensor position $\mathbf{r}$, the S_k^j are not. When their nature or structure is defined, they are invariant characteristics of the object.

The nature of the modal response function $\mathbf{G}_k(\mathbf{r})$ just depends on the nature of the response parameters S_k^j that are applied. For example, if the S_k^j are equivalent charges at the object location, then the $\mathbf{G}_k(\mathbf{r})$ will just be the appropriate Green function for that source type, see e.g. (11). The S_k^j can be used thereafter for repeated forward modeling, requiring only that one decompose any excitation at hand, *i.e.* obtain the applicable b_j for each sensor-object configuration. As we shall see, having obtained the characteristic S_k^j for candidate objects, one can use them within fast forward models during optimization to determine whether recorded signals are most likely to have been produced by one of the candidates. Alternatively, one can infer the S_k^j for unknown targets and then use those parameters themselves as discriminators.

The fundamental problem in formulating the SEA approach in EMI resides in the requirement that one produce some appropriate basis $\mathbf{H}_j^{\mathrm{PR}}(\mathbf{r})$ for decomposing the primary fields. In the EMI/MQS regime, there are no plane waves nor, for that matter, any waves at all. One solution is to apply basic solutions of the Laplace equation in the spheroidal coordinate system,

$$\begin{aligned}\mathbf{H}^{\mathrm{PR}}(\eta,\xi,\phi) &= -\sum_{m,n} b_{m,n}\nabla\left(P_n^m(\eta)P_n^m(\xi)\begin{Bmatrix}\sin m\phi\\ \cos m\phi\end{Bmatrix}\right)\\ &= -\sum_j b_j\nabla\psi_j^{\mathrm{PR}}(\eta,\xi,\phi),\end{aligned} \tag{16}$$

where P_n^m is the associated Legendre function of the first kind of order m and degree n, and j denotes admissible combinations of m and n ([1] and references therein). We frequently choose prolate spheroidal systems, with origin at the (possibly hypothetical) object location, because UXO typically have elongated, rotationally symmetric shapes, requiring few terms in the series. Figure 6 shows example magnetic field lines for some of the lowest modes in the series in (16).

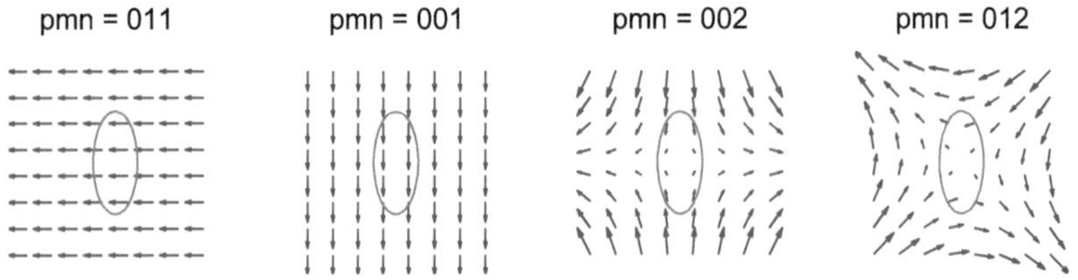

Fig. 6 Any primary field can be considered as the superposition of a set of predefined spheroidal excitation modes $\mathbf{H}_j^{\mathrm{PR}}(\mathbf{r})$. Here we see the magnetic field lines corresponding to some of the most fundamental modes.

Instead of using responses by equivalent sources (e.g. charges), one can construct the response to each fundamental excitation using an analytic spheroidal function series, similar to the one expressing the primary field:

$$\begin{aligned}\mathbf{H}_j^s(\eta,\xi,\phi) &= -\sum_{m,n} B_{m,n}^j \nabla \left(P_n^m(\eta) Q_n^m(\xi) \begin{Bmatrix} \sin m\phi \\ \cos m\phi \end{Bmatrix} \right) \\ &= -\sum_k B_k^j \nabla \psi_j^s(\eta,\xi,\phi),\end{aligned} \tag{17}$$

where Q_n^m is the associated Legendre function of the second kind of order m and degree n. In this case, the S_k^j in (15) are just the coefficients in the series in (17), i.e. the B_k^j. While this formulation presents its own difficulties relative to a source-based response parameterization, it is distinguished by the fact that the B_k^j are unique. That is, one can show that they are characteristics of the object, regardless of excitation or manner of observation, and any object possesses one and only one set of them in a chosen coordinate system [7, 8].

A crucial feature of EMI SEA decomposition using spheroidal potential functions is that very few of them are required. Figure 7 shows averages over instrument position of the b_j values obtained experimentally for a particular UXO, using a relatively small FD sensor. The sensor-UXO separation was about one to two characteristic lengths relative to both sensor and target. Also, this sensor produces rather nonuniform primary fields relative to most other instruments. Even so, only about four modes dominate the primary field distribution around the UXO. Figure 8 shows a test in which the UXO response predicted by the SEA is compared to measurements, using either four or eight excitation modes. The measurements proceed in sweeps along the grid lines, producing peaks in the signal as the sensor moves past the object. The plots on the right illustrate the character of the overall results. The eight-mode formulation produces slightly more accurate results than the four-mode one; however, the difference is not great and probably not justified on the basis of cost vs. benefit. The reader is referred to the literature [8, 12, 19, 21, 24, 25] for

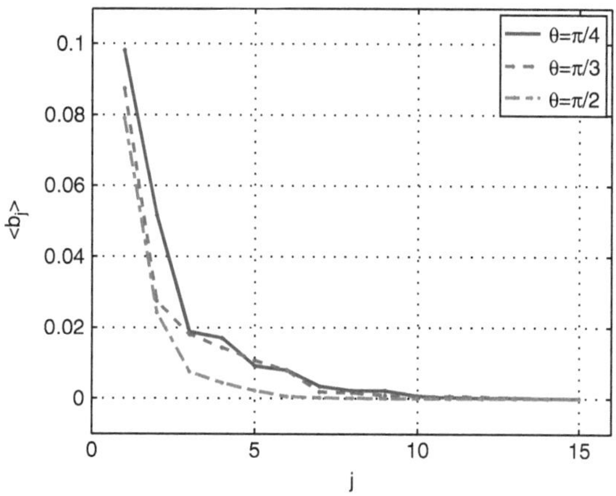

Fig. 7 The most significant coefficients in the decomposition of an example EMI primary field.

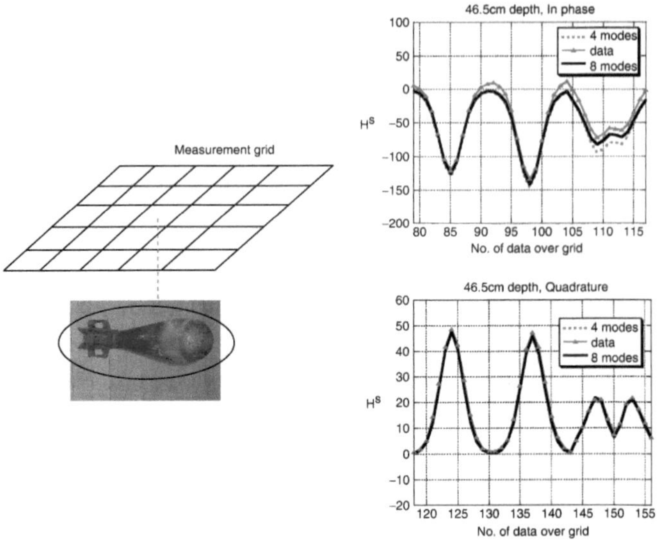

Fig. 8 SEA parameters for a model of the UXO are obtained from controlled measurements over the grid, then used (*plots on right*) to predict the signal at other elevations of the sensor. The results using 8 excitation modes show only slight benefits from inclusion of the higher-order coefficients.

Object	Material	Axis ($2a$)	Axis ($2b$)	$e = b/a$
S2	Steel	30 mm	182 mm	6
S3	Steel	30 mm	90 mm	3
S4	Steel	15 mm	90 mm	6
S7	Steel	30 mm	30 mm	1
A2	Aluminum	30 mm	91 mm	3
A3	Aluminum	15 mm	91 mm	6
C1	S4; S7			
C2	A3; S7			
U1	UXO of Figure 8, mainly steel			

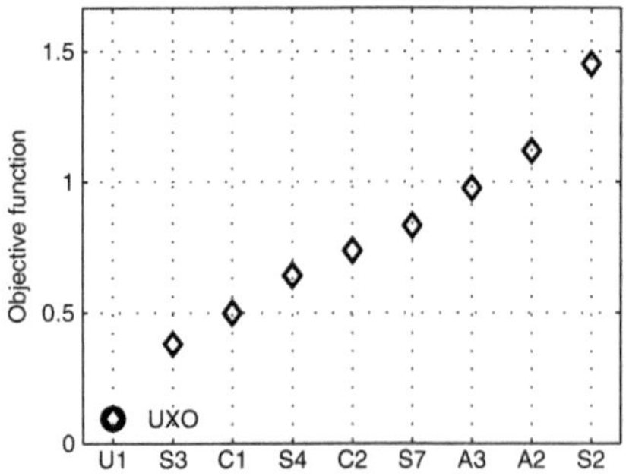

Fig. 9 In a signal pattern matching test over the grid, optimization using the SEA model of the mortar indicates correctly that the model is capable of producing the best match when the UXO in fact produced the data, as opposed to the other objects. The other items produce lowest mismatches that are roughly an order of magnitude worse [24].

discussion of alternatives in pursuing the modal parameters. Issues include control of possible ill-conditioning when one seeks to use higher modes or those only marginally supported by the data.

The SEA model of the same UXO was also used in an optimization over position and orientation to determine the best match (lowest mismatch) it can produce with each recorded signal from a collection of objects. These consisted of machined metal spheroids, combinations of spheroids, and the UXO (Figure 9). The SEA rendering of the UXO indeed produces the lowest mismatch to the actual UXO signal.

Further developments of the SEA include the Normalized Surface Magnetic Charge (NSMC) formulation [18]. The NSMC uses a particularly simple breakdown of the primary field together with synthesis of responses via connected equivalent source distributions. The integral of the source mechanisms itself furnishes a distinctive characterization of the object.

4 Support Vector Machines and Their Application

A possible way to avoid time-consuming nonlinear searches during UXO discrimination can be to perform "before the fact" inversion. One can run a trustworthy model very many times to generate artificial data representative of the expected parameter space. An algorithm could then take those results, make sense of them by weighing the available empirical evidence without any reference to the underlying model, and apply this knowledge to make predictions about unseen cases. In this section we describe one such method, the Support Vector Machine (SVM) [2, 10]. We describe how an SVM can perform binary classification, a task to which most classification and regression problems can be reduced, and then show the results of some SVM experiments related to UXO discrimination.

The "examples" from which an SVM learns to classify are a set $\{\mathbf{x}_i\}$ of n-dimensional vectors. In the UXO problem these can be raw measured fields or distilled parameters – dipole moments or spheroidal expansion coefficients, for example – expected to contain evidence of the character of an object. Depending on the classification we want to make we assign a yes/no attribute to each point: examples belonging to the desired class have $y_i = 1$ and the others $y_i = -1$. SVMs carry out the classification by finding a linear surface, a hyperplane, that divides the parameter space into two distinct regions, each of which hopefully contains points from only one of the categories (Figure 10). During the learning process the machine readjusts the hyperplane parameters to accommodate every training vector until it

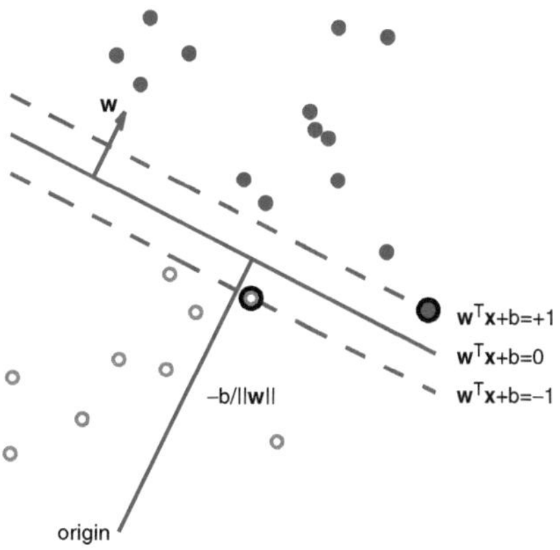

Fig. 10 Support vector classification. The weight vector is perpendicular to the separating hyperplane. The negative of the bias divided by the norm of the weight is the separation between the hyperplane and the origin. The support vectors are circled.

reaches an optimal compromise. At that point, only those examples whose removal would significantly change the locus of the hyperplane suffice to specify a predicting function. These points with high information content are the *support vectors* that give the method its name.

Most data sets are not linearly separable in the space they occupy, and even bona fide separable sets may be corrupted into nonseparability by noise. On the other hand, it should be possible to make any set separable by projecting it into a space of high enough dimensionality. The separating surface would be flat by construction in the new space but could be curved – even multiply connected – in the original. However, there must be a means to limit the *capacity* of the machine, its ability to classify any data set without error: a machine with too much capacity is like a model with too many adjustable parameters in its tendency to overfit data and noise and concentrate on details rather than on essentials. We must be willing to tolerate some mistakes if we want to generalize well, and the SVM algorithm incorporates this in a transparent way [3].

A hyperplane in n dimensions is completely described by the equation

$$\mathbf{w}^{\mathrm{T}}\mathbf{x}+b=0, \tag{18}$$

where the *weight* vector $\mathbf{w}$ is perpendicular to it and the scalar *bias* b is proportional to its separation from the origin (see Figure 10). Knowing $\mathbf{w}$ and b the machine classifies any subsequent example $\mathbf{z}$ by evaluating $f(\mathbf{z}) = sgn(\mathbf{w}^{\mathrm{T}}\mathbf{z}+b)$.

Statistical learning theory [22] proves that the hyperplane that minimizes a properly defined "expected generalization error" for a given set of points is that with the smallest norm [2]. An SVM sets out to solve the constrained minimization problem

$$\begin{aligned} \min_{\mathbf{w},b} \quad & \frac{1}{2}\mathbf{w}^{\mathrm{T}}\mathbf{w} \\ \text{s.t.} \quad & (\mathbf{x}_i^{\mathrm{T}}\mathbf{w}+b)\,y_i \geq 1. \end{aligned} \tag{19}$$

To prevent overfitting, we relax the constraints by introducing *slack variables* that measure how far a point strays into the "wrong" side:

$$\begin{aligned} \min_{\mathbf{w},b} \quad & \frac{1}{2}\mathbf{w}^{\mathrm{T}}\mathbf{w}+C\sum_i \xi_i \\ \text{s.t.} \quad & (\mathbf{x}_i^{\mathrm{T}}\mathbf{w}+b)\,y_i \geq 1-\xi_i, \;\; \xi_i \geq 0. \end{aligned} \tag{20}$$

Note that we also penalize the objective function for each misclassified example; the proportionality constant C is the capacity we referred to above.

It is more convenient to solve this problem in its *dual* formulation. To paraphrase an introductory calculus problem, instead of finding the rectangle with minimum perimeter when given its area we will fix the perimeter and look for the rectangle with maximum area. Both problems have the same answer, but the second involves a simpler constraint and is easier to solve; this advantage is amplified in the multidimensional problem (19), whose inequality constraints become equalities [10]:

$$\begin{array}{ll} \max_{\boldsymbol{\alpha}} & \sum_i \alpha_i - \frac{1}{2}\sum_{i,j} \alpha_i y_i \mathbf{x}_i^{\mathrm{T}} \mathbf{x}_j y_j \alpha_j \\ \text{s.t.} & \sum_i \alpha_i y_i = 0, \qquad 0 \leq \alpha_i \leq C. \end{array} \tag{21}$$

The solution to the new problem is a vector of Lagrange multipliers α_i, each of which in a sense measures the information content of its corresponding point. Only a small fraction of the examples, the support vectors, have nonzero α_i. Note that in this formulation the capacity limits the amount of knowledge that an example can store; problematic points are eventually "sacrificed" in the interest of good generalization. The symmetric convex quadratic programming problem (21) has no local minima, and that, along with the sparsity of the solution, make the SVM a viable and attractive classifier. The weight is given by $\mathbf{w} = \sum_i \alpha_i y_i \mathbf{x}_i = \sum_{i \in \mathrm{SV}} \alpha_i y_i \mathbf{x}_i$, the bias can be computed by applying the Karush-Kuhn-Tucker conditions to the support vectors [3, 10], and the machine predicts new cases using

$$f(\mathbf{x}) = \mathrm{sgn}\Big(\sum_{i \in \mathrm{SV}} \alpha_i y_i \mathbf{x}_i^{\mathrm{T}} \mathbf{x} + b \Big). \tag{22}$$

Having protected the generalization ability of the machine we are ready to increase the dimensionality of the space. We use a device that follows from the realization that in both (21) and (22) the data enter the problem only in the form of scalar products. It is then possible to have a nonlinear separating surface while still keeping the linearity of the machine by substituting

$$\mathbf{x}_1^{\mathrm{T}} \mathbf{x}_2 \rightarrow K(\mathbf{x}_1, \mathbf{x}_2) = \phi(\mathbf{x}_1)^{\mathrm{T}} \phi(\mathbf{x}_2) \tag{23}$$

for some mapping $\phi(\mathbf{x})$. A function K that can be so expressed is called a *kernel*. There is no one-to-one correspondence between mapping and kernel, and, more important, it is not necessary to know ϕ to find K. The mapping may be into a space of hundreds of thousands of dimensions, yet to gain access to it we only need the much smaller and simpler kernel.

Some kernels stretch out the examples into the added dimensions in such a way that gaps open up between the examples which permit a flat separating surface to pass through. For example, Figure 11 shows the effect of applying to a nonseparable

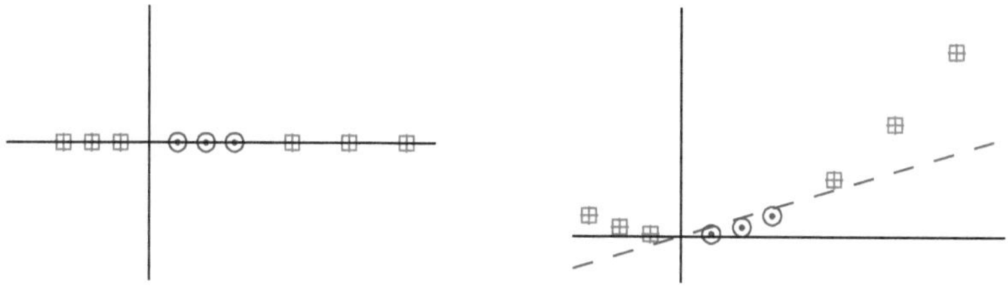

Fig. 11 Applying a polynomial kernel to a nonseparable set of points projects it into a higher-dimensional space and results in a linearly separable distribution. *Left*: original data in X, with two classes corresponding to the circles and squares. *Right*: data projected into $(X,Y,Z) = (x\sqrt{2}, x^2, 1)$, shown on the X-Y plane.

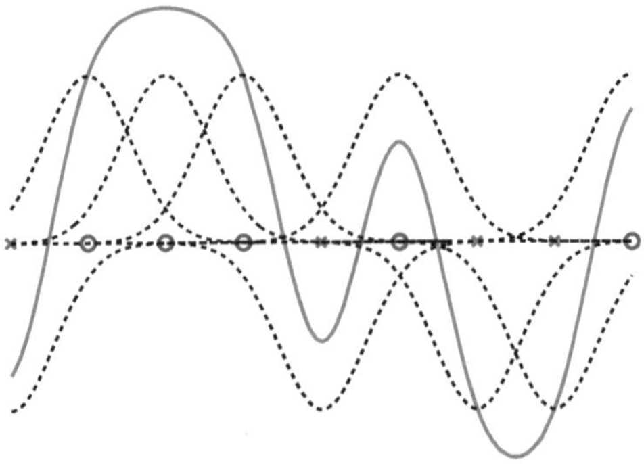

Fig. 12 Two classes of points interspersed, not linearly separable, with $y_i = +1$ for the circles and $y_i = -1$ for the crosses. Dashed lines show the Gaussian (RBF) potential functions around each point; the solid curve is the sum of those functions, which produces the class separator.

one-dimensional set the mapping $x \to \phi(x) = [x\sqrt{2}\ \ x^2\ \ 1]^{\mathrm{T}}$, which corresponds to the polynomial kernel $K(\mathbf{x}_1, \mathbf{x}_2) = (\mathbf{x}_1{}^{\mathrm{T}}\mathbf{x}_2 + 1)^2$. A very popular alternative, the radial basis function (RBF) kernel

$$K(\mathbf{x}_1, \mathbf{x}_2) = \exp(-(\mathbf{x}_1 - \mathbf{x}_2)^{\mathrm{T}}(\mathbf{x}_1 - \mathbf{x}_2)/2\sigma^2), \tag{24}$$

surrounds every point with a (usually Gaussian) surface that resembles a potential function in the sense that it "repels" the separating surface, as shown in Figure 12. The explicit mapping is not known, though in any case in the hyperspace implied by the kernel the separating surface will become flat. The Gaussian width σ is an adjustable parameter; the RBF kernel is found to work best when σ is on the order of the average separation between points.

One view of the RBF kernel, as well as other alternatives, is that it contains some measure of proximity or similarity between two vectors $\mathbf{x}_1$ and $\mathbf{x}_2$. The function attains a maximum when $\mathbf{x}_1 = \mathbf{x}_2$ and declines as the points become more distant or dissimilar. (In the case of the polynomial kernel the similarity involves parallelism rather than closeness.) This provides some measure by which the system can determine whether new cases are similar to (*i.e.*, in the same class as) others.

The SVM principle can also be applied to regression problems, where $y_i \in \mathbb{R}$ instead of $\{-1, 1\}$. The machine creates a surface with a surrounding tube of adjustable width and wiggles it until a given loss function reaches a minimum. The usual choice for this loss function assigns no penalty to points that rest inside the tube and penalizes outliers linearly. The resulting optimization problem is similar to (20): the objective function to be minimized has to strike a balance between fitting accuracy (measured by the capacity and slack variables) and model simplicity (measured by the norm of the weight vector) [10].

Figure 13 shows the results of an SVM classification experiment using synthetic dipole parameters. We choose a range of spheroid diameters (from 1.5 to 15 cm) and a range of elongations (from 0.5 to 4.5) representative of UXO and generate 1,000 spheroids with random diameters and elongations and with conductivities and permeabilities representative of steel and aluminum. We then compute their induced dipole moments under unit axial and transverse uniform primary field excitation using analytic solutions of the EMI equations [1]. Then for given values of the diam-

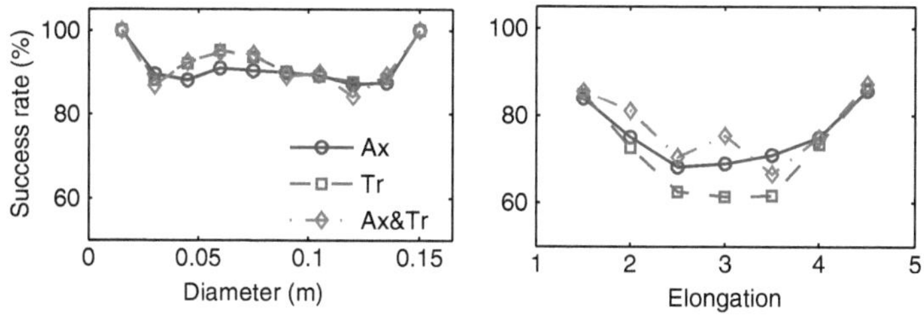

Fig. 13 SVM classification example using dipole moments.

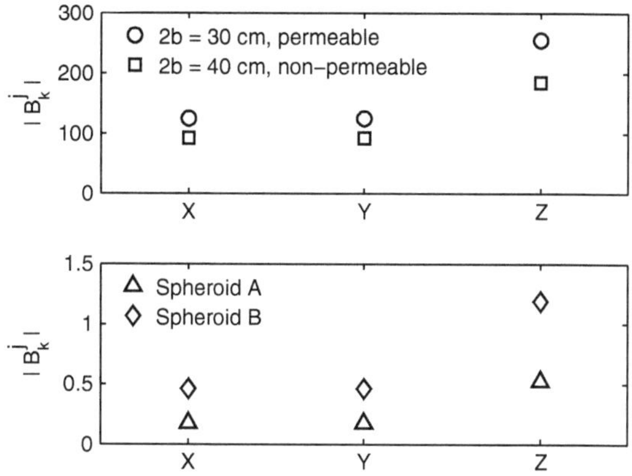

Fig. 14 *Top*: Low-frequency modal response magnitudes B_k^j corresponding to dipole responses to uniform excitation fields in the corresponding directions, for a permeable and a non-permeable spheroid. The smaller object produces larger values. *Bottom*: Same, but for two spheroids of the same material and with the same volume. With different elongation ratios (1.5 and 2) they produce disparate dipole responses at 2 kHz.

eter we train an open-source SVM implementation [17] with 200 examples, telling the machine which ones are larger than the given diameter and which ones are not. After that, we test the SVM predictions on the 800 remaining examples. The figure shows the success rate (defined as the number of correct predictions divided by the total number of tests) as a function of increasing cutoff diameter. The second panel displays the results of repeating the experiment using elongation as the classification parameter. Classification is imperfect, but the results are encouraging [11].

A key discrimination quantity of interest to field workers is the size or volume of an unseen object. To motivate use of the SVM, as well as SEA parameters, we note first that the basic dipole parameters do not necessarily correspond in a simple way to volume. Larger objects do not always produce larger dipole values, especially for composite objects observed over a broad band. Dipole responses can be extracted from within the lowest orders of excitation and response in the complete set of B_k^j. Figure 14 shows that both different materials and also different object proportions can reverse the intuitive ordering of the responses in terms of dipole magnitudes. Even while, by contrast, the (reasonably truncated) full set of B_k^j expresses all possible response behaviors of the objects, there is again no simple, intuitively evident correspondence between the parameter magnitudes that allows ready inference of volume [25]. We need a tool such as the SVM. As a test, an SVM was trained on sets of B_k^j modal response parameters for spheroids of different shapes, materials, and volumes. These had been sorted into "small" and "large" classes based on a chosen volume cutoff. When the trained algorithm was applied to 200 new ("unknown") cases it classified them as shown in Table 1 [25]. These excellent results suggest that the limitations on classification in Figure 13 are due to the shortcomings of the dipole parameters for classification, not to the SVM itself.

The type of analysis just described becomes problematic when we go to the field because it is difficult to obtain these intrinsic object characteristics from measured data. To find the dipole moments – not entirely consistent discriminators anyway – we first have to determine the target's location and orientation, which results precisely in the nonlinear searches that we want to avoid. The spheroidal coefficients are better discriminators but are nontrivial to determine, even when the location and orientation are known exactly. On the other hand, practitioners in areas like handwriting recognition [2] routinely exploit the statistical, model-independent character of the SVM algorithm and apply it to "raw" data that has not been distilled into simpler or unifying parameters. Encouraged by their results, we have employed the SEA [21] as a dependable and accurate model to generate synthetic secondary fields for a collection of UXO at known depths and then used SVM regression to extract unknown depths for other instances. Figure 15 shows a frequency-domain example. A similar approach has been tried on measured data, with reasonable success [13].

Table 1 SVM classification of spheroids based on their response coefficients.

	Predicted large	Predicted small
True large	99	1
True small	0	100

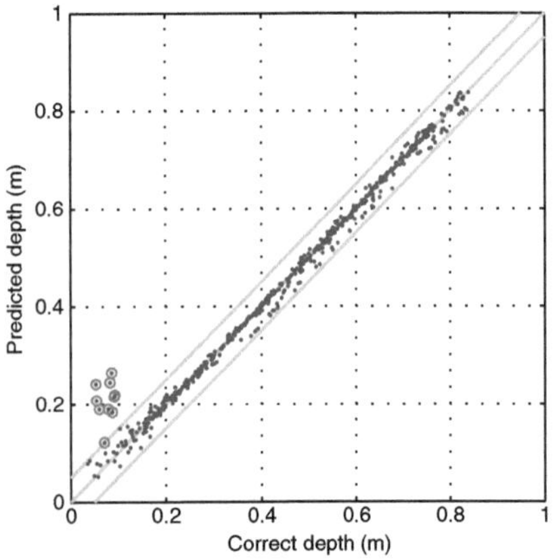

Fig. 15 SVM regression for depth. The machine uses 800 training examples obtained with the SEA and takes 600 tests, all with normalized inphase and quadrature data at $f = 390$ Hz and 14 spatial points. Less than 3% of the tests (*highlighted with circles*) have their depth misjudged outside the 5-cm range shown by dim gray lines.

5 Clutter Reduction by Upward Continuation

In most realistic situations, subsurface UXO discrimination is clutter-limited. Physical clutter causes signal clutter, which lowers the SNR to a point where signals are unintelligible by any means of processing. This applies particularly to GPR but afflicts EMI as well. Wherever ordnance has failed to explode other ordnance probably has exploded, leaving fragments of metal in or on the soil. As a general, order-of-magnitude rule of thumb, the magnetic response of a metal body is proportional to its volume. However, fragments are often shallower than a UXO, and thereby nearer the sensor. The $1/r^3$ or $1/r^6$ factors for signal decay cited above can therefore make clutter signals quite strong relative to those from UXO, even while the fragments are inherently much weaker scatterers. If one could observe the scene from a greater elevation, the ratio of distances to UXO and to clutter would become similar, eliminating this problem. While actual sensor elevation is impractical – if only because the overall level of signal might well diminish to the level of the background – the same advantages can sometimes be obtained by computational upward continuation of fields that are measured near the ground.

Our strategy here will be to obtain magnetic-field data over a grid near the surface at some elevation z_m. The data are then used to infer a sheet of equivalent sources over the ground surface that reproduces the measured values. In particular, given the $\hat{\mathbf{e}}$ component of $\mathbf{H}$ over a surface, we use the gradient of (11) to solve for q_m over a lower surface S_0:

$$H_e(\mathbf{r}) = \int_{S_0} dS'\, q_m(\mathbf{r}') \frac{\hat{\mathbf{e}} \cdot \hat{\mathbf{R}}}{4\pi R^2}, \tag{25}$$

where $\mathbf{R} = \mathbf{r} - \mathbf{r}'$, $\mathbf{r}$ is at elevation z_m, and $\mathbf{r}'$ is on the ground surface. For computational purposes (25) is discretized to form a matrix equation. In this instance, in contrast to assumptions heretofore, let us assume that the object is responding to a single broad primary field while the receivers sample the resulting single secondary field at diverse points (the system depicted in Figure 16, where the survey field is surrounded by a single large transmitter loop [5]). In effect, H_e forms a boundary condition on the field above a plane at z_m. The q_m solution produces fields through

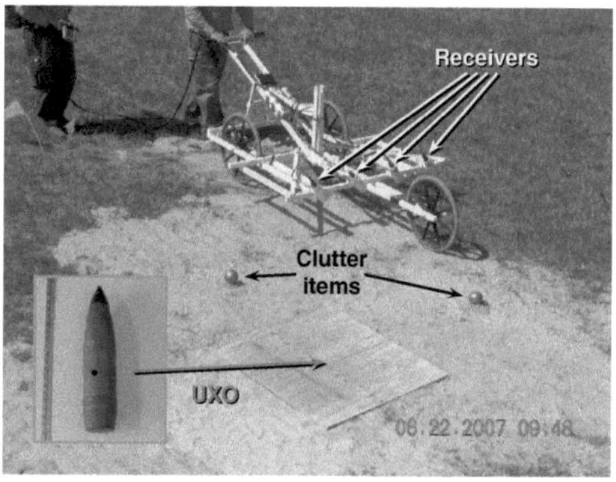

Fig. 16 Field setup, UXO, clutter, and sensor.

the integral that satisfy that boundary condition and that also produce an **H** field above the boundary that will satisfy all applicable governing equations. Therefore, having obtained the q_m from the boundary data via (25), we use it to predict data that would be obtained at a higher elevation. This approach was tested in the field using a buried UXO and shallower pieces of clutter (shotputs, Figure 16).

Figures 17 and 18 show results from this scheme. In the top row, the calculated field on the data plane at z_m based on the q_m solution at $z = 0$ closely reproduces the data. Note that the clutter signals are substantially stronger than those from the UXO. In the bottom row, the signal calculated by continuation to a higher elevation once again matches the corresponding data quite well. While all signals have declined in magnitude, the UXO now produces a response roughly twice that of the clutter. Further, the picture from upward continuation is not afflicted with noise characteristic of the weaker measurements from greater elevation. Projecting the secondary field further upward (Figure 18) shows further and further decrease in the contribution of the clutter and greater prominence of the UXO. This illustration is chosen because the clutter signals are clearly visible alongside the UXO, a situation

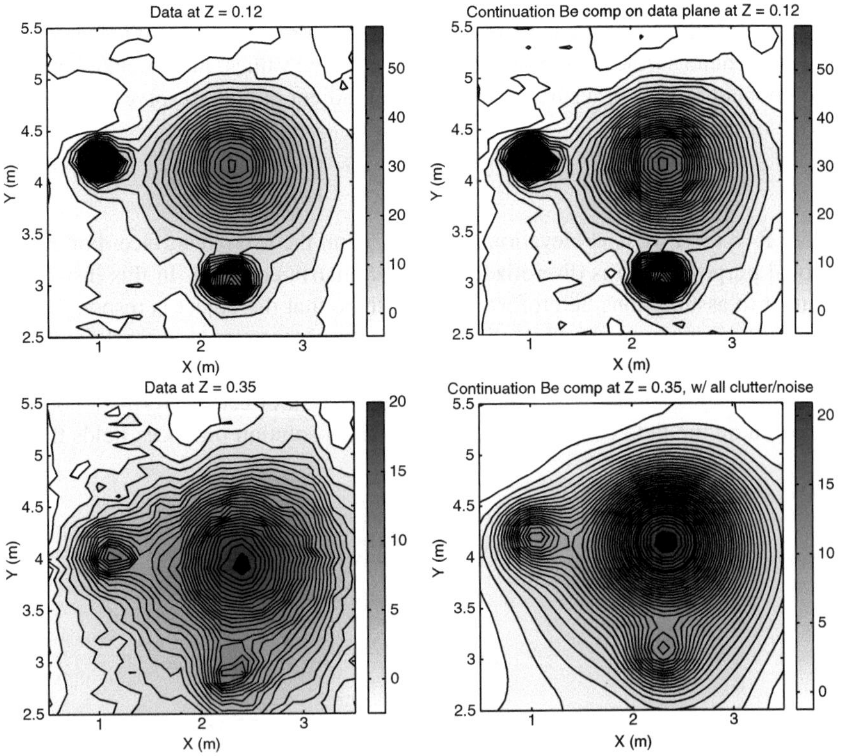

Fig. 17 Case with clutter produced by two shotputs and a 105-mm projectile at the center. *Left*: data. *Right*: Continuation results. *Top row*: Initial elevation. *Bottom*: Signal for sensor at an elevation 23 cm higher.

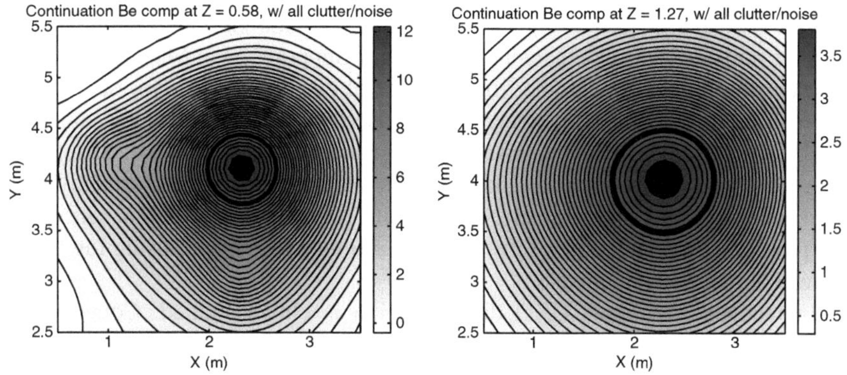

Fig. 18 Computed signals for a 105 mm UXO plus two surface clutter items, with data continued beyond the 35-cm elevation of the previous figure to 58, 81, and 127 cm (*top, middle, bottom*).

one might readily deal with otherwise in practice. However, further investigations show this same sort of clutter suppression in the continued fields even when a clutter signal much stronger than the UXO is mixed within it, from an object directly above it [15].

6 Summary

To sense subsurface UXO one must detect the metal they contain using electromagnetic devices. Discrimination is difficult because low frequencies and long wavelengths are required for ground penetration and reasonable SNR, which limits resolution. Electromagnetic induction sensors have the advantage of being UWB (10s of Hz up to 10s or perhaps 100s of kHz) with the ground essentially transparent to magnetic fields over the entirety of the band. Within metal objects, the fields operate by diffusion. Over the relevant soil and air distances ($<$10 m), the fields can be expressed entirely in terms of a scalar magnetic potential. In our parameter space and scale of observation, there are no waves in the EMI regime, hence no true reflections, diffractions, refractions, or resonances in the wave sense.

One can produce rigorous and fast forward models for EMI responses by chosen metal objects, for known sensors, using the Standardized Excitation Approach. This can be done without solving in detail for the responses within the object. Instead one infers the object's exterior responses to a basis of defined fundamental excitations. Any complete excitation can be expressed as a superposition of the fundamental excitations; therefore any complete response is a corresponding superposition of responses to those excitations. Once the SEA model for an object is defined, its execution is very fast.

For discrimination or classification, the SEA provides highly realistic possible signals in the course of optimizations to match patterns in data from unknown objects. The method can also provide large numbers of high fidelity training examples for statistical learning machines, such as the Support Vector Machine. The SVM operates by implicitly mapping from an original data space to a hyperspace. In the latter the originally intermixed classes for discrimination are separated quite simply by a (hyper)plane. Having trained SVMs using the SEA or analytical solutions, we show some success in using the method to classify objects geometrically, based on their dipole moment parameters; to classify them for size based on sets of their unique SEA parameters themselves; and to estimate their depth using raw signals instead of distilled parameters.

In realistic situations, UXO discrimination is clutter limited. Computational upward continuation shows promise as a physics-based method of EMI clutter suppression. A sheet of equivalent sources is inferred from data at one elevation. These can then predict the signal at greater elevations, at which clutter influences fade.

Acknowledgements We gratefully acknowledge contributions to the work reported here by Dr. Benjamin E. Barrowes, Prof. Xudong Chen, Ms. Xiang-Xiang Cheng, Mr. Steven M. Griffin, Prof. Shah A. Haider, Mr. Lynn Helms, Prof. Keith D. Paulsen, Ms. Irma Shamatava, Prof. Fridon Shubitidze, Dr. Keli Sun, Dr. Bae-Ian Wu, Dr. Beijia Zhang, and the late Prof. Jin-Au Kong.

References

1. Barrowes, B.E., O'Neill, K., Grzegorczyk, T.M., Chen, X., Kong, J.A.: Broadband analytical magneto-quasistatic electromagnetic induction solution for a conducting and permeable spheroid. IEEE Trans. Geosci. Remote Sensing **42**, 2479–2489 (2004)
2. Boser, B.E., Guyon, I.M., Vapnik, V.N.: A training algorithm for optimal margin classifiers. In: D. Haussler (ed.) Proceedings of the 5th Annual ACM Workshop on Computational Learning Theory, pp. 144–152. ACM Press, New York (1992)
3. Burges, C.J.C.: A tutorial on Support Vector Machines for pattern recognition. Data Min. Knowl. Disc. **2**, 121–167 (1998)
4. Butler, D.K.: Implications of magnetic backgrounds for unexploded ordnance detection. J. Appl. Geophys. **54**, 111–125 (2003)
5. Cattach, M.K., Stanley, J.M., Lee, S.J., Boyd, G.W.: Sub-Audio-Magnetics (SAM): a high-resolution technique for simultaneously mapping electrical and magnetic properties. Explor. Geophys. **24**, 387–400 (1993). See also http://www.g-tek.biz/library.html
6. Chen, C.C., Higgins, M.B., O'Neill, K., Detsch, R.: Ultrawide-bandwidth fully-polarimetric ground penetrating radar classification of subsurface unexploded ordnance. IEEE Trans. Geosci. Remote Sensing **39**, 1221–1230 (2001)
7. Chen, X., O'Neill, K., Barrowes, B.E., Grzegorczyk, T.M., Kong, J.A.: Application of a spheroidal mode approach with differential evolution in inversion of magnetoquasistatic data for UXO discrimination. Inv. Prob. **20**, 27–40 (2004)
8. Chen, X., O'Neill, K., Grzegorczyk, T.M., Kong, J.A.: Spheroidal mode approach for the characterization of metallic objects using electromagnetic induction. IEEE Trans. Geosci. Remote Sensing **45**, 697–706 (2007)
9. Cheng, D.K.: Field and Wave Electromagnetics. Addison-Wesley, Reading, MA (1989)
10. Cristianini, N., Shawe-Taylor, J.: An Introduction to Support Vector Machines and Other Kernel-Based Learning Methods. Cambridge University Press, Cambridge (2000)
11. Fernández, J.P., Barrowes, B., O'Neill, K., Paulsen, K., Shamatava, I., Shubitidze, F., Sun, K.: Evaluation of SVM classification of metallic objects based on a magnetic-dipole representation. In: J.T. Broach, R.S. Harmon, J.H. Holloway Jr. (eds.) Detection and Remediation

Technologies for Mines and Minelike Targets XI, *Proceedings of SPIE*, vol. 6217, pp. 6217–03. Bellingham, WA (2006)
12. Fernández, J.P., Barrowes, B., O'Neill, K., Shamatava, I., Shubitidze, F., Sun, K.: A data-derived time-domain SEA for UXO identification using the MPV sensor. In: R.S. Harmon, J.T. Broach, J.H. Holloway Jr. (eds.) Detection and Sensing of Mines, Explosive Objects, and Obscured Targets XIII, *Proceedings of SPIE*, vol. 6953, pp. 6953–1H. Bellingham, WA (2008)
13. Fernández, J.P., Sun, K., Barrowes, B., O'Neill, K., Shamatava, I., Shubitidze, F., Paulsen, K.: Inferring the location of buried UXO using a Support Vector Machine. In: R.S. Harmon, J.T. Broach, J.H. Holloway Jr. (eds.) Detection and Remediation Technologies for Mines and Minelike Targets XII, *Proceedings of SPIE*, vol. 6553, pp. 6553–0B. Bellingham, WA (2007)
14. O'Neill, K.: Ultra-wideband, fully polarimetric ground penetrating radar for UXO discrimination. ESTCP Final Technical Report, Project 199902. www.estcp.org (2005)
15. O'Neill, K.: Processing for clutter evasion in UXO discrimination. SERDP Project MM-1590 Final Technical Report. www.serdp.org (2008)
16. Pasion, L.R., Oldenburg, D.W.: A discrimination algorithm for UXO using time domain electromagnetics. J. Environ. Eng. Geophys. **6**, 91–102 (2001)
17. Rüping, S.: mySVM-Manual. University of Dortmund, Lehrstuhl Informatik 8 (http://www-ai.cs.uni-dortmund.de/SOFTWARE/MYSVM/) (2000)
18. Shubitidze, F., O'Neill, K., Barrowes, B., Shamatava, I., Sun, K., Fernández, J.P., Paulsen, K.: Application of the normalized surface magnetic charge model to UXO discrimination in cases with overlapping signals. J. Appl. Geophys. **61**, 292–303 (2007)
19. Shubitidze, F., O'Neill, K., Shamatava, I., Sun, K., Paulsen, K.D.: Fast and accurate calculation of physically complete EMI response by a heterogeneous metallic object. IEEE Trans. Geosci. Remote Sensing **43**, 1736–1750 (2005)
20. Stratton, J.A.: Electromagnetic Theory. McGraw-Hill, New York (1941)
21. Sun, K., O'Neill, K., Shubitidze, F., Shamatava, I., Paulsen, K.: Fast data-derived fundamental spheroidal excitation models with application to UXO discrimination. IEEE Trans. Geosci. Remote Sensing **43**, 2573–2583 (2005)
22. Vapnik, V.N.: An overview of statistical learning theory. IEEE Trans. Neural Netw. **10**, 988–999 (1999)
23. Wait, J.R.: A conducting sphere in a time varying magnetic field. Geophysics **16**, 666–672 (1951)
24. Zhang, B.: Classification, identification, and modeling of unexploded ordnance in realistic environments. Ph.D. thesis, Massachusetts Institute of Technology (2008)
25. Zhang, B., O'Neill, K., Kong, J.A., Grzegorczyk, T.M.: Support vector machine and neural network classification of metallic objects using coefficients of the spheroidal MQS response modes. IEEE Trans. Geosci. Remote Sensing **46**, 159–171 (2008)
26. Zhang, B., O'Neill, K., Kong, J.A.: Absolute calibration of EMI measurements and application to soil magnetic susceptibility inference. J. Environ. Eng. Geophys. **13**, 223–235 (2008)

Some Advances in UWB GPR

Gennadiy Pochanin*

Abstract A principle of operation and arrangement of UWB antenna systems with frequency independent electromagnetic decoupling is discussed. The peculiar design of the antenna makes it possible to use it in two different modes: horizontal scanning mode and accurate definition of local object location mode. The technique for automatic local objects detection on GPR images is considered. It is based on the Hough transform for detection of hyperbolic curves. Estimation of the accuracy of the objects′ measured coordinates and evaluation of the detection probability have been performed for the case of automatic interpretation of GPR sounding results.

Keywords: Ground penetrating radar, GPR, impulse signal, transmitting-receiving antennas, decoupling, Hough transform, detection probability, false alarm probability

1 Introduction

A wide variety of ground penetrating radars (GPR) is considered as possible equipment for mine and UXO detection. There are many reports and scientific papers discussing different achievement in this area (e.g. [10–12] and others).

However, in practice, the power budget of GPR leaves much to be desired when experiments on GPR sounding are carried out. The large power budget of GPR means deeper sounding, higher resolution, higher detection probability and lower false alarm probability. Taking into account characteristics of an existing short pulse generator, there is no problem driving a radiating antenna by a power pulse. However, GPR is a short range radar and it demands that the receiving antenna be close to the transmitting antenna. Widely known antennas, like "bow-ties", wideband

*A.Ya. Usikov Institute for Radiophysics and Electronics of NAS of Ukraine,
Akad. Proskury St. 12 Kharkov, 61085, Ukraine, e-mail: gpp@ire.kharkov.ua
Tel.: 38(057)7203470
Fax: 38(057)3152105

dipoles and TEM horns when used as transmitting–receiving GPR antennas are electromagnetically coupled. This means that power pulses are induced in the receiving antenna when the transmitting antenna is excited. It is this coupling phenomena that limits the power budget of GPR. Quite good GPR antenna systems provide decoupling which is about -30 dB.

To overcome this problem the author with his colleagues suggested a way to achieve full frequency independent electromagnetic decoupling, as described in [9]. Details of the transmitting–receiving (TR) antenna design and operational principles are given in Section 2.

Usually UXO and land mines are quite small objects. Under these conditions it is possible to analyze them as local objects. Section 3 considers an approach providing automatic detection of local objects.

It is well known that the local objects with small dimensions form hyperbolic curves in a GPR image. The Hough transform is an effective technique for automatically searching for curves in binary images [2]. This technique is applied to detection of hyperbolic curves in the GPR images as described in [1, 8]. The theory of the Hough transform for fast and precise detection of local objects and for determination of soil properties has been stated in the papers [4, 6]. This method is tested using simulated and experimental data. Relations between the object detection probability and the false alarm probability, and the accuracy of determination of the objects' coordinates are obtained [5].

2 High Decoupled Antenna for UWB Pulse GPR

2.1 Principle of operation and arrangement

2.1.1 Principle of operation

Two dipoles which are placed symmetrically with respect to the YZ plane and antiphased excited by $G1$ and $G2$ (Figure 1) generate an electromagnetic field with only Ex and Hy components in the YZ plane. This means that if we place a plane conductor there, the pair of radiating dipoles does not induce any current in this conductor. Thus, receiving antenna in the YZ plane does not receive the electromagnetic field generated by the two dipoles of the transmitting antenna. There is an absolute mutual compensation of electromagnetic fields Ey, Ez, Hx and Hz generated by the transmitting dipoles. Moreover, this holds independently of the exciting signal waveform.

Very high frequency independent electromagnetic decoupling between the TR modules is possible if a single dipole is the radiating antenna and the receiving antenna is a pair of dipoles placed symmetrically with respect to the YOZ plane. It is only necessary to connect the outputs of the receiving dipoles in an appropriate way (Figure 2).

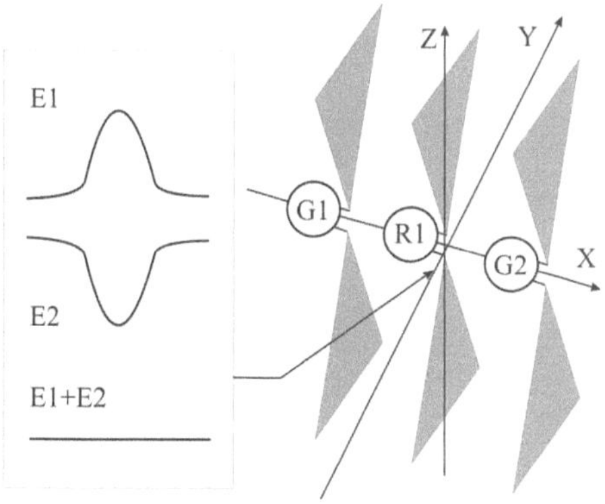

Fig. 1 Electric fields compensation.

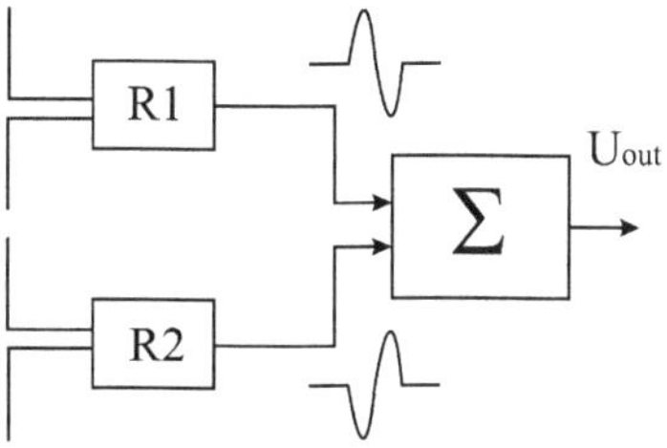

Fig. 2 Summation of signal from the outputs of the receiving module.

EMFs with the same waveform and amplitude are induced in the receiving dipoles under the influence of the radiated electromagnetic field. Subtracting the signals from the outputs of the receiving dipoles, in the summing unit, we achieve a minimal signal at the antenna output. As a result, we have TR antennas decoupling.

Since the signal at the receiving antenna output is the difference of signals received by its two elements; the receiving module is a low-cut filter. The lowest working frequency depends on the relative delay between the signals received by the elements of the receiving module.

2.1.2 Arrangement

The antenna system (Figure 3) consists of a bow-tie transmitting antenna on the middle plate and a pair of receiving bow-ties, one above and one below the middle plate. The distance between the antenna elements of the receiving module is 160 mm. Thus, it effectively receives the electromagnetic field which arrives from the direction of the X axis, and the typical rise time is less than 0.5 ns. A high voltage short pulse generator is used to drive the radiating antenna. The principle of pulse forming by a drift step-recovery diode [7] provides generation of 450 V in amplitude, 0.5 ns in rise time, and 25 kHz in repetition rate pulses.

Fig. 3 The antenna system arrangement.

Under these driving signal parameters and absent reflecting objects near the antenna system (Figure 3) the pulse amplitude at the output of the receiving antenna is less than 3 mV. This implies that the decoupling value is better than -103 dB.

2.2 Modes of operation

There is one more advantage of the antenna system. Its radiation pattern is the product of those of the transmitting and receiving modules. Thus it has only two peaks along the perpendicular to the main plate (in Figure 3). The pattern has nulls in the bow-ties' plane in any direction. It is unresponsive to clutter coming from objects and other sources of electromagnetic radiation situated in the antenna symmetry plane.

GPR with the described antenna system is able to work in two modes:

1. Horizontal scanning mode
2. Accurate definition of local object location mode

Mode 1 is commonly used when the antenna system moves on the ground. The antenna pattern has two peaks in both the nadir and zenith directions, and a null in the horizontal plane. In fact, using this GPR system in horizontal scanning mode is similar to the usual GPR technique. The only advantage is in the power budget, owing to higher decoupling in comparison with conventional bow-tie GPR antenna systems.

Mode 2. In order to provide accurate definition of local object location mode it is necessary to rotate the antenna system around the Z axis (Figure 1) and to perform sounding moving the antenna along the X axis in Figure 1.

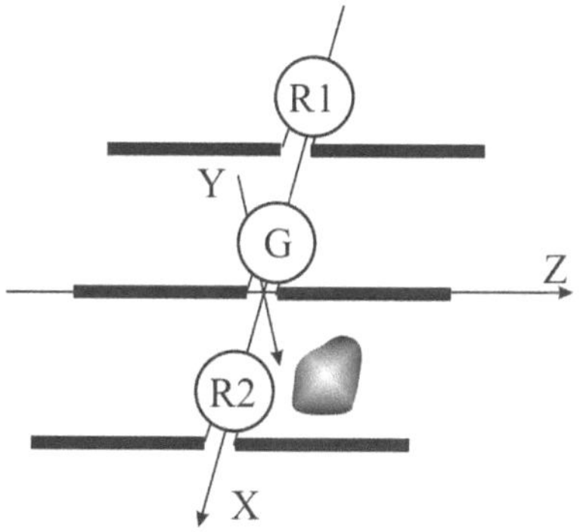

Fig. 4 Accurate definition of local object location.

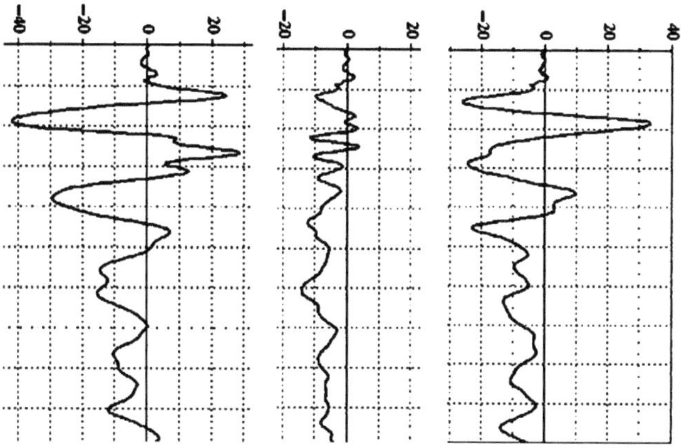

Fig. 5 Output signals.

If the antenna system moves over a local object (Figure 4), the received signal changes its waveform (Figure 5).

When the object is located far from the antenna system, the output signal amplitude is very small. As the distance between the antenna and object decreases the output amplitude increases. It reaches its maximum when one of the elements of the receiving antenna is over the object (Figure 5).

The amplitude goes to zero when the object is in the antenna symmetry (YZ) plane. At further antenna displacement the signal amplitude increase again. It changes its polarity and reaches its maximum when the other element of the receiving antenna goes directly over the object. Thus, if during movement the signal amplitude at the output of the antenna system goes through zero and changes polarity, it means that a local object is in the ground and the location of the object corresponds to the location of the antenna symmetry plane where the output signal was minimal.

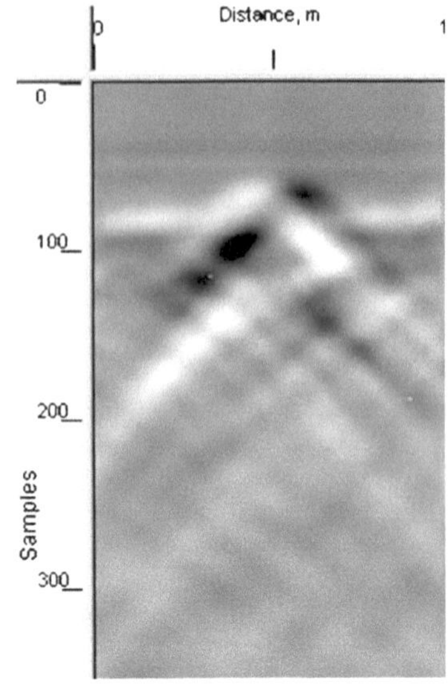

Fig. 6 Experimental profile.

Figure 6 shows results of a test of the antenna working in accurate definition of local object location mode. The initial GPR profile, without applying any data processing procedures, is shown. It corresponds to a section of the sounded path 1 m in length.

Accuracy of the object's horizontal coordinate measurement is about 2 cm. It should be noted that, in contrast to horizontal scanning mode, the accurate definition of local object location mode provides high horizontal resolution at shallow depth.

3 Automatic Object Detection with GPR Images Containing a Response from a Local Object

3.1 Use of the Hough transform for detection of GPR hyperbolic curves

The Hough transform associates the original binary image of the profile (the so-called "space of signals") with another image (the Hough space) where a set of hyperbolic curves that cross at one point with coordinates x_0', y_0' (position of a local object in the Hough space) corresponds to one hyperbola in the space of signals. In other words, one pixel that is a component of the source hyperbola drawn in Figure 7 as a dashed curve is the vertex of the hyperbola in the Hough space.

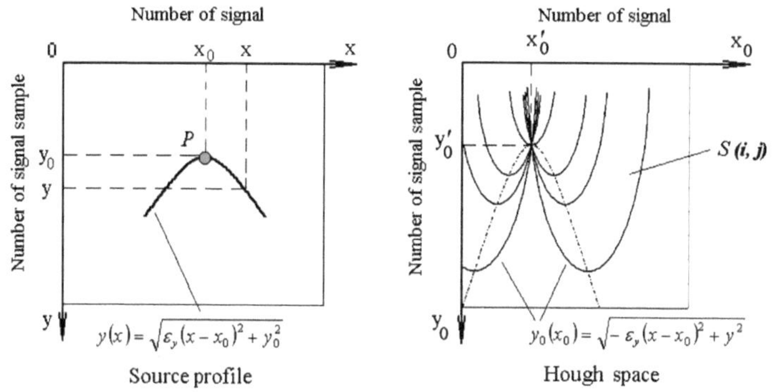

Fig. 7 The hyperbolic curve in the binary image of the profile and the Hough space.

In the "classical" case [2] for HT calculation the Hough space should be divided by a rectangular mesh into collecting elements $S(i, j)$ of fixed size. The number of black points in the original binary image that lie on the curve $y_0(x_0)$ is calculated for every collecting element. Thus the spatial accuracy depends on the size of a collecting element. Then maximal values of S are calculated as a function of three variables y_0, x_0 and ε, and a collecting element with the highest value corresponds to three parameters defining the detected hyperbola in the original binary image.

The standard HT requires long-term computations. The authors have suggested a way to reduce computation time using the following algorithm:

- The Hough space should be divided into collecting elements – 1×1 pixels in size.
- One hyperbola $y_0(x_0)$ in the Hough space should be plotted for every black point of the whole original image.
- Coordinates of each point of this hyperbola should be calculated, and the accumulator corresponding to these coordinates should be increased by one.

So, if several hyperbolas fall within one element, the accumulator grows according to the number of hyperbolas.

It is obvious that the Hough space should be calculated and plotted only once. All points in the original image are already taken into consideration. Thus, this reduces the calculation time considerably. Moreover, this technique precisely determines coordinates of the hyperbola vertices because the element size is originally 1×1 pixels.

3.2 Hough space at different permittivity values

Consider the Hough space of a single hyperbola when the value of ε used in calculations differs from the correct value. The Hough space has been imaged for a test

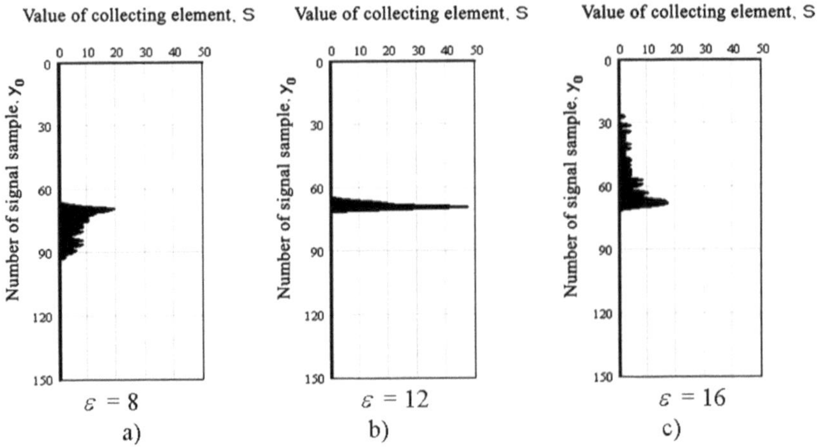

Fig. 8 Histograms of collecting elements along the coordinate y_0.

hyperbola similar to those shown in Figure 7 when x_0 and y_0 are fixed and ε takes several different values. The correct value of ε is 12.

The simulation results are shown in Figure 8. Histograms of collecting elements along the coordinate y_0, i.e., vertical profiles of the accumulator in collecting elements (VPACE), form a cross-section of the Hough space with the plane along the straight line $x_0 = x_0'$. It is perpendicular to the plane x_0y_0.

One can see that at the exact matching of the actual value ε the VPACE looks like a peak at y_0' (Figure 8b). Depending on the value of mismatch and its sign (greater or less than the actual ε) it shows changes with depth either from zero to fast growing and then slow drop, or slow growing then peak and the fast drop to zero.

Thus, when ε changes there are phenomena typical for focusing. Therefore, when a specified value of permittivity does not correspond to its actual value, there is a defocusing in the Hough space. The defocusing pattern depends on the difference between the calculated and actual values of ε.

An algorithm for adaptive selection of ε (Figure 9b) has been developed based on the behavior of $S(\varepsilon)$.

This algorithm has been tested using several simulated and experimental GPR profiles. Two percent error for permittivity estimation can be achieved for the geometrically generated curves. The error increases to 5–10% for simulated images and to 12% for experimental data because of the presence of clutter. Error in ε calculation corresponds to error in calculation of coordinates of the local object location.

Thus, the algorithm is applicable for automatic GPR data processing and automatic detection of local objects in GPR profile. It allows minimizing the influence of the human factor on data processing, and calculates the object coordinates with accuracy which is not worse than one pixel in the GPR image.

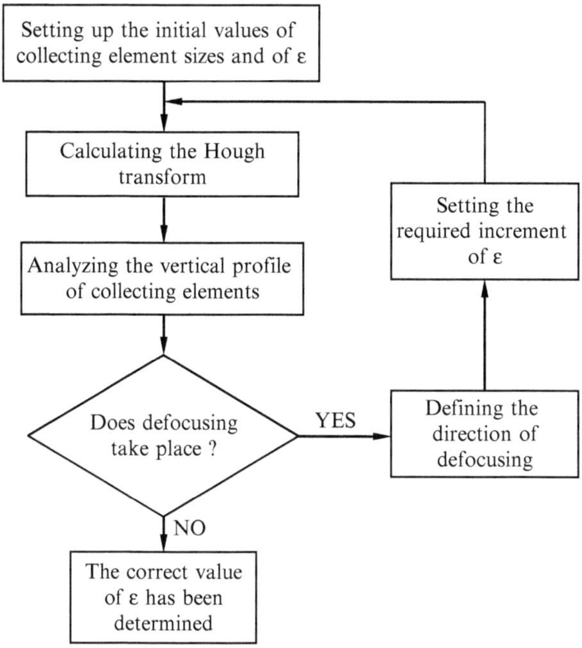

Fig. 9 The algorithm for adaptive selection of ε.

3.3 Performance of automatic object detection method

Usually object detection cannot be done unambiguously. If we deal with a binary classifier, which divides a set of detected targets into two classes – real (the positive instances) and false alarms (the negative instances), the ensemble of metrics in [3] is used for performance measurement.

The second problem that has to be solved while searching for objects is determination of coordinates of the object's location and estimation of errors in these coordinates. It is possible to simulate the necessary profiles using the finite-difference time domain method (FDTD) software.

The automatic object detection method has been tested with

- Geometrically simulated local objects images
- FDTD simulated local objects images
- Experimental data with several local objects [5]

In the first item the classifier marked all true positive instances correctly while processing the image simulated using hyperbola tracing, though several false alarms appeared. In the FDTD examples some problems with classification occurred due to the presence of clutter and to the non-ideal shape of the hyperbolic curves. Comparison between coordinates of the detected peaks and coordinates of the hyperbola vertices shows that the X coordinates of all the vertices in the first example have been determined with zero error, while the determined coordinates of vertices for the second example have a maximal displacement of two pixels from the source values. This means that it is possibile to determine the horizontal coordinates of the object with zero error. The absolute error in the determination of the Y coordinate of the hyperbola vertices did not exceed three pixels for all images.

Experimental data analysis shows that if the separation threshold is selected so that it provides detection of all objects, the false alarm probability will be 91.6%. If the threshold is selected so that the false alarm probability equals 0, then only 60% of all objects (three of five) are detected.

Examples of GPR images with only one clearly visible hyperbola in the binary image were analyzed as well. Here, if the threshold yields 100% detection probability then the false alarm probability will be 33.3%. This is less than in the previous example.

Thus, the possibility to detect subsurface objects in a GPR image automatically and to find their coordinates accurately has been shown using examples of simulated and experimental GPR images. The developed method based on the Hough transform allows this. It has been demonstrated that 100% object detection probability is achievable, and at the same time the false alarm probability is minimal. Nevertheless, it is necessary to optimize the criteria for choosing the optimal threshold for separating the Hough space peaks. This requires more complete statistical data and enough simulated and experimental GPR images.

4 Summary

Two ideas regarding an antenna system with high and frequency independent transmitting–receiving antennas decoupling, and the Hough transform for automatic detection of local objects in GPR images, were discussed. Improved GPR performance was shown.

Acknowledgment The author is grateful to Mr. M.M. Golovko for his contribution to the investigation on automatic object detection with GPR images containing a response from a local object.

References

1. L. Capineri, P. Grande, and J. A. G. Temple. Advanced image-processing technique for real-time interpretation of ground-penetrating radar images. In *International Journal of Imaging Systems and Technology* (9):51–59, 1998.
2. K. S. Fu, R. C. Gonzalez, and C. S. G. Lee. Robotics: Control, Sensing, Vision and Intelligence. New York: McGraw-Hill, 1987.
3. E. Gelenbe and T. Kocak. Area-based results for mine detection. In *IEEE Transactions on Geoscience and Remote Sensing* (38):12–24, Jan., 2000.
4. M. M. Golovko. The automatic determination of soil permittivity using the response from a subsurface local object. In *Proceedings of the 2nd International Conference on "Ultra-*

wideband and ultrashort impulse signals" UWBUSIS-2004 Sevastopol, Ukraine, Sept. 19–22, 2004, 248–250.
5. M. M. Golovko. The evaluation of performances of automatic method for the object detection in GPR images. In *Proceedings of the 5th International Symposium on "Image and Signal Processing and Analysis"* Istambul, Turkey, Sept. 27–29, 2007, 476–481.
6. M. M. Golovko and G. P. Pochanin. Application of the Hough transform for automatic detection of objects in georadar profiles. In *Elektromagnitnye volny i elektronnye sistemy (Electromagnetic waves and electronics systems, in Russian)* 9(9–10):22–30, 2004.
7. V. M. Tuchkevich and V. M. Grekhov. New principles of high power commutation with semiconductor devices. Leningrad: Nauka, 1988.
8. T. Kaneko. Radar image processing for locating underground linear objects. In *IEICE Transactions* E74(10):3452–3458, 1991.
9. Yu. A. Kopylov, S. A. Masalov, and G. P. Pochanin. Method for decoupling between transmitting and receiving modules of antenna system. Patent UA 81652. Jan. 25, 2008.
10. www.sic.rma.ac.be/Publications
11. www.g-p-r.com/biblio.htm
12. www.tudelft.nl

Operational Research Approach to Decision Making

Oleg Pokrovsky*

Abstract The decision making (DM) problem is of great practical value in many areas of human activities. Most widely used DM methods are based on probabilistic approaches. The well-known Bayesian theorem for a conditional probability density function (PDF) is a background for such techniques. It is needed due to some uncertainty in many parameters entered in any model which describes the functioning of many real systems or objects. Uncertainty in our knowledge might be expressed in an alternative form. We offer to employ appropriate confidence intervals for model parameters instead of a relevant PDF. Thus one can formulate a prior uncertainty in model parameters by means of a set of linear constraints. The related cost or goal function should be defined at a corresponding set of parameters. That leads us to stating the problem in terms of operational research or mathematical linear programming. It is more convenient to formulate such optimization problems for discreet or Boolean variables. A review of relevant problem statements and numerical techniques are presented as well as many examples.

Keywords: Decision making, Bayesian theory, linear and integer programming, optimal design

1 Introduction

Decision theory is a theory about decisions. The subject is not a very unified one. To the contrary, there are many different ways to theorize about decisions, and therefore also many different research traditions. This chapter attempts to reflect some of the diversity of the subject. Its emphasis lies on the mathematical aspects of decision theory. Decision theory focuses on how we use our freedom. In the situations treated by decision theorists, there are options to choose between, and we choose

*Main Geophysical Observatory, Karbyshev str.7, St. Petersburg, 194021, Russian Federation, e-mail: pokrov@main.mgo.rssi.ru

in a non-random way. Our choices, in these situations, are goal-directed activities. Hence, decision theory is concerned with goal-directed behaviour in the presence of options. We do not decide continuously. In the history of almost any activity, there are periods in which most of the decision-making is made, and other periods in which most of the implementation takes place. Decision theory tries to throw light, in various ways, on the former type of period. Decision makers divide the decision process into the following five steps:

- Identification of the problem
- Obtaining necessary information
- Production of possible solutions
- Evaluation of such solutions
- Selection of a strategy for performance

The set of above issues is sequential in the sense that they divide decision processes into parts that always come in the same order or sequence. This approach might be criticized. Some empirical material indicates that the "stages" are performed in parallel rather than in sequence. A more realistic model should allow the various parts of the decision process to come in different order in different decisions.

2 Bayesian Decision Theory

Bayesian decision theory is based on the statistical inference in which evidence or observations are used to update or to newly infer the probability that a hypothesis may be true. The name "Bayesian" comes from the frequent use of Bayes' theorem in the inference process. Bayes' theorem was derived from the work of the Reverend Thomas Bayes. Bayesian inference uses aspects of the scientific method, which involves collecting evidence that is meant to be consistent or inconsistent with a given hypothesis. As evidence accumulates, the degree of belief in a hypothesis ought to change. With enough evidence, it should become very high or very low. Thus, proponents of Bayesian inference say that it can be used to discriminate between conflicting hypotheses: hypotheses with very high support should be accepted as true and those with very low support should be rejected as false. However, detractors say that this inference method may be biased due to initial beliefs that one needs to hold before any evidence is ever collected. Bayesian inference uses a numerical estimate of the degree of belief in a hypothesis before evidence has been observed and calculates a numerical estimate of the degree of belief in the hypothesis after evidence has been observed. Bayesian inference usually relies on degrees of belief, or subjective probabilities, in the induction process and does not necessarily claim to provide an objective method of induction. Nonetheless, some Bayesian statisticians believe probabilities can have an objective value and therefore Bayesian inference can provide an objective method of induction.

$$P(H/E) = \frac{P(E/H)P(H)}{P(E)} \tag{1}$$

where:

- H represents a specific hypothesis, which may or may not be some null hypothesis.
- P(H) is called the prior probability of H that was inferred before new evidence, E, became available.
- P(E/H) is called the conditional probability of seeing the evidence E if the hypothesis H happens to be true. It is also called a likelihood function when it is considered as a function of H for fixed E.
- P(E) is called the marginal probability of E: the a priori probability of witnessing the new evidence E under all possible hypotheses. It can be calculated as the sum of the product of all probabilities of any complete set of mutually exclusive hypotheses and corresponding conditional probabilities:

$$P(E) = \sum_i P(E/H_i)P(H_i). \tag{2}$$

- P(H/E) is called the posterior probability of H given E.

The factor P(E/H)/P(E) represents the impact that the evidence has on the belief in the hypothesis. If it is likely that the evidence E would be observed when the hypothesis under consideration is true, but unlikely that E would have been the outcome of the observation, then this factor will be large. Multiplying the prior probability of the hypothesis by this factor would result in a larger posterior probability of the hypothesis given the evidence. Conversely, if it is unlikely that the evidence E would be observed if the hypothesis under consideration is true, but a priori likely that E would be observed, then the factor would reduce the posterior probability for H. Under Bayesian inference, Bayes' theorem therefore measures how much new evidence should alter a belief in a hypothesis.

Bayesian statisticians argue that even when people have very different prior subjective probabilities, new evidence from repeated observations will tend to bring their posterior subjective probabilities closer together. However, others argue that when people hold widely different prior subjective probabilities their posterior subjective probabilities may never converge even with repeated collection of evidence. These critics argue that worldviews, which are completely different initially, can remain completely different over time despite a large accumulation of evidence.

Thus, one applies Bayes theorem (see (1) and (2)), multiplying the prior by the likelihood function and then normalizing, to get the posterior probability distribution, which is the conditional distribution of the uncertain quantity given the data. A prior is often the purely subjective assessment of an experienced expert. Some will choose a conjugate prior when they can, to make calculation of the posterior distribution easier. In decision theory, a Bayes estimator is an estimator or decision rule that maximizes the posterior expected value of a utility function or minimizes the posterior expected value of a loss function (also called posterior expected loss). Unfortunately, there are many decision making examples where Bayes theory fails due to difficulties in determining the prior probability distribution. Standard statistical practice ignores model uncertainty. Data analysts typically select a model from some class of models and then proceed as if the selected model had generated the data. This approach ignores the uncertainty in model selection, leading to over-confident inferences and decisions that are more risky than one thinks they are. Bayesian model averaging (BMA) provides a coherent mechanism for accounting for this model uncertainty. Several methods for implementing BMA have recently emerged [6, 13, 14]. Nonetheless, the BMA approach cannot solve the decision problem entirely [17].

3 Decision Under Severe Uncertainty

It is common to make uncertain decisions [1]. What can be done to make good (or at least the best possible) decisions under conditions of uncertainty? Info-gap robustness analysis evaluates each feasible decision by asking: how much deviation from an estimate of a parameter value, function, or set, is permitted and yet "guarantee" acceptable performance? In everyday terms, the "robustness" of a decision is set by the size of deviation from an estimate that still leads to performance within requirements when using that decision. It is sometimes difficult to judge how much robustness is needed or sufficient. However, according to info-gap theory, the ranking of feasible decisions in terms of their degree of robustness is independent of such judgments. To this end, the following questions must be addressed:

- What are the characteristics of decision problems that are subject to severe uncertainty?
- What difficulties arise in the modelling and solution of such problems?
- What type of robustness is sought?
- How does info-gap theory address these issues?

In what way is info-gap decision theory similar to and/or different from other theories for decision under uncertainty? Two important points need to be elucidated in this regard at the outset:

- Considering the severity of the uncertainty that info-gap was designed to tackle, it is essential to clarify the difficulties posed by severe uncertainty.
- Since info-gap is a non-probabilistic method that seeks to maximize robustness to uncertainty, it is imperative to compare it to the single most important "non-probabilistic" model in classical decision theory, namely Wald's maximin paradigm.

The maximin rule tells us to rank alternatives by their worst possible outcomes: we are to adopt the alternative the worst outcome of which is superior to the worst outcome of the others. After all, this paradigm has dominated the scene in classical decision theory for well over 60 years. So, first let us clarify the assumptions that are implied by severe uncertainty:

1. A parameter λ, whose true value is subject to severe uncertainty
2. A region of uncertainty Δ, where the true value of λ lies
3. An estimate $\tilde{\lambda}$ of the true value of λ

Two remarks should be made with account to above assumption. First, The region of uncertainty is relatively large. Second, the estimate is a poor approximation of the true value of λ. Info-gap decision theory is radically different from all current theories of decision under uncertainty. The difference originates in the modeling of uncertainty as an information gap rather than as a probability. In general, info-gap's robustness model is a mathematical representation of a local worst-case analysis in the neighborhood of a given estimate of the true value of the parameter of interest. Under severe uncertainty the estimate is assumed to be a poor indication of the true value of the parameter and is likely to be substantially wrong. The fundamental question therefore is: given the severity of the uncertainty, the local nature of the analysis and the poor quality of the estimate, how meaningful and useful are the results generated by the analysis, and how sound is the methodology as a whole? The robust optimization literature (see [2]; Kouvelis and Yu, 1997) provides methods and techniques that take a global approach to robustness analysis. These methods directly address decision under severe uncertainty, and have been used for this purpose for more than 30 years. Wald's Maximin model is the main instrument used by these methods. The principal difference between the Maximin model employed by info-gap and the various Maximin models employed by robust optimization methods is in the manner in which the total region of uncertainty is incorporated in the robustness model. Info-gap takes a local approach that concentrates on the immediate neighborhood of the estimate. In sharp contrast, robust optimization methods set out to incorporate in the analysis the entire region of uncertainty, or at least an adequate representation thereof. In fact, some of these methods do not even use an estimate. The info-gap's robustness model is an instance of the generic Maximin model. Therefore, it is instructive to examine the mathematical programming (MP) formats of these generic models [4, 16, 10].

4 Linear programming

A Linear Programming (LP) problem is a special case of a Mathematical Programming problem [3, 8]. From an analytical perspective, a mathematical program tries to identify an extreme (i.e., minimum or maximum) point of a function $f(x_1,x_2,...,x_n)$, which furthermore satisfies a set of constraints, e.g., $g(x_1,x_2,...,x_n) \geq b$. Linear programming is the specialization of mathematical programming to the case where both function f, to be called the objective function, and the problem constraints g are linear. From an applications perspective, mathematical (and therefore, linear) programming is an optimization tool, which allows the rationalization of many managerial and/or technological decisions required by contemporary techno-socio-economic applications. An important factor for the applicability of the mathematical programming methodology in various application contexts is the computational tractability of the resulting analytical models. Under the advent of modern computing technology, this tractability requirement translates to the existence of effective and efficient algorithmic procedures able to provide a systematic and fast solution to these models. For Linear Programming problems, the Simplex algorithm provides a powerful computational tool, able to provide fast solutions to very large-scale applications, sometimes including hundreds of thousands of variables (i.e., decision factors). In fact, the Simplex algorithm was one of the first Mathematical Programming algorithms to be developed [3], and its subsequent successful implementation in a series of applications significantly contributed to the acceptance of the broader field of Operations Research as a scientific approach to decision making.

4.1 Illustrative example

Let us consider a simple example of the MP problem formulation [8]. Assume that a company produces two types of products P_1 and P_2. Production of these products is supported by two workstations W_1 and W_2, with each station visited by both product types. If workstation W_1 is dedicated completely to the production of product type P_1, it can process 40 units per day, while if it is dedicated to the production of product P_2, it can process 60 units per day. Similarly, workstation W_2 can produce daily 50 units of product P_1 and 50 units of product P_2, assuming that it is dedicated completely to the production of the corresponding product. If the company's profit by disposing one unit of product P_1 is \$200 and that of disposing one unit of P_2 is \$400, and assuming that the company can dispose its entire production, how many units of each product should the company produce on a daily basis to maximize its profit?

First notice that this problem is an optimization problem. Our objective is to maximize the company's profit, which under the problem assumptions is equivalent to maximizing the company's daily profit. Furthermore, we are going to maximize the company profit by adjusting the levels of the daily production for the two items P_1 and P_2. Therefore, these daily production levels are the control/decision factors, the values of which we are asked to determine. In the analytical formulation of the problem the role of these factors is captured by modeling them as the problem *decision variables*:

- X_1 = number of units of product P_1 to be produced daily
- X_2 = number of units of product P_2 to be produced daily

In view of the above discussion, the problem objective can be expressed analytically as:

$$f(X_1, X_2) = 200X_1 + 400X_2. \tag{3}$$

Equation (3) will be called the objective function of the problem, and the coefficients 200 and 400, which multiply the decision variables in it, will be called the objective function coefficients.

Furthermore, any *decision* regarding the daily production levels for items P_1 and P_2, in order to be realizable in the company's operational context, must observe the production capacity of the two workstations W_1 and W_2. Hence, our next step in the problem formulation seeks to introduce these technological constraints. Let's focus first on the *constraint*, which expresses the finite production capacity of workstation W_1. Regarding this constraint, we know that one day's work dedicated to the production of item P_1 can result in 40 units of that item, while the same period dedicated to the production of item P_2 will provide 60 units of it. Assuming that production of one unit of product type $P_i (i = 1,2)$, requires a constant amount of processing time $T_{1i} (i = 1,2)$ at workstation W_1, it follows that: $T_{11} = \frac{1}{40}$ and $T_{12} = \frac{1}{60}$. Under the further assumption that the combined production of both items has no side-effects, i.e., does not impose any additional requirements for production capacity of the workstation (e.g., zero set-up times), the total capacity (in terms of time length) required for producing X_1 units of product P_1 and X_2 units of product P_2 is equal to $\frac{1}{40}X_1 + \frac{1}{60}X_2$. Hence, the technological constraint imposing the condition that our total daily processing requirements for workstation W_1 should not exceed its production capacity, is analytically expressed by:

$$\frac{1}{40}X_1 + \frac{1}{60}X_2 \leq 1. \tag{4}$$

Notice that in Equation (4) time is measured in days.

Following the same line of reasoning (and under similar assumptions), the constraint expressing the finite processing capacity of workstation W_2 is given by:

$$\frac{1}{50}X_1 + \frac{1}{50}X_2 \leq 1. \tag{5}$$

Constraints (4) and (5) are known as the technological constraints of the problem. In particular, the coefficients of the variables $X_i (i = 1,2)$, $\frac{1}{T_{ij}} (i, j = 1,2)$, are known as the technological coefficients of the problem formulation, while the values on the right-hand-side of the two inequalities define the right-hand side vector of the constraints. Finally, to the above constraints we must add the requirement that any permissible value for variables $X_i (i = 1,2)$ must be nonnegative since these values express production levels. These constraints are known as the variable sign restrictions. Combining Equations (3) to (5), the analytical formulation of our problem is as follows:

$$max\{f(X_1, X_2)\} = max\{200X_1 + 400X_2\} \tag{6}$$

$$\begin{aligned}
&\frac{1}{40}X_1 + \frac{1}{60}X_2 \leq 1 \\
&\frac{1}{50}X_1 + \frac{1}{50}X_2 \leq 1 \\
&X_i \geq 0 (i = 1,2).
\end{aligned}$$

4.2 The general "linear programming" formulation

Generalizing formulation (6), the general form for a Linear Programming problem is as follows [5]:

Linear Objective Function (LOF) maximization:

$$max\{f(X_1,X_2,...,X_n)\} = max\{\sum c_i X_i\} \tag{7}$$

under *Linear Constraints (LC)*:

$$\sum_j a_{ij} X_j \begin{matrix} \leq \\ or \\ = \\ or \\ \geq \end{matrix} b_i (i = 1,...,m). \tag{8}$$

The LC (8) might be used in important particular cases, when variables signs are prescribed:

$$(X_j \geq 0), or (X_j \leq 0). \tag{9}$$

We conclude our discussion on the general LP formulation by formally defining the solution search space and optimality. Specifically, we shall define as the feasible region of the LP of Equations (6) to (8), the entire set of vectors $\mathbf{X} = (X_1,...,X_n)^T$ that satisfy the LC of (8) and the sign restrictions of (9). An optimal solution to the problem is any feasible vector that further satisfies the optimality requirements expressed by (7)–(9). Introducing integrality requirements for some of the variables in an LP formulation turns the problem to one belonging in the class of (Mixed) Integer Programming (MIP) or Integer Programming (IP).

4.3 Graphical LP's interpretation

In this section, we consider a solution approach for LP problems, which is based on a geometrical representation of the feasible region and the objective function [5]. In particular, the space to be considered is the n-dimensional space with each dimension defined by one of the LP variables (X_1, X_2). Thus we present an illustration for the two-variable case.

We start our investigation regarding the geometrical representation of two-var linear constraints by considering first constraints of the *equality type*, i.e.,

$$a_1 X_1 + a_2 X_2 = b. \tag{10}$$

Assuming $a_2 \neq 0$, this equation corresponds to a straight line with slope $s = \frac{-a_1}{a_2}$ and intercept $d = \frac{b}{a_2}$. In the special case a_2=*0* the solution space (locus) of Equation (10) is a straight line perpendicular to the X_1-axis, intersecting it at the point $(\frac{b}{a_1};0)$.

Notice that the presence of an equality constraint restricts the dimensionality of the feasible solution space by one degree of freedom, i.e., it turns it from a planar area to a line segment.

Consider the *inequality constraint*:

$$a_1X_1 + a_2X_2 \overset{\leq}{\underset{\geq}{\overset{or}{\underset{or}{=}}}} b. \tag{11}$$

The solution space of this constraint is one of the closed *half-planes* defined by Equation (11). To show this, let us consider a point (X_1, X_2) , which satisfies Equation (11) as equality, and another point $(X_1^{'}, X_2^{'})$ for which Equation (11) is also valid. For any such pair of points, it holds that:

$$a_1(X_1^{'} - X_1) + a_2(X_2^{'} - X_2) \overset{\leq}{\underset{\geq}{\overset{or}{\underset{or}{=}}}} 0. \tag{12}$$

Let us consider the left side of (12) as the inner (dot) product of the two vectors $\mathbf{a} = (a_1, a_2)^T$and $\Delta\mathbf{X} = ((X_1^{'} - X_1), (X_2^{'} - X_2))^T$. It is equal to $\left|\left|\Delta\mathbf{X}\right|\right|\left|\left|\mathbf{a}\right|\right|\cos(\Delta\mathbf{X}, \mathbf{a})$. In this case a line $a_1X_1 + a_2X_2 = b$ can be defined by the point (X_1, X_2) and the set of points $(X_1^{'}, X_2^{'})$ such that vector **a** is at right angles with vector $\Delta\mathbf{X}$. Furthermore, the set of points that satisfy the inequality parts of Equation (12) have the vector forming an acute (obtuse) angle with vector **a**, and therefore they are "above" ("below") the line. Hence, the set of points satisfying each of the two inequalities implied by Equation (11) is given by one of the two half-planes the boundary of which is defined by the corresponding equality constraint. Figure 1 summarizes the above discussion.

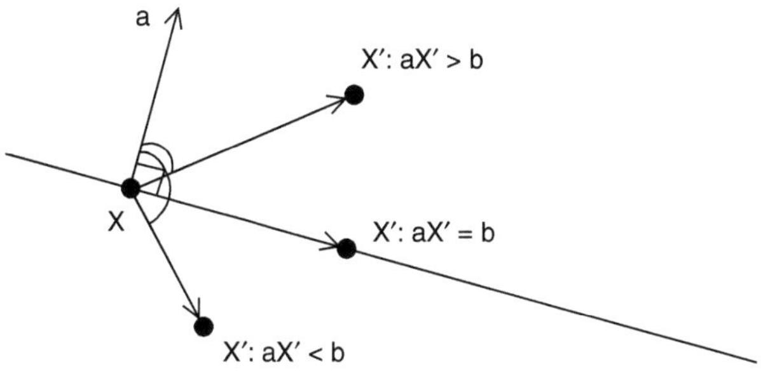

Fig. 1 Half-planes: the feasible region of a linear inequality.

An easy way to determine the half-plane depicting the solution space of a linear inequality is to draw the line depicting the solution space of the corresponding equality constraint and then test whether the point (0, 0) satisfies the inequality. In case of a positive answer, the solution space is the half-space containing the origin, otherwise, it is the other one.

From the above discussion, it follows that the feasible region for the prototype LP of Equation (6) is the shaded area in the following figure:

The next step is a maximization (minimization) of the objective function. The most typical way to represent a two-variable function $c_1X_1 + c_2X_2$ is to perceive it as a surface in an (orthogonal) three-dimensional space, where two of the dimensions correspond to the independent variables X_1 and X_2 , while the third dimension provides the function value for any pair (X_1, X_2). However, in the context of our discussion, we are interested in expressing the information contained in the two-var LP objective function $c_1X_1 + c_2X_2$ in the Cartesian plane defined by the two independent variables X_1 and X_2. For this purpose, we shall use the concept of contour plots. Contour plots depict a function by identifying the set of points (X_1, X_2) that correspond to a constant value of the function $(c_1X_1 + c_2X_2) = a$, for any given range of a's. The plot obtained for any fixed value of a is a contour of the function. Studying the structure of a contour is expected to identify some patterns that essentially depict some useful properties of the function. In the case of LP's, the linearity of the objective function implies that any contour of it will be of the type:

$$(c_1X_1 + c_2X_2) = a \tag{13}$$

i.e., a straight line. For a maximization (minimization) problem, this line will be called an isoprofit (isocost) line. Assuming that $c_2 \neq 0$, Equation (13) can be rewritten as:

$$X_2 = -\frac{c_1}{c_2}X_1 + \frac{a}{c_2}$$

which implies that by changing the value of a , the resulting isoprofit/isocost lines have constant slope and varying intercept, i.e, they are parallel to each other (which makes sense, since by the definition of this concept, isoprofit/isocost lines cannot intersect). Hence, if we continuously increase a from some initial value a_o, the corresponding isoprofit lines can be obtained by "sliding" the isprofit line corresponding to $(c_1X_1 + c_2X_2) = a_o$ parallel to itself, in the direction of increasing or decreasing intercepts, depending on whether c_2 is positive or negative. The " sliding motion" suggests a way for identifying the optimal values for, let's say, a max LP problem. The underlying idea is to keep "sliding" the isoprofit line$(c_1X_1 + c_2X_2) = a_o$ in the direction of increasing a's until we cross the boundary of the LP feasible region. The implementation of this idea on the LP of Equation (6) (see also Figure 2) is depicted in Figure 3.

From Figure 3, it follows that the optimal daily production levels for the protoype LP are given by the coordinates of the point corresponding to the intersection of line $\frac{1}{50}X_1 + \frac{1}{50}X_2 = 0$ with the X$_2$-axis, i.e., $X_1^{opt} = 0, X_2^{opt} = 50$. The maximal daily profit is $200 * 0 + 400 * 50 = 20{,}000\$$. Notice that the optimal point is one of the

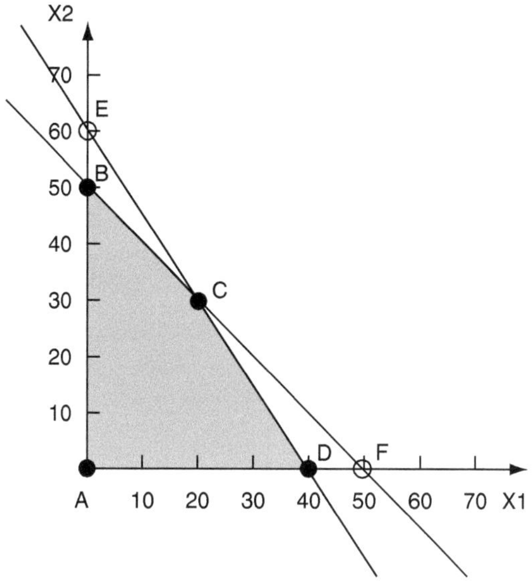

Fig. 2 The feasible region of the example LP considered in 3.1.

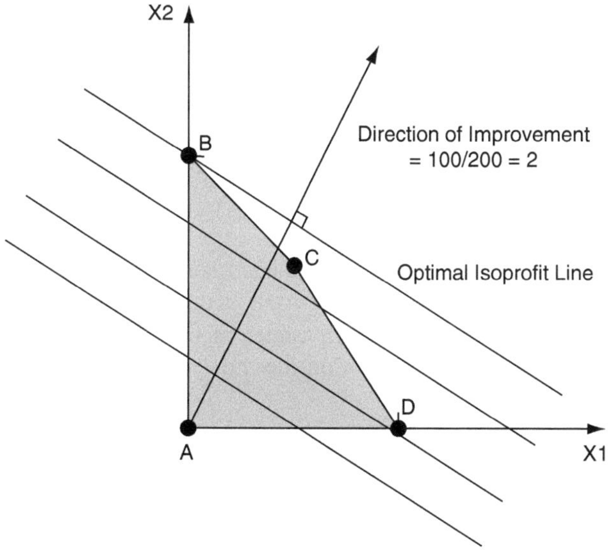

Fig. 3 Graphical solution of the example LP (6).

"corner" points of the feasible region depicted in Figure 3. Can you argue that for the geometry of the feasible region for two-var LP's described above, if there is a bounded optimal solution, then there will be one which corresponds to one of the corner points? (This argument is developed for the broader context of n-var LP's in the next section.)

Fig. 4 An infeasible LP.

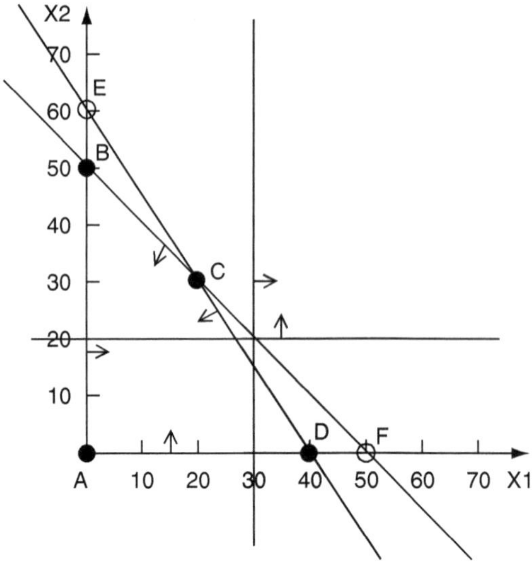

There are two fail options related to LP problem solution. First is absence of any solution, when the feasible region is empty. Consider again the original example (6), modified by the additional requirements (imposed by the company's marketing department) that the daily production of product X_1 must be at least 30 units, and that of product X_2 should exceed 20 units. These requirements introduce two new constraints into the problem formulation, i.e., $X_1 \geq 30, X_2 \geq 20$. Attempting to plot the feasible region for this new problem, we get Figure 4, which indicates that there are no points in the (X_1, X_2)-plane that satisfy all constraints, and therefore our problem is infeasible (over-constrained).

A second particular option is an unbounded solution. In the LP's considered above, the feasible region (if not empty) was a bounded area of the plane. For this kind of problems it is obvious that all values of the LP objective function (and therefore the optimal) are bounded. Consider however the following modified LP problem:

$$max\{2X_1 - X_2\}$$

under constraints:

$$X_1 - X_2 < 1$$
$$2X_1 + X_2 > 6$$
$$X_1 \geq 0, X_2 \geq 0.$$

The feasible region and the direction of improvement for the isoprofit lines for this problem are given in Figure 5. It is easy to see that the feasible region of this problem is unbounded, and furthermore, the orientation of the isoprofit lines is such that no matter how far we "slide" these lines in the direction of increasing the objective function, they will always share some points with the feasible region. There-

Fig. 5 An unbounded LP.

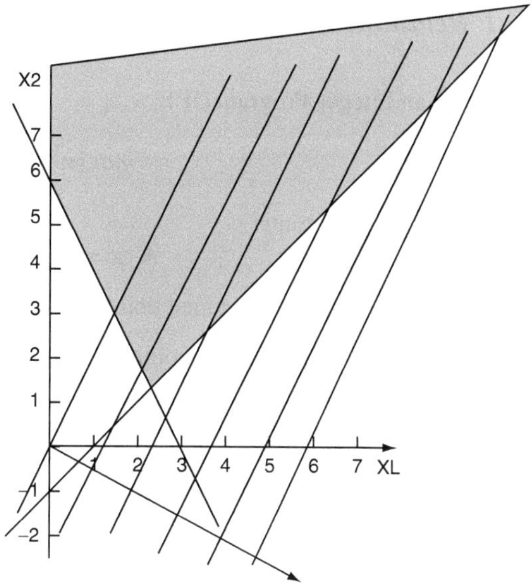

fore, this is an example of a (two-var) LP whose objective function can take arbitrarily large values. Such an LP is characterized as unbounded. Notice, however, that even though an unbounded feasible region is a necessary condition for an LP to be unbounded, it is not sufficient; to convince yourself, try to graphically identify the optimal solution for the above LP in the case that the objective function is changed to:

$$max\{2X_1 - X_2\} = -X_2.$$

Summarizing the above discussion, we have shown that a two-var LP can either:

- Have a unique optimal solution which corresponds to a "corner" point of the feasible region or
- Have many optimal solutions that correspond to an entire "edge" of the feasible region or
- Be unbounded, or be infeasible

5 Integer Programming

The use of integer variables in production when only integral quantities can be produced is the most obvious use of integer programs [7, 9]. In this section, we will look at some less obvious ones. The text also goes through a number of them (some are repeated here).

5.1 Relationship to linear programming

Given is an Integer Program (IP):

$$max\{\mathbf{c}^{\mathbf{T}} \cdot \mathbf{x}\}$$

subject to constraints:

$$\mathbf{A} \cdot \mathbf{x} = \mathbf{b}; \mathbf{x} \geq \mathbf{0}.$$

Since (LP) is less constrained than (IP), the following are immediate:

If (IP) is a minimization, the optimal objective value for (LP) is less than or equal to the optimal objective for (IP).

If (IP) is a maximization, the optimal objective value for (LP) is greater than or equal to that of (IP).

If (LP) is infeasible, then so is (IP).

If (LP) is optimized by integer variables, then that solution is feasible and optimal for (IP).

If the objective function coefficients are integer, then for minimization, the optimal objective for (IP) is greater than or equal to the "round up" of the optimal objective for (LP).

For maximization, the optimal objective for (IP) is less than or equal to the "round down" of the optimal objective for (LP). So solving (LP) does give some information: it gives a bound on the optimal value and, if we are lucky, may give the optimal solution to IP. We saw, however, that rounding the solution of LP will not in general give the optimal solution of (IP). In fact, for some problems it is difficult to round and even get a feasible solution.

5.2 Capital budgeting

Let us consider one example of IP having a practical value [7]. Suppose we wish to invest \$14,000. We have identified four investment opportunities. Investment 1 requires an investment of \$5,000 and has a present value (a time-discounted value) of \$8,000; investment 2 requires \$7,000 and has a value of \$11,000; investment 3 requires \$4,000 and has a value of \$6,000; and investment 4 requires \$3,000 and has a value of \$4,000. Into which investments should we place our money so as to maximize our total present value?

Our first step is to decide on our variables. This can be much more difficult in integer programming because there are very clever ways to use integrality restrictions. In this case, we will use a (0-1) variable $x_i(i = 1,..,4)$ for each investment. If x_i is 1 then we will make investment i. If it is 0, we will not make the investment. This leads to the 0-1 IP problem:

$$max\{8x_1 + 11x_2 + 6x_3 + 4x_4\}$$

subject to constraints:

$$5x_1 + 7x_2 + 4x_3 + 3x_4 \leq 14$$
$$x_i \in \{0;1\}, (i = 1,\ldots,4).$$

Now, a straightforward decision suggests that investment 1 is the best choice. In fact, ignoring integrality constraints, the optimal linear programming solution is $(x_1 = 1;\ x_2 = 1;\ x_3 = 0.5;\ x_4 = 0)$for a objective value of \$22,000. Unfortunately, this solution is not integral. Rounding down to 0 gives a feasible solution with a value of \$19,000. There is a better integer solution $(x_1 = 0;\ x_2 = 1;\ x_3 = 1;\ x_4 = 1)$, however, for an objective value of \$21,000. This example shows that rounding does not necessarily give an optimal value.

5.3 Branch and bound method

We discuss the branch and bound method by means of the simple IP example considered above. Our IP problem is as following:

$$max\{z\} = max\{8x_1 + 11x_2 + 6x_3 + 4x_4\}$$

subject to constraints:

$$5x_1 + 7x_2 + 4x_3 + 3x_4 \leq 14$$
$$x_i \in \{0;1\}, (i = 1,\ldots,4).$$

The linear relaxation solution is $(x_1 = 1; x_2 = 1; x_3 = 0.5; x_4 = 0)$ with an objective function value of 22. We know that no integer solution will have value more than 22. Unfortunately, since x_3 is not integer, we do not have an integer solution yet. We want to force it to be integral. To do so, we branch on x_3, creating two new problems. In one, we will add the constraint $x_3 = 0$. In the other, we add the constraint $x_3 = 1$. This is illustrated in Figure 6.

Note that any optimal solution to the overall problem must be feasible to one of the subproblems. If we solve the LP by linear relaxations of the subproblems, we get the following solutions:

$$x_3 = 0;\ z = 21.65; for x_1 = 1, x_2 = 1, x_3 = 0, x_4 = 0.667$$
$$x_3 = 1;\ z = 21.85; for x_1 = 1, x_2 = 0.714, x_3 = 1, x_4 = 0.$$

At this point we know that the optimal integer solution is no more than 21.85, but we still do not have any feasible integer solution. So, we will take a subproblem and branch on one of its variables. In general, we will choose the subproblem as follows:

- We will choose an active subproblem, which so far only means one we have not chosen before.

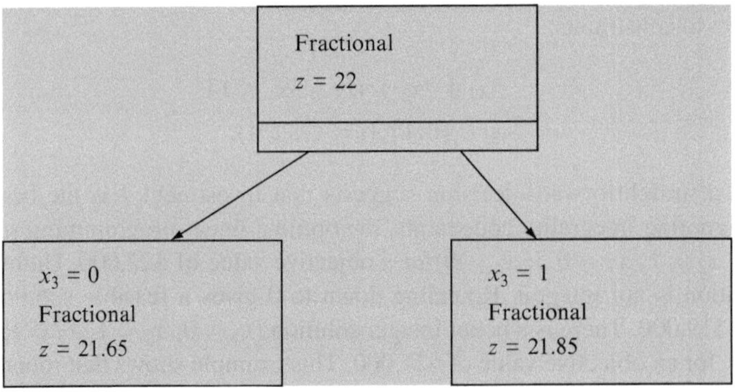

Fig. 6 First branching.

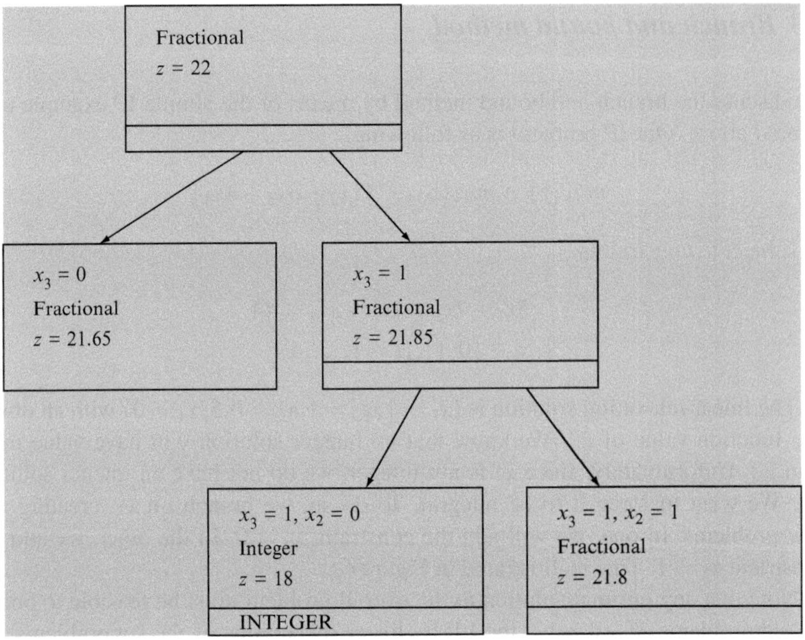

Fig. 7 Second branching.

- We will choose the subproblem with the highest solution value (for maximization; lowest for minimization).

In this case, we will choose the subproblem with $x_3 = 1$, and branch on x_2. After solving the resulting subproblems, we have the branch and bound tree in Figure 7.

The solutions are:

$$x_3 = 1;\, z = 18;\, for x_1 = 1, x_2 = 0, x_3 = 1, x_4 = 1$$
$$x_3 = 1;\, z = 21.8;\, for x_1 = 0.6, x_2 = 1, x_3 = 1, x_4 = 0.$$

We now have a feasible integer solution $x_3 = 1, x_2 = 0$ with objective value 18. Furthermore, since the IP problem gave an integer solution, no further branching on that problem is necessary. It is not active due to the integrality of solution. There are still active subproblems that might give values more than 18. Using our rules, we will branch again to get Figure 8.

The solutions are:

$$x_3 = 1, x_1 = 0, x_2 = 1; z = 21; for x_1 = 0, x_2 = 1, x_3 = 1, x_4 = 1$$
$$x_3 = 1, x_1 = 1, x_2 = 1; infeasible.$$

Our best integer solution now has objective value 21. The subproblem that generates that is not active due to integrality of the solution. The other subproblem generated is not active due to infeasibility. There is still a subproblem that is active. It is the subproblem with solution value 21.65. By our "round-down" result, there is no better solution for this subproblem than 21. But we already have a solution with

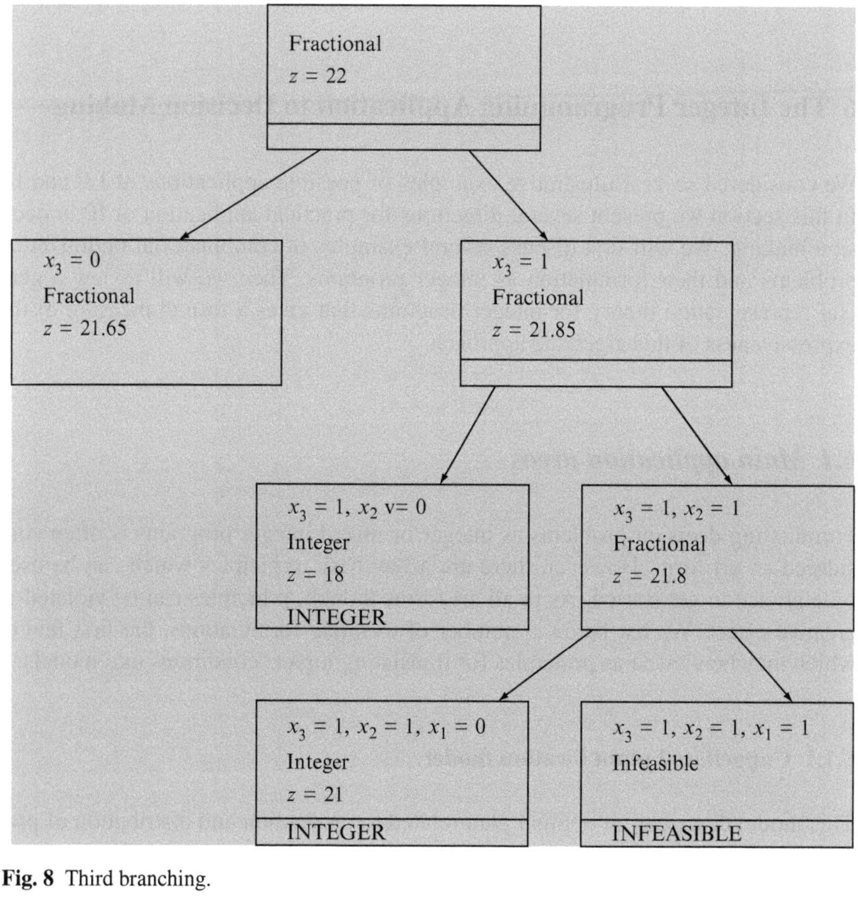

Fig. 8 Third branching.

value 21. It is not useful to search for another such solution. We can fathom this subproblem based on the above bounding argument and mark it not active. There are no longer any active subproblems, so the optimal solution value is 21.

We have seen all parts of the branch and bound algorithm. The essence of the algorithm is as follows:

- Solve the linear relaxation of the problem. If the solution is integral, then we are done. Otherwise create two new subproblems by branching on a fractional variable.
- A subproblem is not active when any of the following occurs:

  1. You used the subproblem to branch on.
  2. All variables in the solution are integer.
  3. The subproblem is infeasible.
  4. You can fathom the subproblem by a bounding argument.

Choose an active subproblem and branch on a fractional variable. Repeat until there are no active subproblems.

6 The Integer Programming Application to Decision Making

We considered several illustrative examples of possible applications of LP and IP. In this section we present several directions for practical application of IP in decision making. We will first discuss several examples of combinatorial optimization problems and their formulation as integer programs. Then we will review a general representation theory for integer programs that gives a formal measure of the expressiveness of this algebraic approach.

6.1 Main application areas

Formulating decision problems as integer or mixed integer programs is often considered an art form. However, there are a few basic principles which can be used by a novice to get started. As in all art forms though, principles can be violated to creative effect. We list below a number of example formulations, the first few of which may be viewed as principles for translating logical conditions into models.

6.1.1 Capacitated plant location model

This model describes an optimal plan related to production and distribution of produced wares in accordance to demand sites. Let us introduce the following input parameters:

$i = \{1,...,m\}$ – possible locations for plants
$j = \{1,...,n\}$ – demand sites
k_i – a capacity of plant i; if opened
f_i – fixed cost of opening plant i
c_{ij} – per unit production cost at i plus transportation cost from plant i to site j
d_j – a demand at location j

Our task is to choose the plant locations so as to minimize total cost and meet all demands. This task might be formulated as the IP problem:

$$min\{\sum_j \sum_i c_{ij}x_{ij} + \sum_i f_i y_i\}$$

subject to constraints:

$$\sum_i x_{ij} \geq d_j; (j = 1,...,n)$$
$$\sum_j x_{ij} \leq k_i y_i; (i = 1,...,m)$$
$$x_{ij} \geq 0$$
$$y_i = \{0; 1\}.$$

to satisfy demands.

If the demand d_j is less than the capacity k_i for some "*ij*" combination, it is useful to add the constraint

$$x_{ij} \leq d_j y_i$$

to improve the quality of the linear programming relaxation.

6.1.2 Traveling salesman problem

A recurring theme in IP is that the same decision problem can be formulated in several different ways. Principles for sorting out the better ones have been the subject of some discourse [9]. We now illustrate this with the well known traveling salesman problem. Given a complete directed graph with distance c_{ij} of arc (i; j), we are to find the minimum length tour beginning at node 1 and visiting each node of this graph exactly once before returning to the start node 1. This task might be formulated as the IP problem:

$$min\{\sum_j \sum_i c_{ij}x_{ij}\}$$

subject to constraints:

$$\sum_i x_{ij} = 1, (j = 1,...,n)$$
$$\sum_j x_{ij} = 1, (i = 1,...,n)$$

$$\sum_j \sum_i c_{ij} x_{ij} \geq 1$$
$$x_{ij} = \{0;1\}.$$

6.1.3 Covering and packing problems

A wide variety of location and scheduling problems can be formulated as set covering or set packing or set partitioning problems. The three different types of covering, partitioning and packing problems can be succinctly stated as follows: Given

(a) A finite set of elements $M = \{1,...,m\}$ and
(b) A family F of subsets of M with each member $F_j = \{1,...,n\}$ having a profit (or cost) c_j associated with it

find a collection, S, of the members of F that maximizes the profit (or minimizes the cost) while ensuring that every element of M is in:

(P1): at most one member of S (set packing problem)
(P2): at least one member of S (set covering problem)
(P3): exactly one member of S (set partitioning problem).

The three problems (P1), (P2) and (P3) can be formulated as integer linear programs as follows: Let $\boldsymbol{A}$ denotes the m*n matrix where

$$A_{ij} = \left\{ \begin{array}{c} 1, if\, element``i"belongsto\, F_j \\ 0, otherwise \end{array} \right\}.$$

The decision variables are $x_j (j = 1,...,n)$, where

$$x_j = \left\{ \begin{array}{c} 1, if\ F_j\, is\, chosen \\ 0, otherwise \end{array} \right\}.$$

The set packing problem is (P1)

$$max\{\mathbf{c^T} \cdot \mathbf{x}\}$$

subject to constraints:

$$\mathbf{A} \cdot \mathbf{x} \leq \mathbf{e}_m;$$
$$x_i = \{0;1\}$$

where $\mathbf{e}_m$ is an m–dimensional column vector of "1"s. The set covering problem (P2) is (P1) with less than or equal to constraints replaced by greater than or equal to constraints and the objective is to minimize rather than maximize. The set partitioning problem (P3) is (P1) with the constraints written as equalities. The set partitioning problem can be converted to a set packing problem or a set covering problem (see [9]) using standard transformations. If the right hand side vector $\mathbf{e}_m$ is replaced by a non-negative integer vector $\mathbf{b}$, (P1) is referred to as the generalized set packing problem. The airline crew scheduling problem is a classic example of the set partitioning or the set covering problem. Each element of M corresponds to a flight segment. Each subset F_j corresponds to an acceptable set of flight segments of a crew. The problem is to cover, at minimum cost, each flight segment exactly once. This is a set partitioning problem.

6.2 Environmental application

6.2.1 Multi-user consortium

Requirements for weather forecast products can vary significantly and are typically oriented to the needs of specific user groups. Nonetheless, in many respects the requirements are rather similar, such as a common need for information on basic variables such as temperature, humidity, and precipitation (mean, maximum, minimum). On other hand, it is hard to imagine that every user could provide their own forecast product because of substantial costs of both inputs and model development/maintenance. In the case of a specified forecast some additional observations might be required to increase prescribed reliability or probability. Therefore, it is more rational to select a set of a few forecast models and observing systems, which respond to the correct extent to an optimal set of requirements generated by a multi-user economical and mathematical model. A consortium of multi-users will get benefits of mathematically optimal decisions under minimal costs. User investments in a weather forecast system should be proportional to their expected benefits derived from the early warning of short-term weather fluctuations or extreme events. Under such circumstances a consortium of multi-users approach would be more likely to derive benefits from the mathematically optimal decisions for minimum investment. The meteorological community is interested in such an approach in order to reduce the number of observing programs and forecasting models [11, 12].

6.2.2 Elementary statement of problem

Let us assume that there are *n users* of climate forecasting data with their *n benefits* of early warning: $c_i(i=1,...,n)$ $(i=1,..,n)$. These users are interested to forecast *m specific meteorological events* numerated as $j=1,\ldots m$. The potential usefulness of them varies and is described by the matrix of coefficients $\mathbf{A}=\{a_{ij}\}$. Each magnitude a_{ij} can be considered as the expense of the *i-th user* for the *j-th meteorological event* delivered by some forecast model. The *minimum expected efficiency* for the *i-th user* is bounded by b_i^{min} . Let us introduce the decision maker variable:

$$x_i = \left\{ \begin{matrix} 1, if\, user\text{``}i\text{''} adopts\, forecast\, data \\ 0, otherwise \end{matrix} \right\}.$$

Now we come to formulation of the optimization problem for $\{x_i\}$:

$$max\left\{\sum_i c_i x_i\right\} \tag{14}$$

subject to constraints:

$$\sum_j a_{ij} x_j \geq b_i^{min}. \tag{15}$$

Another interpretation of the coefficients and a more complex method to derive them is possible. A generalization to the forecast multi-model case is evident.

6.2.3 Illustrative example

Let us consider multi-user decision making for many meteorological events. We used the European Center for Medium Range Weather Forecasting (ECMWF) Ensemble Prediction System (EPS) forecast for the T850 (air temperature field at the standard level of 850 mb) anomaly, Europe, Jan–Feb, 1998 (Figure 9) (see details in [15]) with n = 3 (number of users), m = 4 (number of meteorological events). The matrix of EPS forecast relative economic values are presented in Table 1, the minimal efficiency for each user in Table 2. In the case of equal importance of users we came to the optimal solution x_{opt} for (14) constrained by (15). This solution

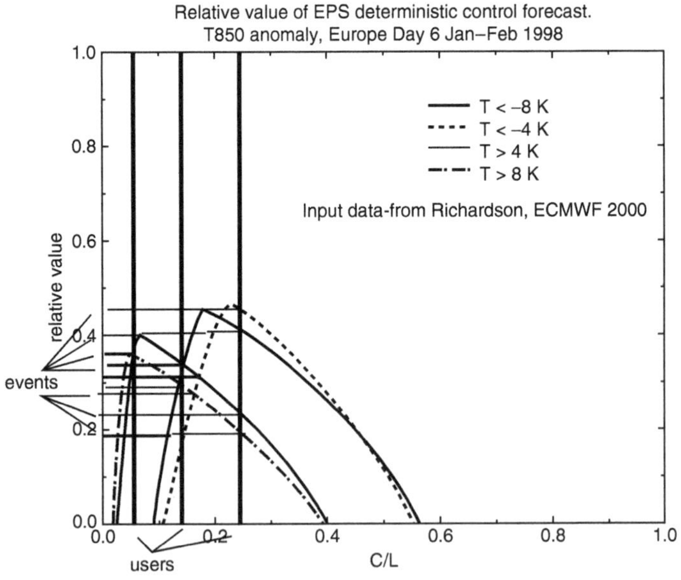

Fig. 9 Ensemble Prediction System forecast relative values (usefullness) responded to multi-user and multi-event case [15].

Table 1 Matrix of constraints $\mathbf{A} = \{a_{ij}\}$.

Users	(T<-8K)	(T<-4K)	(T> +4K)	(T> +8K)
1	0.40	0.36	0	0
2	0.32	0.29	0.32	0.19
3	0.22	0.19	0.41	0.46

Table 2 Constraint vector of minimal efficiencies -$\mathbf{b}_{min}$.

Users	b_i^{min}
1	0.1
2	0.2
3	0.3

Table 3 Optimal decision x_{opt} in the case of priority user "3":$\mathbf{c} = (0.5, 0.5, 1)^T$.

Users	x_{opt}
1	2.26
2	0.36
3	1.99

Table 4 Optimal decision x_{opt} in the case of priority user "3":$c = (0.5, 0.5, 1)^T$.

Users	x_{opt}
1	2.26
2	0.36
3	1.99

shows that the EPS forecasting system has prior importance for *user "2"*. The least contribution is related to *user "3"* . Let us now enhance the a priori importance of *user "3"* by changing values of the target function (1) from $\mathbf{c} = (1,1,1)^T$ to $\mathbf{c} = (0.5, 0.5, 1)^T$(Tables 3 and 4). Even in this case *user "3"* remains at second place after *user "1"*. It is interesting to note that the output for *user "1"* is its insensitivity with account to a priory weights.

7 Conclusion

An approach based on MP and IP finds a wide application area in many branches of economical sciences. It can be used in decision making related to multidimensional target functions constrained by many linear cost restrictions. This chapter indicates that similar problems arising in many important practical areas might be efficiently solved by the described approach.

References

1. Ben-Haim, Y. (2001) Information-Gap Theory: Decisions Under Severe Uncertainty, Academic, London.
2. Ben-Tal, A., Boyd, S., Nemirovski, A. (2006) Extending the Scope of Robust Optimization: Comprehensive Robust Counterparts of Uncertain Problems. Mathematical Programming, 107, 63–89.
3. Dantzig G.B. (1949) Programming of interdependent activities, Mathematical model. Econometrica, 17, 3, 200–211.
4. Ecker J.G. and Kupferschmid,M. (1988) Introduction to Operations Research, Wiley, New York.
5. Gass S.I. (1958) Linear Programming (Methods and Applications), McGraw-Hill, Toronto/London.
6. George, E. I. (1999) Discussion of Model averaging and model search strategies by M. Clyde. In Bayesian Statistics 6 (J. M. Bernardo, J. O. Berger, A. P. Dawid and A. F. M. Smith, eds.), 157–185. Oxford University Press.
7. Gomory R.E. (1963) An algorithm for integer solution to linear programming, McGraw-Hill, Toronto/London.
8. Kantorovich L.V. (1966) Mathematical models and methods of optimal economical planning. Novosibirsk, Nauka, Moscow, 256 p. (in Russian).
9. Korbut, A.A., and Yu.Yu. Finkelstein (1969) Discreet programming. Nauka, Moscow, 302 p. (in Russian).
10. Kouvelis P. and G. Yu (1997) Robust Discrete Optimization and Its Applications, Kluwer, Dordrecht.
11. Pokrovsky O.M. (2005) Development of integrated "climate forecast-multi-user" model to provide the optimal solutions for environment decision makers. – Proceedings of the Seventh International Conference for Oil and Gas Resources Development of the Russian Arctic and CIS Continental Shelf, St. Petersburg, 13–15 September, 2005, AMAP, Oslo, Norway, September 2005, pp. 661–663.
12. Pokrovsky O.M. (2006) Multi-user consortium approach to multi-model weather forecasting system based on integer programming techniques – Proceedings of the Second THORPEX International Scientific Symposium (Landshut, 3–8 December, 2006), WMO/TD, No. 1355, pp. 234–235.
13. Raftery, A. E. (1996a) Approximate Bayes factors and accounting from model uncertainty in generalised linear models. Biometrika 83, 251–266.
14. Raftery, A. E. (1996b) Hypothesis testing and model selection. In Markov Chain Monte Carlo in Practice (W. R. Gilks and D. Spiegelhalter, eds.) 163–188. Chapman & Hall, London.
15. Richardson R. (2000) Skill and relative economic value of the ECMWF Ensemble Prediction System. - Quarterly Journal of Royal Meteorological Society, 126, pp. 649–668.
16. Thie, P. (1988) An Introduction to Linear Programming and Game Theory. Wiley, New York.
17. Weakliem, D. L. (1999) A critique of the Bayesian information criterion for model selection. Sociological Methods and Research, 27, 359–297.

Recent Advances in Space-Variant Deblurring and Image Stabilization

Michal Šorel*,1, Filip Šroubek* and Jan Flusser*

Abstract The blur caused by camera motion is a serious problem in many areas of optical imaging such as remote sensing, aerial reconnaissance or digital photography. As a rule, this problem occurs when low ambient light conditions prevent an imaging system from using sufficiently short exposure times, resulting in a blurred image due to the relative motion between a scene and the imaging system. For example, the cameras attached to airplanes and helicopters are blurred by the forward motion of the aircraft and vibrations. Similarly when taking photographs by hand under dim lighting conditions, camera shake leads to objectionable blur. Producers of imaging systems introduce compensation mechanisms such as gyroscope gimbals in the case of aerial sensing or optical image stabilization systems in the case of digital cameras. These solutions partially remove the blur at the expense of higher cost, weight and energy consumption. Recent advances in image processing make it possible to remove the blur in software. This chapter reviews the image processing techniques we can use for this purpose, discusses the achievable performance and presents some promising results achieved by the authors.

Keywords: Camera shake, image stabilization, image registration, space-variant restoration, deblurring, blind deconvolution, point spread function, regularization

1 Introduction

The blur caused by sensor motion is a serious problem in a large number of applications from remote sensing to landmine detection to amateur photography. In general, this problem occurs if the time needed to capture an image is so long that the imaging system moves relative to the scene.

*,1 Institute of Information Theory and Automation of the ASCR, Pod Vodarenskou vezi 4, Praha 8, CZ-18208, Czech Republic, e-mail: {sorel,sroubekf,flusser}@utia.cas.cz

J. Byrnes (ed.), *Unexploded Ordnance Detection and Mitigation,*

An example application in landmine detection is the general survey of minefields in the aftermath of military conflicts using visible light or infrared cameras. The cameras attached to airplanes and helicopters are blurred by the forward motion of the aircraft and vibrations. While the vibrations can be dumped to some extent using gyroscope stabilizers, there is no simple way to do the same with the forward movement. A similar problem arises in the case of cameras attached to moving vehicles. For example, thermal infrared cameras attached to armoured vehicles can be used to detect anti-personnel and anti-tank mines on roads and tracks.

Similarly, when taking photographs under low light conditions, the camera needs a long exposure time to gather enough light to form the image, which leads to objectionable blur. To mitigate this problem, producers of digital cameras introduced two types of hardware solutions. The technically simpler one is to increase the sensitivity of a camera (ISO) by amplifying the signal from the sensor, which permits faster shutter speed. Unfortunately, especially in the case of compacts, this results in a decrease of image quality because of more noise. Optical image stabilization (OIS) systems, containing either a moving image sensor or an optical element to counteract camera motion, are technologically more demanding. They help to remove blur without increasing noise level but at the expense of higher cost, weight and energy consumption.

A system removing the blur in software would be an elegant solution to the problem. In this chapter we give an overview to possible approaches to this problem. The algorithms are explained in connection with photography but the results can be applied to other cases such as aerial reconnaissance and infrared imaging as well.

We start with an outline of approaches. Then, in Section 3 we describe a mathematical model of blurring. For each approach (Sections 4–7), we summarize its strong and weak points and present a typical state-of-the-art method. Section 8 summarizes results and indicates the potential of individual approaches.

2 Overview of Approaches

An obvious way to avoid camera motion blur is to take a sequence of underexposed images so that the exposure time is short enough to prevent blurring. After registration, the whole sequence can be summed to get the original sharp image with a reasonable noise level. In Section 4 we briefly discuss why this idea turns out to be impractical for more than a few images. In the rest of this chapter, we discuss situations where we already have a blurred image (or a sequence of images) and wish to remove the blur.

To simplify the problem, the blur is usually assumed to be homogenous in the whole image. In this case the blur can be modeled by convolution. That is why the reverse problem to find the sharp image is called *deconvolution*. If the PSF is not known, which is the case in most real situations, the problem is called *blind deconvolution*.

While non-blind deconvolution problems can be easily solved, solutions of blind deconvolution problems from a single image are highly ambiguous. To find a stable solution some additional knowledge is required. This case is treated in Section 5. The most common approach is regularization, applied both on the image and blur. Regularization terms mathematically describe *a priori* knowledge and play the same role as prior distributions in stochastic models. For the present, probably the best published blind deconvolution methods are those of Fergus et al. [2] and coming soon [9].

Another approach, extensively studied in past years, is to use multiple images capturing the same scene but blurred in a different way (Section 6). The camera takes two or more successive images and each exhibits different blurring due to the basically random motion of the photographer's hand or, for example, aircraft vibrations. Multiple images permit estimation of the blurs without any prior knowledge of their shape, which is hardly possible in single image blind deconvolution [10].

One particular multi-image setup attracted considerable attention only recently. Taking images with two different exposure times (long and short) results in a pair of images, in which one is sharp but underexposed and another is correctly exposed but blurred. Instead of the underexposed image we can equivalently take an image with high ISO. Both can be easily achieved in continuous shooting mode by exposure and ISO bracketing functions of DSLR cameras. For Canon compact cameras these functions can be written in the scripting language implemented within the scope of the CHDK project (http://chdk.wikia.com/wiki/CHDK).

To estimate the sharp image, two different ideas were proposed in the literature. The first adjusts the contrast of the underexposed image to match the histogram of the blurred one [7]. However, this technique is applicable only if the difference between exposure times is small. The second way [5, 11] uses the image pair to estimate the blur and then deconvolves the blurred image. This path was followed by [15], where the authors show an effective way to suppress ringing artifacts produced by Richardson-Lucy deconvolution. In Section 7 we give an example of an algorithm of this type proposed by the authors of this chapter. To be applicable even for wide angle lenses, we consider space-variant blur.

3 Blur Model

It is well known that homogenous blurring can be described by *convolution*

$$\mathbf{z} = \mathbf{u} * \mathbf{h}\,[x,y] = \int \mathbf{u}(x-s, y-t)\mathbf{h}(s,t)\,ds dt, \tag{1}$$

where $\mathbf{u}$ is an original image, $\mathbf{h}$ is called the *convolution kernel* or *point-spread function* (PSF) and $\mathbf{z}$ is the blurred image. In our case of camera motion blur the PSF is a plane curve given by an apparent motion of each pixel during the exposure.

If the focal length of the lens is short or camera motion contains a significant rotational component about the optical axis, this simple model is not valid. The blur is then different in different parts of the image and is a complex function of camera motion and depth of scene [14]. We can see an example in Figure 5, where the image was divided into 49 (7×7) rectangles and convolution kernels were estimated within these subimages (by a method described in Section 7). Notice for example the difference between the upper left and right kernels.

Nevertheless, this spatially varying blur can be described by a more general linear operation

$$\mathbf{z} = \mathbf{u} *_v \mathbf{h}\ [x,y] = \int \mathbf{u}(x-s, y-t)\mathbf{h}(x-s, y-t; s,t)\ ds dt, \tag{2}$$

where $\mathbf{h}$ is again called the *point-spread function* as in the case of convolution. Note that convolution is a special case, with the function h independent of coordinates x and y, that is $\mathbf{h}(x,y;s,t) = \mathbf{h}(s,t)$. We can look at (2) as convolution with a kernel that changes with its position in the image, and speak about *space-variant convolution*. The subscript v distinguishes from ordinary space-invariant convolution, denoted by asterisk.

Because the rotational component of camera motion is usually dominant, the blur is independent of depth and the PSF changes in a continuous gradual way. Therefore the blur can be considered locally constant and can be locally approximated by convolution. This property can be used to efficiently estimate even the space-variant PSF, as described in Section 7.

4 Summing of Underexposed Images

At first sight, the idea to sum a sequence of underexposed images seems to be very attractive. It is a well known property of shot (Poisson) noise that an image taken with an exposure time t has the same level of noise as the sum of N images each taken with time t/N. So, apparently, the only problem we must solve is to register images with sufficient precision. There exist many fast image registration methods and, without doubt, one of them could be used in this case. Registration is made easier also by the fact that the difference between images is not large as the images are taken quickly one after another.

Unfortunately, for the present, there is a serious problem that limits the use of this idea in practice. Images taken by present day digital cameras are huge and it takes a lot of time to read them out from sensor to camera memory. For consumer level DSLRs it typically takes about $1/3$ of second, for compacts even more. For example, imagine that we want to replace one $1/4$ s image by a sequence of 16 images taken with exposure time $1/60$ s, which corresponds to the use of ISO 1,600 instead of ISO 100. Now the camera needs $16 \times 1/3$, or more than 5 s. For many situations this is simply too long.

To summarize, on one hand this approach is computationally simple and can potentially be implemented inside a camera. On the other hand, to be useful for really low lighting conditions, the read-out time will have to be significantly shortened.

In the rest of this chapter we will treat blurred images, which is less demanding with respect to read-out time and can actually be used with present day cameras. On the other hand, deblurring is computationally more time consuming and assumes postprocessing on the photographer's personal computer.

5 Single-Image Blind Deconvolution

There has been a considerable effort in the image processing community in the last three decades to find a reliable algorithm for single image blind deconvolution. For a long time, the problem seemed too difficult to be solved for complex blur kernels. Proposed algorithms usually worked only for special cases such as astronomical images with uniform (black) background. There was no reliable result applicable to natural scenes.

Only recently, in 2006, Rob Fergus et al. [2] proposed an interesting Bayesian method with very impressive results. Another method of this kind should appear at SIGGRAPH 2008 [9]. The authors claim even better results than [2] with much simpler and faster computation. In this chapter we briefly describe the method [2].

The method assumes a simple convolution model of blurring

$$\mathbf{z} = \mathbf{u} * \mathbf{h} + \mathbf{n}, \tag{3}$$

where $\mathbf{n}$ is an independent Gaussian zero mean noise.

The basic idea is to estimate the *a posteriori* probability distribution of the gradient of the original image and of the blur kernel

$$p(\mathbf{u}, \nabla\mathbf{h} | \nabla\mathbf{z}) = p(\nabla\mathbf{z} | \nabla\mathbf{u}, \mathbf{h}) p(\nabla\mathbf{u}) p(\mathbf{h}), \tag{4}$$

using knowledge of independent prior distributions of the image gradient $p(\nabla\mathbf{u})$ and of the kernel $p(\mathbf{h})$. The likelihood $p(\nabla\mathbf{z}|\nabla\mathbf{u},\mathbf{h})$ is considered Gaussian with mean $\nabla\mathbf{u} * \mathbf{h}$ and an unknown variance. After estimation of the full posterior distribution $p(\mathbf{u}, \nabla\mathbf{h}|\nabla\mathbf{z})$, it computes the kernel with maximal marginal probability. Finally, the original image is restored by the classical Richardson-Lucy algorithm. This final phase could obviously be replaced by an arbitrary non-blind deconvolution method.

The algorithm is quite complex. It approximates the full posterior distribution by the product $p(\mathbf{u}|\nabla\mathbf{z})p(\nabla\mathbf{h}|\nabla\mathbf{z})$ in the sense of Kullback-Leibler distance, which can be efficiently computed by the variational scheme described in [6] for cartoon images. The image gradient prior is considered in the form of a Gaussian mixture. In a similar way, the prior on kernel values is expressed as a mixture of exponential distributions, which reflects the fact that most kernel values for motion blur are zero. Both types of priors are learned from a typical natural image.

Figure 1 shows an example of an image restored by this method. We can see that the convolution kernel is recovered surprisingly well. Some artifacts appear because there are no smoothing constraints in the algorithm. Another problem is the high number of artifacts produced by non-blind deconvolution in the final phase of the

(a) Blurred image, 800 × 600 pixels

(b) Deblurred image

(c) Estimated PSF, 35 × 35 elements

Fig. 1 Example results of single-image blind deconvolution provided by Fergus et al. [2].

algorithm. A typical example is the well known ringing effect. New papers [9,15,16] seem to deal with this problem successfully.

Bringing this all together, there are reliable methods for estimating the blur kernel and subsequent restoration from a single blurred image. The main problem is the need for user assistance to choose a suitable part of the image for kernel inference.

6 Multi-image Blind Deconvolution

In this approach we use multiple images (Fig. 2a) capturing the same scene but blurred in a different way. We can easily take such a sequence using continuous shooting modes of present day cameras. Multiple images permit one to estimate the blurs (Fig. 3) without any prior knowledge of their shape.

(a) Blurred input images, 1024×768 pixels

(b) Deconvolution from only first image (c) Result of multi-image deconvolution

Fig. 2 Example results achieved by multi image blind deconvolution algorithm [10].

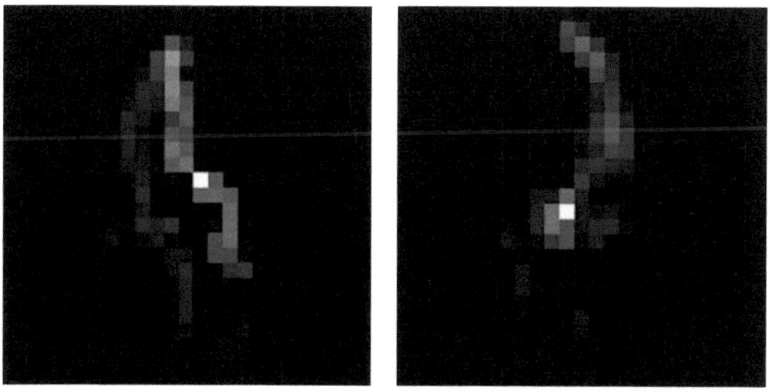

Fig. 3 Convolution kernels corresponding to images in Figure 2a.

Mathematically, the situation is described as convolution of the original image $\mathbf{u}$ with P convolution kernels $\mathbf{h}_p$

$$\mathbf{z}_p = \mathbf{u} * \mathbf{h}_p + \mathbf{n}_p, \quad p = 1,..,P. \tag{5}$$

In this section, we describe one of the best working multi-image deblurring algorithms [10].

As in the single image situation, the algorithm can be viewed as a MAP (maximum a posteriori) estimate of distributions of the sharp image and the blur kernels. It is equivalent to minimization of the functional

$$E(\mathbf{u},\mathbf{h}_1,...,\mathbf{h}_p) = \frac{1}{2}\sum_{p=1}^{P} \|\mathbf{u}*\mathbf{h}_p - \mathbf{z}_p\|^2 + \lambda_u Q(\mathbf{u}) + \lambda_h \sum_{i\neq j} R(\mathbf{h}_i,\mathbf{h}_j) \qquad (6)$$

with respect to the latent image $\mathbf{u}$ and blur kernels $\mathbf{h}_1,...,\mathbf{h}_p$. The first term of (6), called the *error term*, is a measure of the difference between input blurred images $\mathbf{z}_p$ and the original image $\mathbf{u}$ blurred by kernels $\mathbf{h}_k$. The size of the difference is measured by the L_2 norm $\|.\|$. The inner part of the error term is nothing more than the matrix of errors at the individual points of image p, which should be close to zero for the correct image and kernel. Note that kernels $\mathbf{h}_p$ incorporate a possible shift of the camera between the images.

The role of regularization terms

$$Q(\mathbf{u}) = \int |\nabla \mathbf{u}| \qquad (7)$$

and

$$R(\mathbf{h}_i,\mathbf{h}_j) = \|\mathbf{z}_j * \mathbf{h}_i - \mathbf{z}_i * \mathbf{h}_j\| \qquad (8)$$

is to make the problem well-posed and incorporate prior knowledge about the solution [12].

Thus, $Q(\mathbf{u})$ is an image regularization term which can be chosen to properly represent the expected character of the image function. For the majority of images a good choice is total variation (7), where $\nabla \mathbf{u}$ denotes the gradient of $\mathbf{u}$. The size of the gradient is integrated over the whole area of the image. Very good anisotropic denoising properties of the total variation were shown by Rudin et al. [8]. A reason why total variation works so well for real images is that it favors piecewise constant functions. In real images object edges create sharp steps that appear as discontinuities in the intensity function. For a more detailed discussion of image regularization, see [1, 10, 13]. The kernel regularization term is a constraint useful for kernels of limited support.

The functional (6) is minimized by alternating minimization in the subspaces corresponding to the image and the blur kernels.

The main problem of the multi-image approach is speed. For this reason, it is practically impossible to generalize this approach to space-variant blur. As a result, this approach can be applied mainly for tele-lens photos if the rotational component of camera motion about the optical axis is negligible. In general, it usually works for the central section of an arbitrary blurred image.

7 Restoration from a Pair of Blurred and Noisy Images

The idea to use two images with two different exposure times appeared only recently [5, 11, 15]. Most algorithms of this group first estimate the blur from the image pair and then deconvolve the blurred image. The main problem of the deconvolution phase is suppression of ringing artifacts. A method of handling this problem for the Richardson-Lucy algorithm was proposed in [15, 16].

None of the aforementioned methods are general enough to be applicable to full uncropped photos. The reason is that the blur is not constant throughout the image, especially in the case of lenses with shorter focal length (<50 mm). In addition, it often happens that camera motion has a considerable rotational component about the optical axis and then the blur is space-variant, even for tele-lenses. Another effect modifying blurs is lens distortion. All these effects are accentuated in regions close to image borders. Therefore a space-variant approach is necessary for artifact-free results.

Space-variant restoration was already considered in astronomy and microscopy but there is almost no work applicable in photography. Only recently, in [14], is space-variant blur considered for a camera moving without rotation, but this assumption does not correspond to the real trajectory of a handheld camera.

In the following paragraphs we describe a state-of-the-art algorithm proposed by the authors. To avoid ringing effects we use a constrained least squares method with total variation regularization. To be applicable even for wide angle lenses, we consider space-variant blur.

7.1 Algorithm

For input the algorithm requires a pair of images, one of them blurred and another noisy but sharp. The algorithm works in three phases:

1. Robust image registration
2. Estimation of convolution kernels on a grid of windows followed by an adjustment at places where estimation failed
3. Restoration of the sharp image

In the first step, we need a robust registration procedure working with precision significantly better than the considered size of blur kernels. We can assume that the change of camera position is negligible with respect to scene distance (very short baseline) and consequently it can be approximated by a projective transform independent of scene depth. Experiments have also shown that misalignments due to lens distortion do not harm the algorithm because they are compensated by the shift of the corresponding part of the space-variant PSF. For the purpose of this algorithm, we apply the standard RANSAC [3,4] approach to estimate the homography matrix. Then we transform the blurred image accordingly. The transformed image will be denoted by $\mathbf{z}^T$.

In the second step of the algorithm we make use of the fact that the blur can be locally approximated by convolution. We do not estimate the blur kernels in all pixels. Instead, we divide the image into rectangular windows (a 7×7 grid in our example in Figure 6) and estimate only a small set of kernels $\mathbf{h}_{i,j}$ $(i, j = 1.7$ in our example in Figure 5). The estimated kernels are assigned to centers of the windows where they were computed. In the rest of the image, the PSF $\mathbf{h}$ is approximated by bilinear interpolation from blur kernels in four adjacent windows.

Thus, we estimate blur kernels on a grid of windows, where the blur can be approximated by convolution

$$\mathbf{z}^T_{i,j} = (\mathbf{u}_{i,j} - \mathbf{n}_{i,j}) * \mathbf{h}_{i,j} = \mathbf{u}_{i,j} * \mathbf{h}_{i,j} - \mathbf{n}_{i,j} * \mathbf{h}_{i,j}, \tag{9}$$

where $\mathbf{z}^T_{i,j}$ is a section of the transformed blurred image $\mathbf{z}^T$, $\mathbf{u}_{i,j}$ the corresponding part of the noisy image, $\mathbf{h}_{i,j}$ the locally valid convolution kernel and $\mathbf{n}_{i,j}$ an independent Gaussian noise contained in the noisy image.

We estimate the solution of this problem in a least squares sense as

$$\mathbf{h}_{i,j} = \arg\min_{\mathbf{k}} \|\mathbf{u}_{i,j} * \mathbf{k} - \mathbf{z}^T_{i,j}\|^2 + \alpha \|\nabla \mathbf{k}\|^2, \qquad \mathbf{k}(s,t) \geq 0, \tag{10}$$

where $\mathbf{h}_{i,j}(s,t)$ is an estimate of $h(x_0, y_0, s, t)$, (x_0, y_0) being the center of the current window $\mathbf{z}_{i,j}$, and $\|.\|$ is the L_2 norm. Regularization helps reduce the noise arising from the imprecise model.

The kernel estimation procedure (10) can fail. Such kernels must be identified, removed and replaced by the average of adjacent (valid) kernels. There are basically two reasons why kernel estimation fails. Therefore we need two different measures to decide which kernel is wrong. To identify textureless regions we compute entropy of the kernels and take those with entropy above some threshold. The other, more serious case of failure is pixel saturation, that is pixel values above the sensor range. This situation can be identified by computing the sum of kernel values, which should be close to one for valid kernels. Therefore, we simply remove kernels whose sum is too different from unity, again above some threshold.

For the restoration step, we use an energy minimization approach with total variation as an image regularization term, which belongs to the category of constrained least squares estimators [14]. Notice that it has the same form as (6). Total variation behaves satisfactorily for most photographs since it removes noise efficiently while not oversmoothing edges. It also helps to some extent to suppress artifacts caused by pixel saturation.

The restoration phase of the proposed algorithm can be described as minimization of the functional

$$E(\mathbf{u}) = \frac{1}{2} \|\mathbf{u} *_v h - \mathbf{z}\|^2 + \lambda \int |\nabla \mathbf{u}| \tag{11}$$

with respect to the unknown sharp image $\mathbf{u}$, where the second term is the total variation of the image.

Its derivative can be written as

$$\partial E(\mathbf{u}) = (\mathbf{u} *_v h - \mathbf{z}) \circledast_v h - \lambda \operatorname{div}\left(\frac{\nabla \mathbf{u}}{|\nabla \mathbf{u}|}\right), \tag{12}$$

where $\circledast_v$ is the operator adjoint to space-variant convolution

$$\mathbf{u} \circledast_v \mathbf{h}\,[x,y] = \int \mathbf{u}(x-s,y-t)\mathbf{h}(x,y;-s,-t)\,dsdt. \tag{13}$$

To minimize functional (11) we used a half-quadratic iterative approach, reducing this problem to a sequence of linear subproblems [14].

Alternatively, to speed up the restoration step, we could use a variant of the Richardson-Lucy algorithm, similar to methods [15,16]. Figures 4–6 show an example of a real image restored by this method.

In our opinion, this is the best of the three deblurring approaches. It is quite fast and reliable. Because of its stability it can be used to estimate the space-variant PSF, which makes it more applicable for a much larger range of situations. Another plus is that it can be used to segment moving objects, which is hardly possible from one image.

Fig. 4 Image of a shopping center taken in an evening with shutter speed 1/2 s (*left*), results of our algorithm with PSF adjustment (*right*). Close-ups are shown in Figure 6.

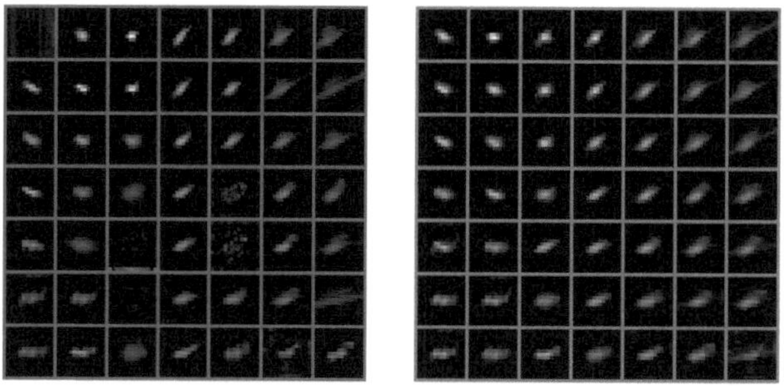

Fig. 5 Fourty-nine convolution kernels estimated in the shopping center image (*left*). Notice the wrong kernels at places of low-contrast texture (*upper left corner*) and pixel saturation (lights inside the building). Adjusted kernels on the right.

Fig. 6 Details of the shopping center image. From left to right – the blurred image, noisy image, result of deconvolution and our result.

8 Summary

In this chapter, we reviewed approaches to software image stabilization in the sense of removing blur caused by camera motion (Table 1).

The first possibility is to avoid blur from the beginning by taking a sequence of underexposed images. This idea is impractical because of the time needed for sensor read-out. We followed with the description of a deblurring algorithm from a single image. Although there are usable algorithms for this case, the main disadvan-

Table 1 Summary of approaches to image stabilization.

Approach	Speed	Quality	Main problem
Multiple underexposed images	High	High	Slow read-out, precise registration
Single-image deconvolution	Slow/medium	Medium	Homogenous blur only
Multi-image deconvolution	Slow	Medium/high	Slow computation
One blurred and one noisy image	Medium	Medium	More artifacts than multi-image deconvolution

tages are speed and difficulties with the segmentation of moving objects. The third approach was deconvolution from a sequence of blurred images. The main disadvantage of existing algorithms from this category is speed. They are even slower than single image deconvolution methods.

The last and, in our opinion, most advantageous approach is to use a pair of images, one blurred and one underexposed. Its main assets are relative speed, reliability, ability to deal with space-variant blur and the potential to segment moving objects.

Acknowledgements This work has been supported by the Czech Ministry of Education under the project No. 1*M*0572 (Research Center DAR) and the Czech Science Foundation under the project No. GACR 102/08/1593.

References

1. M. R. Banham and A. K. Katsaggelos. Digital image restoration. *IEEE Signal Process. Mag.*, 14(2):24–41, March 1997.
2. R. Fergus, B. Singh, A. Hertzmann, S. T. Roweis, and W. T. Freeman. Removing camera shake from a single photograph. *ACM Trans. Graphics, SIGGRAPH 2006 Conf. Proc., Boston, MA*, 25:787–794, 2006.
3. M. A. Fischler and R. C. Bolles. Random sample consensus: A paradigm for model fitting with applications to image analysis and automated cartography. *Commun. ACM*, 24(6):381–395, 1981.
4. R. Hartley and A. Zisserman. *Multiple View Geometry in Computer Vision*. Cambridge University, Cambridge, 2nd edition, 2003.
5. S. H. Lim and D. A. Silverstein. Method for deblurring an image. US Patent Application, Pub. No. US2006/0187308 A1, August 24 2006.
6. J. Miskin and D. MacKay. Ensemble learning for blind image separation and deconvolution. In M. Girolami (Ed.), *Advances in independent component analysis*. Berlin: Springer-Verlag, (pp. 123–141), 2000.
7. Q. R. Razligh and N. Kehtarnavaz. Image blur reduction for cell-phone cameras via adaptive tonal correction. pages I: 113–116, 2007.
8. L. I. Rudin, S. Osher, and E. Fatemi. Nonlinear total variation based noise removal algorithms. *Physica D*, 60:259–268, 1992.

9. Qi Shan, J. Jia, and A. Agarwala. High-quality motion deblurring from a single image. *ACM Trans. Graphics (SIGGRAPH)*, 27(3) 2008.
10. F. Šroubek and J. Flusser. Multichannel blind deconvolution of spatially misaligned images. *IEEE Trans. Image Process.*, 14(7):874–883, July 2005.
11. M. Tico, M. Trimeche, and M. Vehvilainen. Motion blur identification based on differently exposed images. In *Proc. IEEE Int. Conf. Image Process.*, pp. 2021–2024, 2006.
12. A. Tikhonov and V. Arsenin. *Solution of Ill-Posed Problems*. Wiley, New York, 1977.
13. D. Tschumperlé and R. Deriche. Vector-valued image regularization with pdes: A common framework for different applications. *IEEE Trans. Pattern Anal. Mach. Intell.*, 27(4):506–517, 2005.
14. M. Šorel and J. Flusser. Space-variant restoration of images degraded by camera motion blur. *IEEE Trans. Image Process.*, 17(2):105–116, February 2008.
15. L. Yuan, J. Sun, L. Quan, and H.-Y. Shum. Image deblurring with blurred/noisy image pairs. In *SIGGRAPH '07: ACM SIGGRAPH 2007 papers*, p. 1, ACM, New York, 2007.
16. L. Yuan, J. Sun, L. Quan, and H.-Y. Shum. Progressive inter-scale and intra-scale non-blind image deconvolution. *ACM Trans. Graphics (SIGGRAPH)*, 2008.

UXO Detection Techniques Using Sonar and Radar

Edmund J. Sullivan*

Abstract Several approaches to the detection of Unexploded Ordnance (UXO) in the ground are discussed. Methods exploiting the coupling of sound into the earth are shown to have promise. These approaches can use both linear and non-linear phenomena as clues. Also discussed is the potential of a ground penetrating radar method that is based on a nonlinear phenomenon.

Keywords: Sonar, acoustic, detection, radar, mines, nonlinear, Laser Doppler Velocimetry, speckle noise

1 Introduction

The general problem of detecting and identifying buried objects has grown more difficult with time. During the second world war, the detection of buried mines was reasonably successful since most mines were metal and thus could be detected with reasonable success by any of several types of metal detectors [1]. The sophistication of mines has increased since then, however. Anti-personnel mines are much smaller than anti-tank mines and many of today's mines are nonmetallic. Other techniques have been tried, such as biological (use of dogs, rats and bees), and infrared techniques. For an overview of these approaches, see reference [1].

A large part of the problem is that, when trying to detect a UXO from, say, a moving vehicle, time is of the essence, so that the issue of false alarms becomes paramount. There has been some recent work which seeks to overcome this problem. In this chapter we will concentrate mainly on seismo-acoustic (SA) methods. Also, a short discussion on a Ground-Penetrating Radar (GPR) method is given.

---

*Prometheus Inc., Newport, Rhode Island, 02840, USA, e-mail: ed@prometheus-us.com

J. Byrnes (ed.), *Unexploded Ordnance Detection and Mitigation*,

2 Detection and Identification

The signal processing practitioner usually separates problems into detection, estimation and identification. The simplest detection problem, the one we will be addressing here, is the binary hypothesis test. That is, we wish to indicate the presence or absence of a target with a reasonably high probability of success. The next level is the estimation problem. For the UXO problem, the estimation of its location is not a major issue since, upon detection, its location is fairly well known. For our purposes, identification is the main issue, since without some means to obtain an approximation of the nature of the contact, the false alarm problem becomes a major issue, since in the case of a UXO buried in a roadway, there likely will be objects of similar size (rocks, discontinuities, etc.) in the same region.

3 Acoustic Methods

The speed of sound for compressional waves in soil is on the order of 200–300 m/s, as compared to 343 m/s in air. This means that to resolve a target with a characteristic size on the order of, say 0.25 m, would require wavelengths on the order of 0.1 m or less. Thus, frequencies on the order of 2 kHz would be necessary. As it turns out, these frequencies are already too high to be of any use, as will be seen in the following.

3.1 Acoustic properties of soils

Sound propagation in porous media is well described by the theory developed by Biot [2, 3]. This theory predicts that shear waves and two types of compressional waves are supported in such solids. Of the two compressional waves, sometimes referred to as "fast" and "slow" compressional waves, the slow wave is rapidly attenuated, as is the shear wave. The speeds of the fast and slow waves are actually quite close to each other, differing by only a few 10 s of c/s. There seems to be no general agreement as to which of these compressional waves plays the major role in acoustic UXO detection methods.

Since the sound speeds in soil are significantly less than in air, any sound coupled into the ground can be assumed to refracted downward. Also, since the sound coupling into the soil cannot realistically be thought of as occurring at a well-defined interface, the concepts of reflection and transmission coefficients cannot always be considered to be a realistic model. The phenomenon is usually referred to as *seismo acoustic* or SA coupling, where the interface is viewed as a region of interaction [7]. Generally speaking, it is not the coupling that hinders SA UXO detection methods, but the absorption and false alarm problem.

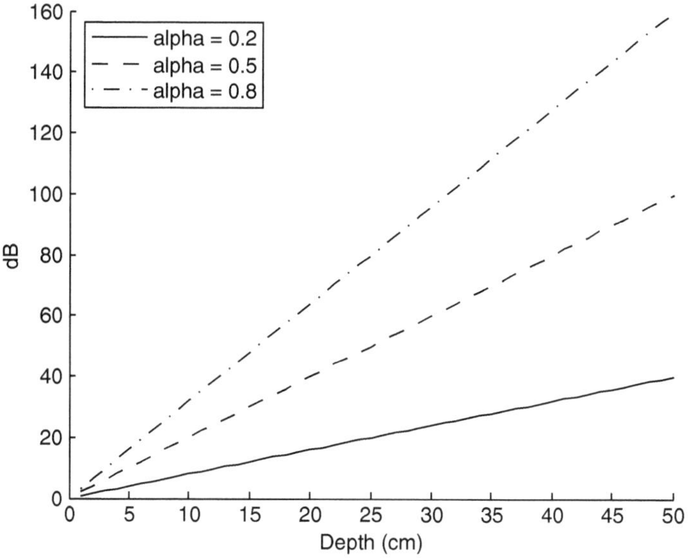

Fig. 1 Absorption loss in soil. Alpha has units of dB/m/kHz.

A study of the behavior of sound waves is soils was carried out by Oelze et al. [4]. They studied six soil compositions with clay content ranging from 2% to 38%, silt content ranging from 1% to 82%, sand content ranging from 2% to 97%, and organic matter ranging from 0.1% to 11.7%. Soils were classified as "loose" to "dense" and water content from dry to saturated. As might be expected, the results varied over wide ranges.

Attenuation coefficients α determined over frequencies of 2–6 kHz ranged from 0.12 to 0.96 dB/m/kHz. Lower attenuation tended to be in loose dry samples. Propagation speeds ranged from 86 to 260 m/s.

The two-way attenuation loss can now be estimated. Figure 1 shows the loss as a function of depth at 2 kHz for α values of 0.2, 0.5, and 0.8. Here, it can be seen that to expect to detect a buried object at this frequency, at a depth of more than a few tens of centimeters, is unlikely. To complicate the problem, there will likely be a great deal of clutter, leading to a high false alarm rate.

3.2 The nonlinear approach

There has been some experimental work done in the field of nonlinear detection. In 2002, Donskoy et al. [5] demonstrated that they could detect the nonlinear response of a buried mine-like object by detecting its sum and difference frequencies. The ground was excited with two high-level sound sources, generating acoustic waves in the ground in the region of the objects of interest. The source power levels were on the order of several hundred watts.[1] By using two frequencies, the sum and difference frequencies were detectable by sensing the ground surface vibration with a Laser Doppler Velocimeter (LDV). Interestingly, the nonlinear response is not from the object itself, but arises from the fact that the object is more compliant than the surrounding earth, resulting in a detachment at the interface during the tensile phase of the oscillation.

These results are interesting, since they rely on inducing a resonance in the object, which for anti-personnel mines and anti-tank mines, occurs at frequencies less than a kilohertz, thus ameliorating the absorption problem. In 2004, Korman and Sabatier [6] carried out a series of experiments essentially verifying the work of reference [5] and extending the experiments to include the observation of the effects of nonlinearities on "tuning curves," i.e., the shift of the resonant frequency of the object with amplitude. A major importance of this work is that it shows promise for reducing the false alarm problem, since it is to be expected that the mine will be the most compliant object in the ground.

This work must be considered to still be at the research level, since it is far from being applicable as an operational device.

3.3 The linear approach

Another approach, one that also uses a scanning LDV, does show promise of being applicable as an operational device. In this case, the ground is excited by a broadband high-level sound source, which can excite a resonance of the target. Although such a resonance has a low Q, since the object is in the soil, it appears to be sufficient to permit detection of the reradiation of the object by sensing the surface displacement. In 2001, Sabatier and Xiang [7] published a method in which they drove the ground with a broadband signal with a reasonably flat spectrum between 80 and 300 Hz, with a sound pressure level on the order of 90–130 dB(C)[2] and interrogated the surface with a scanning LDV. By using a correlation detector, they were able to successfully detect VS 2.2 and M21 anti-tank mines at a depth of 7.5 cm. The VS 2.2 is a roughly cylindrical plastic mine with a diameter of 24 cm, and the M21 is a metallic mine with a diameter of 22 cm. In these cases, surface velocities on the order of 50 μm/s at frequencies on the order of 150 Hz were encountered. The laser light has a wavelength of approximately 0.6 μm, so that for vibrational speeds of this order, Equation (6) indicates displacements on the order of 50 nm. Generally speaking, the plastic mines showed a greater response than the metallic ones.

In a later work, Valeau et al. [8] were able to improve the detection performance by using a time-frequency approach which was able to remove much of the speckle effects.

---

[1] It is difficult to translate these numbers into sound pressure level since the sources are in the near field.

[2] Unlike underwater acoustics, where the sound reference level is 1 μPa, the conventional reference in air is 20 μPa, which is approximately at the hearing threshold. C refers to the frequency weighting, which is essentially flat over a band of 63 Hz to 4 kHz.

The importance of this approach is that it holds promise for the development of an operational system, since the scanning LDV allows the processing to be carried out at acceptable speeds, as opposed to the so-called "stop and stare" method, where the LDV is used in a point by point method.

3.4 Principle of the Laser Doppler Velocimeter

The laser Doppler velocimeter is a device that uses the Doppler shift imparted by a moving (vibrating) surface on the reflected energy of an incident laser beam to estimate its instantaneous velocity. The approach used by Sabatier and Xiang [7] is based on the heterodyne method, where the incident beam is modulated by a Bragg cell, sometimes called an Acousto-Optic modulator or A/O modulator, which imparts a frequency shift (usually in the megahertz range) on the optical frequency. This frequency shift plays the role of a carrier frequency which is then frequency modulated by the vibratory motion. For example, if the laser frequency is ω_0 and the modulation frequency is ω_m, then when a beam of amplitude A_i with this frequency is scattered from a surface, and mixed in an interferometer with a reference beam of amplitude A_r and frequency ω_0, the intensity of the sum is given by

$$I_s = |A_i e^{i(\omega_0+\omega_m)t} + A_r e^{i\omega_0 t}|^2 = |e^{i\omega_0 t}|^2 |A_i e^{i\omega_m t} + A_r|^2. \tag{1}$$

Equation (1) now reduces to

$$I_s = I_1 + I_r + 2A_i A_r cos(\omega_m t), \tag{2}$$

with $|A_i|^2 = I_i$ and $|A_r|^2 = I_r$. The result, after removing the DC terms, is simply

$$I_s = A_i A_r cos(\omega_m t). \tag{3}$$

Now suppose the reflecting surface is vibrating with amplitude A_v at radian frequency ω_v. Then there will be a time dependent phase term added to $\omega_m t$ equal to

$$\phi(t) = \left(\frac{2\pi}{\lambda}\right) 2A_v sin(\omega_v t), \tag{4}$$

where λ is the wavelength of the laser light. Thus, Equation (3) becomes

$$I_s = 2A_i A_r cos(\omega_m t + \phi(t)). \tag{5}$$

The output of the LDV photodetector is a current proportional to I_s which can then be demodulated to extract the velocity. That is,

$$v(t) = \frac{\phi'(t)}{4k} = \omega_v A_v cos(\omega_v). \tag{6}$$

The prime indicates the time derivative.

3.4.1 Speckle noise

The LDV suffers from a limitation commonly referred to as "speckle" noise. This is a consequence of the fact that the laser light is highly coherent, so that the phase front of reflected laser light is extremely grainy and non-stationary in time. This can be viewed as a coherent addition of a multiple of spherical wavefronts, arriving from different points on the surface, coherently interfering at the observation point. From a statistical point of view, even though it is deterministic, it can be considered to be a realization of a random process.

If the undulations of the scattering surface have a characteristic deviation greater than the laser wavelength, then the phase structure of the wavefront can be considered to be a zero-mean random process, uniformly distributed from $-\pi$ to π, and its autocorrelation function is sharply peaked with a width on the order of a wavelength. Also, it is reasonable to consider the complex field at an observation point on an observation plane to be complex Gaussian.

In the case of the LDV, the difficulty is that the speckle noise takes the form of random spots that are rapidly moving in the observation plane. These spots have a characteristic size that is strongly dependent on the optical aperture involved. This is a consequence of the fact that highly localized scatterers are not resolvable beyond the capability of the viewing aperture. Thus, the narrower this aperture, the larger the correlation length, and therefore the larger the apparent size of the speckle spots. For a scanning LDV then, the speckle noise emanating from the LDV has a "bursting" type behavior as these speckle spots move past. This noise is difficult to remove. As mentioned above, some progress has been made in dealing with this by Valeau et al. [8] where a space-time representation of the velocity field is used for the detection statistic.

An excellent discussion of speckle is in the book by Barrett and Myers [9].

4 Ground-Penetrating Radar

Ground Penetrating Radar (GPR) techniques have been highly developed in the recent past [10] and have resulted in several commercial devices. It has applications in a number of fields. It is used to make geological measurements, nondestructive testing of large structures and pavements, and locating pipes and other buried objects. It is also extensively used in archaeology.

In spite of its successes however, it has some severe limitations. It performs poorly in any medium that has a high conductivity, such as clayey and moisture laden soils. Also, there is the fact that absorption of electromagnetic energy increases with frequency, whereas high frequency is necessary when resolution of small objects is desired. This means that there are severe depth constraints in such cases. In the case of mine hunting or UXO detection, as with the SA methods, it can suffer from poor detection statistics due to clutter. More information on GPR can be found in reference [10].

4.1 Nonlinear detection

Here we propose a nonlinear approach that may have application to cases where some form of electronic circuitry is contained. As an example, a typical cell phone receives on a carrier on the order of 850 MHz. If we choose a radar signal of this frequency to drive the input, then we should expect that the circuitry itself will have induced currents due to the high field strength. Since these circuits are highly nonlinear, we could expect reradiated frequencies to exhibit spectral components that lie outside of the carrier frequency's band. This means that, even if the reradiated field levels are low, they will have a favorable signal to noise ratio.

In the following example, we consider a clipped sine wave. In Figure 2 we show a 1 kHz unit amplitude sine wave that is clipped to half of its amplitude. Figure 3 depicts the power spectrum of this signal. As can be seen, along with the 1 kHz line, there are several strong lines at odd multiple frequencies.

The exact nature of the nonlinearities and their ability to produce such spectra would most easily be determined by experiment. Clearly, one drawback of this approach is that frequencies of 800–900 kHz, depending on the soil makeup, may not penetrate deeply into the ground. In many cases however, such high frequencies usually can detect at depths on the order of 1–2 ft. For UXO devices buried at such depths, this offers an interesting possibility.

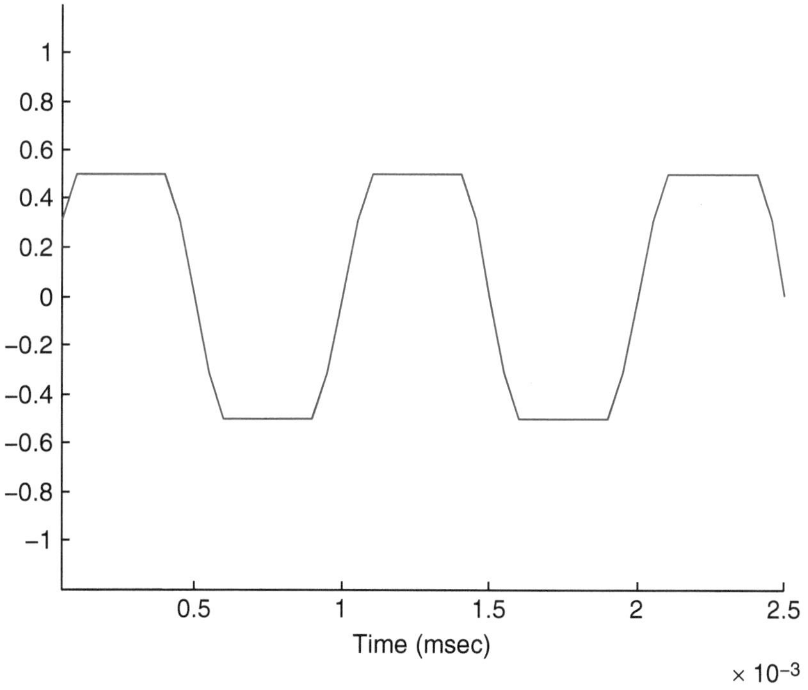

Fig. 2 Sine wave clipped at half amplitude.

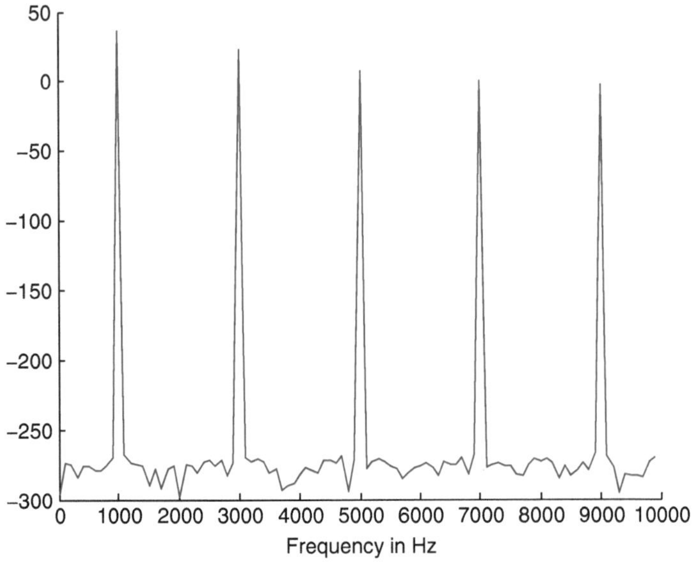

Fig. 3 Spectrum of clipped sine wave.

5 Conclusions

The capabilities of seismo-acoustic coupling and ground penetrating radar have been discussed. Due to the absorption of high-frequency (>1 kHz) sound waves by the soil, direct imaging of a buried object appears to be out of the question.

The exploitation of nonlinear effects shows promise in mitigating the false alarm rate, but they are still at the research stage. The linear approach, which uses SA coupling into the ground to excite the object of interest, shows promise of being closer to an operational system. Here, the surface vibration induced by the vibrating buried object is sensed with a scanning LDV.

The possibility of exploiting nonlinear effects in any electronic circuitry used as a detonator is shown to offer the possibility of detection of the reradiation from such devices when excited by a GPR source.

Acknowledgement The author would like to thank Prof. Ning Xiang of Rensselaer Polytechnic Institute for helpful conversations and suggestions.

References

1. Gooneratne C. P., Mukhopahyay, S. C. and SenGupta, G. (2004) *2nd International Conference on Robots and Agents*, Dec. 13–15, Palmerston North, New Zealand.
2. Biot M. A. (1956) Theory of propagation of elastic waves in a fluid saturated porous solid. I. Low-frequency range. *J. Acoust. Soc. Am.* 28, 168–178.

3. Biot M. A. (1956) Theory of propagation of elastic waves in a fluid saturated porous solid. II. High-frequency range. *J. Acoust. Soc. Am.* 28, 179–191.
4. Oelze M. L., O'Brien W. D. Jr. and Darmody R. G. (2002) Measurement of attenuation and speed of sound in soils, *Soil Sci. Soc. Am. J.* 66, 788–796.
5. Donskoy D., Ekimov A., Sedunov N. and Tsionskiy M. (2002) Nonlinear seismo-acoustic land mine detection and discrimination, *J. Acoust. Soc. Am.* 111, 2705–2714.
6. Korman M. S. and Sabatier J. M. (2004) Nonlinear acoustic techniques for landmine detection, *J. Acoust. Soc. Am.* 116, 3354–3369.
7. Sabatier J. M. and Xiang N. (2001) An investigation of acoustic to seismic coupling to detect buried anti-tank landmines, *IEEE Trans. Geosci. Remote Sens.* 39(6), 1146–1154.
8. Valeau V., Sabatier J. Costley R. D. and Xiang N. (2004) Development of a time-frequency representation for acoustic detection of buried objects, *J. Acoust. Soc. Am.* 116, 5, 2984–2995.
9. Barrett H. H. and Myers K. J., *Foundations of Image Science*, Wiley Series in Pure and Applied Optics. 2004, Chap. 18.
10. Daniels B., Ed. (2004), *Ground Penetrating Radar, 2nd Edition*, The Institution of Engineering Technology.

Subject Index

MIX
Papier aus verantwortungsvollen Quellen
Paper from responsible sources
FSC® C105338

If you have any concerns about our products,
you can contact us on
ProductSafety@springernature.com

In case Publisher is established outside the EU,
the EU authorized representative is:
Springer Nature Customer Service Center GmbH
Europaplatz 3, 69115 Heidelberg, Germany

Printed by Libri Plureos GmbH
in Hamburg, Germany